VETERINARY RADIOLOGY

Basic Principles and Radiographic Positioning

Edited by

A.P. SINGH
MVSc, PhD

JIT SINGH
MVSc, PhD

Department of Surgery and Radiology,
College of Veterinary Sciences,
C.C.S. Haryana Agricultural University, Hisar.

CONSULTANT EDITOR   :   KULDIP SINGH, MVSc, PhD

ILLUSTRATIONS   :   ASHOK KUMAR, MVSc

PHOTOGRAPHY   :   JASPAL SINGH

C B S

CBS Publishers & Distributors Pvt. Ltd.

New Delhi • Bengaluru • Chennai • Kochi • Kolkata • Mumbai
Bhubaneswar • Hyderabad • Jharkhand • Nagpur • Patna • Pune • Uttarakhand • Dhaka

ISBN: 978-81-239-0300-2

First Edition: 1994
Reprint: 1999, 2001, 2004, 2006, 2007, 2009,
         2011, 2012, 2013, 2015, 2017, 2020

Published by **Satish Kumar Jain** and produced by **Varun Jain** for
CBS Publishers & Distributors Pvt. Ltd.,
4819/XI Prahlad Street, 24 Ansari Road, Daryaganj, New Delhi - 110002
delhi@cbspd.com, cbspubs@airtelmail.in • www.cbspd.com
Ph.: 23289259, 23266861, 23266867 • Fax: 011-23243014

Corporate Office: 204 FIE, Industrial Area, Patparganj, Delhi - 110 092
Ph: 49344934 • Fax: 011-49344935
E-mail: publishing@cbspd.com • publicity@cbspd.com

Branches:
• *Bengaluru:* 2975, 17th Cross, K.R. Road, Bansankari 2nd Stage,
  Bengaluru - 70 • Ph: +91-80-26771678/79 • Fax: +91-80-26771680
  E-mail: cbsbng@gmail.com, bangalore@cbspd.com
• *Chennai:* No. 7, Subbaraya Street, Shenoy Nagar, Chennai - 600030
  Ph: +91-44-26681266, 26680620 • Fax: +91-44-42032115
  E-mail: chennai@cbspd.com
• *Kochi:* Ashana House, 39/1904, A.M. Thomas Road, Valanjambalam,
  Ernakulum, Kochi • Ph: +91-484-4059061-65
  Fax: +91-484-4059065 • E-mail: cochin@cbspd.com
• *Kolkata:* 6-B, Ground Floor, Rameshwar Shaw Road, Kolkata - 700014
  Ph: +91-33-22891126/7/8 • E-mail: kolkata@cbspd.com
• *Mumbai:* 83-C, Dr. E. Moses Road, Worli, Mumbai - 400018
  Ph: +91-9833017933, 022-24902340/41 • E-mail: mumbai@cbspd.com

Representatives:

• Hyderabad: 0-9885175004   • Nagpur: 0-9021734563
• Patna: 0-9334159340   • Pune: 0-9623451994
• Jharkhand: 0-9811541605   • Uttarakhand: 0-9716462459

Printed at:
India Binding House, Noida, UP (India)

This Book
is
Dedicated
to
The Memory of

Late Dr. A.K. Bhargava

Formerly Associate Professor of Radiology, College of Veterinary Sciences, Haryana Agricultural University, Hisar; Professor, Division of Experimental Medicine and Surgery, Indian Veterinary Research Institute, Izatnagar.

For
Promoting the Cause of Veterinary Radiology in India

FOREWORD

It gives me immense pleasure to write foreword of a book compiled by my own former students. It is more so because it has been developed at a centre which I myself nursed from its infancy to adulthood. I clearly remember those days when as Professor and Head of the Department of Surgery and Radiology of Haryana Agricultural University, Hisar, I was able to install in the department a 500 mA ceiling suspension X-ray machine in 1972, first of its kind in a veterinary institute in India. The installation of this machine revolutionised large animal practice in the region. The event was a major contributing factor for the development of veterinary radiology in India as others followed suit and radiology in real sense was introduced for research and clinical applications. We in India have marched ahead in veterinary science research but are still struggling to provide primary radiological facilities in the field hospitals. Nevertheless awareness is emerging and many states have introduced the facilities at district headquarters as more and more polyclinics are being established.

Availability of sufficient financial resources and standard textbooks within easy reach have been two important limiting factors for the growth of veterinary science in developing countries. It is high time that need to develop textbooks which suit our own requirements and resources is fully understood. It was with this objective that "Ruminant Surgery" was prepared. The experiment proved a success considering the response that book evoked across the country within a few months of its publication. The compilation of this book on veterinary radiology has thus been on an appropriate time.

This book deals with basic principles, contrast techniques and radiographic positioning. Modern radiology can not be developed or maintained at its present position unless basic principles are fully understood, a subject often given second priority. The editorial of The Veterinary Record, November 30, 1935, Stated, "Success in veterinary radiology is not obtainable by merely purchasing one of the many X-ray installations .... He who employs the apparatus must possess a sound knowledge of X-ray physics, anatomy, pathology and perhaps above all, long experience in the interpretation of X-ray films. Without these qualifications the results obtained are unlikely to enhance the reputation of the would be radiologist." How appropriate are these lines to highlight the importance of understanding the basic principles and to ultimately develop insight for radiographic interpretation.

The authors have presented concise information in simple language amply supported by illustrations, photographs and X-ray plates. The approach should make it easy for the students and clinicians to understand the subject. I am sure undergraduate students of veterinary colleges, post-graduate students, clinicians and teachers will find this book a source of useful information. It should also be of immense value to the radiographers employed in veterinary radiological services and to those who intend to establish new radiological units.

I congratulate the editors and contributors for this tremendous job despite all the odds. Only those who undertake to compile such specialised books in our circumstances can really understand the handicaps involved. All the participants, especially the editors, thus deserve appreciation for a job well done.

R.P.S. TYAGI

Vice-Chancellor,
Himachal Pradesh Krishi Vishwavidalya
Palampur, Himachal Pradesh.

PREFACE

Our basic aim in compiling this textbook is to provide relevant information in a simple and practical style to the veterinary science students and practising veterinarians of the developing countries. There is now an increasing awareness about the importance of radiology as a diagnostic aid even in ruminant practice. Many states of India have started equipping their field veterinary hospitals, especially polyclinics, with modern X-ray machines. Thus veterinary radiology is no more limited to the four walls of the teaching institutes. In order to establish a radiology unit and to obtain quality radiographs, it is essential that basic principles are well understood. There are several excellent books on radiological physics and basic principles but are exhaustive as well as expensive. Most of these books are designed to meet the requirements of medical radiology. Therefore, there was a need to project concise information which is easy to comprehend and is available in an economical edition. Moreover, radiographic positioning protocols for ruminants have not been described in available books of veterinary radiology. This book is made with these objectives in view.

Contributors of this book, though experienced teachers and clinicians, had no formal training in radiological physics and so errors are inevitable. We do not claim any originality as far as basic principles are concerned since a variety of literature was consulted to compile the information. However, every attempt has been made to present the information in a simple way so that matter does not appear to be difficult to understand by an average undergraduate student. A number of illustrations and photographs have been included to make the text meaningful. Major material included in contrast techniques and radiographic positioning is original and produced within the scope of radiographic facilities available. Any material used with permission of various sources has been specified under the figures.

We hope that information provided in this book would enable the students and clinicians to gain a working knowledge of the basic principles of radiology, to establish a radiology unit and to obtain a quality radiograph. We also hope that information provided will be useful to radiographers working in veterinary radiological services who are usually trained in medical institutes in the absence of separate animal health technicians training programmes in most third world countries.

Considering that computers are fast entering day to day life and that these play an important role in a radiology section, a chapter on computers and radiology, incorporating basic information, has also been included. Techniques of alternate imaging described in chapter 9 are hardly used at this stage, except for fluoroscopy, in veterinary practice in developing countries. However, a veterinarian can not justify his training if he is not aware of even basic principles of such techniques which will certainly be used in practice in near future.

We also hope that information contained in this book would help in implementing a uniform syllabus in veterinary radiology teaching across the country for which Veterinary Council of India is emphasising. Wherever appropriate, selected questions have been listed at the end of chapters to help a clinician in testing his acquired knowledge and a student in preparing for an examination.

It was not easy for us to compile this book as we had no finances at our disposal and there were many additional constraints. We do hope that the readers will understand our limitations. Duplication of material was inevitable at a few places in order to maintain clarity in the text. We request the readers to send us their valuable suggestions for improving the book in future.

A.P. SINGH
JIT SINGH

ACKNOWLEDGEMENTS

It was not easy for us to compile this book on basic principles of radiology and radiographic positioning. There were times when we were not even sure of its reality. Nevertheless task could be completed because of continued support of many persons. We thank them all from our heart.

We are grateful to various publishers of standard works and editors of journals who allowed reproduction of illustrations and photographs. Publishers are : W.B. Saunders Co, Philadelphia; Iowa State University Press, Ames; Lea and Febiger, Philadelphia; The C.V. Mosby Co Saint Louis; Kluwer Academic Publishers, Dordrecht, Netherland and American Veterinary Publications. Journals include : Veterinary Radiology and Ultrasound, The Journal of American Veterinary Medical Association, Veterinary Medicine and Modern Veterinary Practice.

We thank Mr. Uri Bargai of Koret University, Israel and Dr. J.N. Kornegay of North Carolina State University, Raleigh, U.S.A., for providing material for inclusion in this book. We also thank M/S Siemens India Pvt Ltd to allow reproduction of photographs of their products.

The material used with permission/courtesy is specified in the legends to figures. But for this support, the contents of this book would have remained incomplete.

It is indeed an honour that foreword of this book has been written by an eminent personality of the profession, Dr. R.P.S. Tyagi, our mentor and guide. His contributions to the veterinary profession in India, especially in the discipline of surgery and radiollogy, are well known. We are grateful to him for sparing his valuable time to go through the contents of this book.

We thank the consultant editor, contributors, Dr. Ashok Kumar and Mr. Jaspal Singh for meeting the deadlines at short notices and for all the help and cooperation. We also thank postgraduate students of our department for the ever ready help. We gratefully acknowledge the help of Mr. D.P. Mangal, artist, in designing the cover of the book and the assistance provided by supporting staff of the department, especially Mr. Mohinder Kumar, Mr. Subhash, Mr. R.D. Sharma, Mr. Rati Ram, Mr. Mahavir and Mr. Parmod during photographic work.

"There is always a lady behind a successful man". We do not claim success but ladies were certainly standing solidly behind us. But for the moral support and cooperation of Mrs Veena A.P. Singh and Mrs. Pamila Jit Singh, this work would not have been completed. We thank Dr. Sudhir Sharma for the help in compiling the subject index. We are grateful to Mr. Pankaj Kalra of Creative Graphics and his staff for a job well done i.e. laser type setting of the manuscript. Last but not the least, we thank C.B.S. Publishers, especially Mr. Satish Jain and Mr. Vinod Jain, for bringing out this publication. Their warming smiles have always been cheering.

A.P. SINGH
JIT SINGH

CONTRIBUTORS

Ashok Kumar, MVSc, Assistant Professor, Department of Veterinary Surgery and Radiology, H.A.U, Hisar, Haryana.

A.P. Bhokre, MVSc, PhD, Professor, Department of Veterinary Surgery and Radiology, M.A.U, Parbhani, Maharashtra.

A.P. Singh, MVSc, PhD, Professor, Department of Veterinary Surgery and Radiology, H.A.U, Hisar, Haryana.

B.A. Moulvi, MVSc, PhD, Assistant Professor, Department of Veterinary Surgery and Radiology, Sher-e-Kashmir University of Agricultural Sciences and Technology, Shuhma, Alusteng, J & K.

B.M. Jani, MVSc, PhD, Hospital Superintendent, Veterinary College, Anand, Gujarat.

Bharat Singh, MVSc. PhD, Associate Professor, Department of Veterinary Surgery and Radiology, College of Veterinary Sciences, Mathura, U.P.

D.B. Patil, MVSc, PhD, Assistant Professor, Department of Veterinary Surgery and Radiology, Veterinary College, Anand, Gujarat.

D. Krishnamurthy, MVSc, PhD, Professor, Department of Veterinary Surgery and Radiology, H.A.U, Hisar, Haryana.

D.M. Makhdoomi, MVSc, Assistant Professor, Department of Veterinary Surgery and Radilogy, Sher-e-Kashmir University of Agricultural Sciences and Technology, Shuhama, Alusteng, J & K.

G.C. Georgie, MVSc, PhD, Professor, Department of Animal Production Physiology, H.A.U., Hisar, Haryana.

Gajraj Singh, MVSc, PhD, Senior Scientist, Division of Experimental Surgery, Indian Veterinary Research Institute, Izatnager, U.P.

Harpal Singh, MVSc, PhD, Dean Postgraduate Studies, G.B. Pant University of Agriculture and Technology, Pantnagar, U.P.

I.S. Chandna, MVSc, PhD, Professor, Department of Veterinary Surgery and Radiology, H.A.U., Hisar, Haryana.

J.R. Jindal, BA, DMR, Senior Radiographer, Department of Veterinary Surgery and Radiology, H.A.U., Hisar, Haryana.

Jit Singh, MVSc, PhD, Professor, Department of Veterinary Surgery and Radiology, H.A.U., Hisar, Haryana.

K.I. Singh, MVSc, Assistant Professor, Department of Veterinary Surgery and Radiology, P.A.U, Ludhiana, Punjab.

K.K. Mirakhur, MVSc, PhD, Associate Professor, Department of Veterinary Surgery and Radiology, P.A.U. Ludhiana, Punjab.

Kuldip Singh, MVSc, PhD, Associate Professor, Department of Veterinary Surgery and Radiology, H.A.U, Hisar, Haryana.

Mohinder Singh, MVSc, Assistant Professor, Department of Veterinary Surgery and Radiology, H.P.K.V, Palampur, Himachal Pradesh.

N.R. Purohit, MVSc, PhD, Assistant Professor, Department of Veterinary Surgery and Radiology, College of Veterinary Sciences, Bikaner, Rajasthan.

N.S. Saini, MVSc, Assistant Professor, Department of Veterinary Surgery and Radiology, P.A.U, Ludhiana, Punjab.

P.K. Peshin, MVSc, PhD, Associate Professor, Department of Veterinary Surgery and Radiology, H.A.U, Hisar, Haryana.

Prem Singh, MVSc, Assistant Professor, Department of Veterinary Surgery and Radiology, H.A.U. Hisar, Haryana.

R.R. Parsania, MVSc, PhD, Professor, Department of Veterinary Surgery and Radiology, Veterinary College, Anand, Gujarat.

Rishi Tayal, MVSc, PhD, Assistant Professor, Department of Veterinary Surgery and Radiology, H.A.U, Hisar, Haryana.

S.K. Chawla, MVSc, PhD, Associate Professor, Department of Veterinary Surgery and Radiology, H.A.U, Hisar, Haryana.

S.M. Behl, MVSc, Assistant Professor, Department of Veterinary Clinic, H.A.U, Hisar, Haryana.

Sukhbir Singh, MVSc, PhD, Assistant Professor, Department of Veterinary Surgery and Radiology, H.A.U, Hisar, Haryana.

S. Thilgar, MVSc, PhD, Associate Professor, Department of Veterinary Surgery and Radiology, Madras Veterinary College, Madras, Tamil Nadu.

T.K. Gahlot, MVSc, PhD, Assistant Professor, Department of Veterinary Surgery and Radiology, College of Veterinary Sciences, Bikaner, Rajasthan.

Trilok Nanda, MVSc, FRVCS (Sweden), Assistant Professor, Department of Veterinary Gynaecology and Obstetrics, H.A.U, Hisar.

V.K. Sobti, MVSc, PhD, Associate Professor, Department of Veterinary Surgery and Radiology, P.A.U, Ludhiana, Punjab.

CONTENTS

CHAPTER

1 GENERAL INTRODUCTION

JIT SINGH
A.P. SINGH

"It is unfortunate that we, the veterinarians, have yet to fully understand the real importance of our profession. If we properly understand our science, we will feel the importance of it, and when we properly apply it, we will understand the importance of our profession. Only after we do it, we can be worthy of our profession and make others understand our importance. Let us become like X-rays to penetrate in depth".

GENERAL TERMINOLOGY

Radiology (Roentgenology): It is that branch of medical science which deals with the diagnostic and therapeutic applications of radiant energy. Radiant energy for this purpose includes x-rays, and beta and gamma radiations.

Veterinary radiology: It is that branch of science which uses radiant energy principally for diagnostic and therapeutic purposes in domestic, zoo and laboratory animals.

Science of radiology: It covers all of the uses of radiant energy in medical and veterinary sciences such as in diagnosis, monitoring and treatment of diseases, and in research programmes.

Radiologist: Any person qualified in medical or veterinary sciences and radiological physics to use radiant energy in the diagnostic, therapeutic and research fields of medicine.

Radiographer: A technically trained person who can obtain quality radiographs for use by a radiologist.

X-Rays: A special type of electromagnetic radiation which has high energy, extremely short wavelength, no mass or charge and travels at the speed of light (detailed properties are discussed in Chapter 2).

Medical X-rays: These X-rays are generated within a vacuum tube consisting of a source of electron and a target. The electrons are accelerated in the tube to travel through the tube at tremendous speed to strike at the target. This electron target interaction generates medical X-rays.

Radiograph: X-rays interact with body tissues with some degree of absorption in the exposed tissues. This photographic record of the extent of penetrability of X-rays through the exposed tissue parts is called a radiograph or ROENTGENOGRAM or SKIAGRAM or simply an X-ray picture.

HISTORICAL PERSPECTIVES

The discovery of X-rays was an accidental product of the work on Crookes-type tubes. Due to the nature and importance of the discovery, rapid progress took place in the production and uses of

X-rays in medical science. Veterinarians also made significant contributions in this regard. Some of the main events only are listed here:

November, 8, 1895	- Wilhem Conrad Roentgen discovered X-rays.
1896	- Lindenthal made first contrast picture of a hand.
	- First oil immersed X-ray tube developed by Trowbridge.
	- Roentgen and colleagues made the first metal-target X-ray tube.
	- First intensifying screen made by Pupin.
	- First Photographic paper developed for recording of X-ray image by Wright.
	- First dental radiograph made by Konig and Morten.
	- First veterinary radiograph, of an enquine foot, published by Paton and Duncan in the March issue of The Veterinary Journal.
	- Papers describing the use of X-rays in veterinary practice published by R.Eberlein and C.Troester of Germany; F.T.G Hobday, V.E.Jhonson and J.A.W. Dollar (of Dollar's Surgery) of England and V.Lemoine of France.
	- A.H. Becquerel discovered radioactivity of uranium.
1897	- J.J. Thomson discovered electron (a negatively charged particle much smaller than atom).
	- X-rays were used to locate bullets in the bodies of soldiers in Greco-Turkish and Sudan-Boer Wars.
1898	- Cannon used X-rays in the form of contrast studies (using bismuth meals) to investigate physiology of the gastrointestinal tract.
1900	- The Roentgen Society of the United States formed.
1901	- On December 10, W.C. Roentgen received first Noble prize in physics for discovery of X-rays.
1902	- G.Holz Knecht developed first dosimeter for radiation therapy.
1903	- Noble prize in Physics awarded to Becquerel (for discovery of radioactivity of uranium), Marie Curie and Pierre Curie (for discovery of radioactivity of radium and polonium).
1905	- Kienbock used strips of silver bromide photographic paper to estimate dosage in radiation therapy.
	- First Roentgenological Congress held at Berlin with veterinarian R. Eberlein being the chairman.
1913	- Gustav Bucky invented grid to remove scatter radiation.
1914	- W.H. Bragg and W.L. Bragg discovered that X-rays could be reflected.
1917	- Self rectifying generators developed for use in X-ray machine.
1918	- Line focus principle was discovered for small focal area to obtain good radiographic detail.
	(In the early years that followed discovery of X-rays, use of a fluoroscopic screen was preferred to visualise an X-ray image because of long exposure time required to obtain a photographic image, poor film emulsions and unpredictable tube performance. The safety precautions were largely ignored and even hands were left unprotected during exposures. This caused permanent radiation injuries to an extent that in some cases grafts or amputation of fingers became necessary. In veterinary practice, X-ray machines were not considered productive because of high costs involved.)
1919	- H.F. Waite constructed oil immersed shock-proof high voltage generators with enclosed collidge tube.
Early 1920's	- Double coated film replaced old glass plate for recording X-ray image.

	- Moving grid invented by Dr. Hollis Potter.
	- Iodine compounds introduced for use as contrast agents.
	- Total reflection, refraction and diffraction of X-rays by ruled grating was shown by Compton and Doan.
1926	- High voltage transformers with valve tube rectification came into general use.
1928	- International recommendations on radiation safety precuations were published.
1930	- Supervoltage single section X-ray tube developed by C.C. Lauriston.
1935	- Substraction technique introduced in radiography by Ziedes de Plantes (a Dutch radiologist).
1937	- Xeroradiography invented by a physicist Chester F. Carlson.
1945	- Gray Schnelle wrote first american book on veterinary radiology.
1950	- Cadmium sulphide crystals were used to detect X-rays by self amplification. It produced 107-fold increase in photoconducting currents.
	- SF_6 gas replaced oil etc as an insulating medium in transformers of portable X-ray machines.
Early 1950's	- X-ray image intensifiers and clinically safe cinfluorography developed.
	- Medical application of ultrasound.
1954	- First meeting of The American Veterinary Radiological Society was held (During this period, W.D. Carlson was a moving force in developing veterinary radiology in the U.S.A).
1957	- The organisation of Educators in Veterinary Radiology (EVRS) formed in the U.S.A.
1958	- Alois Pommer of Australia published exhaustive treatise on veterinary radiotherapy.
1960	- First medical radiographic film with a polyester based used.
	- American Board of Veterinary Radiologists (ABVR) formed.
	- American Veterinary Radiological Society issued its first printed proceedings (now it is a leading world journal "Veterinary Radiology and Ultrasound").
1969	- American Board of Veterinary Radiologists (ABVR) renamed as American College of Veterinary Radiologists (ACVR).
1972	- Rare earth intensifying screens invented..
	- Computerised axial tomography (C.T Scan) developed by G.N. Hounsfield in England.

In comparison to medical radiology, veterinary radiology in India has made slow progress. All veterinary teaching institutes do have facilities for small animal radiography and for radiography of the limbs of large animals. First 500 mA machine suitable for large animal radiography was installed at the College of Veterinary Sciences, Hisar, in 1972. In the early 1970's, late Dr. A.K. Bhargava, then Associate Professor of Radiology at Hisar, propagated the use of radiology in experimental and clinical research and for diagnosis of various disorders, especially in large animals. Credit also goes to Dr. R.P.S. Tyagi, the then Professor and Head of the Department of Veterinary Surgery and Radiology at Hisar, who encouraged the application of radiology in clinical practice. At present, number of institutes have facililties for large animal radiography. However, ultrasound facilities are available only at College of Veterinary Sciences, Ludhiana.

Radiology is being taught at undergraduate and postgraduate level in all veterinary teaching institutes of India. However, there is neither a separate department of veterinary radiology nor a separate postgraduate programme on the subject. Veterinary radiology continues to be a part of the department of veterinary surgery. X-ray film libraries for teaching purposes exist only in a few institutes. There is still a general lack of radiological facilities in field veterinary hospitals though X-ray machines have recently been introduced in polyclinics of a few states of India. Although human X-ray facilities are being used for small animal patients in many cities yet there is no evidence of a close liason with medical personnel. Thus enough scope exists in India to expand the utility of conventional radiographic procedures.

While concluding historical perspectives, a few lines said by Kealy appear befitting, "History would not have been the same without us - without each and every one of us however the humble role we play ....it is this kind of historical perspective - not just recalling what happened on this day or that, or who said this or that - or where - but understanding that we are part of a great design, it is this that gives a meaning to life -..............a sense of satisfaction".

SCOPE AND USES OF VETERINARY RADIOLOGY

It is not feasible at this stage in veterinary practice to depend upon or use all the developed technology of radiology. The cost of equipment for CT-imaging and ultrasound etc has limited the scope of these machines in veterinary practice. Moreover, the emphasis in veterinary practice is different than that in medical practice. Radiology in veterinary practice may not always be used to arrive at a final diagnosis. Nevertheless, radiography remains a good aid in diagnosis. Judicious use of conventional machines, proper positioning of the part to be examined, careful interpretation and genuity of the radiologist, all help to arrive at a reasonable radiographic diagnosis. In practice, radiography should be used only when it is expected to provide more significant information.

Following are the possible uses of radiography in veterinary practice:
   i)   As a diagnostic aid.
   ii) To select methods or techniques of treatment e.g., for fracture repair.
   iii) To detect previously unrecognised lesions.
   iv) To monitor efficacy of a treatment schedule.
   v)   To screen normal animals for morphological evaluation in an attempt to eradicate inherited diseases by selective breeding.
   vi) To determine the age of animals.
   vii) To examine postmortem material.
   viii) For non-destructive examination of archaeological specimens of animal origin.
   ix) As teaching aid in the subject of anatomy.
   x)   In veterinary science research e.g., osteomedullography to evaluate bone healing.

According to Webbon, three approaches are possible for the continued development of radiology without acquiring equipments of prohibitive cost:
   i)   Establishment and maintenance of a close contact with radiologists working in human medicine so that expensive equipment can be made available for use in animals.
   ii) Consolidation of knowledge on the natural history of diseases and significance of minor variations from the normal.
   iii) Application and expansion of diverse uses of radiography in veterinary science, especially for purposes other than diagnosis of diseases in clinical cases e.g., in disease eradication programmes by selective breeding.

DIRECTIONAL TERMS (RADIOGRAPHIC VIEWS)

The nomenclature committee of the American College of Veterinary Radiologists recommended a system of standard nomenclature for radiographic views and these recommendations were accepted by American College of Veterinary Radiologists in 1983. The recommended directional terms are easy to understand and avoid confusion. Standard journals and textbooks have mostly shifted to this new terminology. Therefore, the new terminology is being used in the present text and reader should be familiar with this terminology. These directional terms are based on the principle that each view should be able to indicate the direction that the central ray of the primary beam of X-rays penetrates the body part being examined i.e. from the point of entrance to the point of exit.

These terms replace the previously used terms of anterior, posterior, superior and inferior. These directional terms (Fig.1-1) are described below :

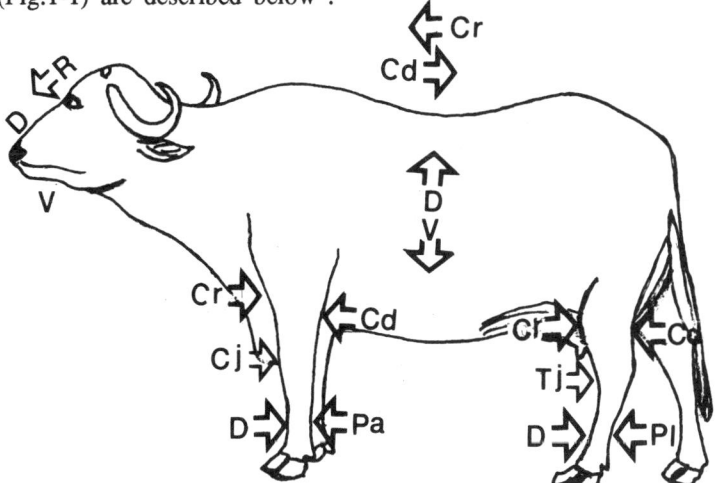

Fig. 1-1 : Directional terms. D = Dorsal, R = Rostral, Cr = Cranial, Cd = Caudal, V = Ventral, Pa = Palmar. Pl = Plantar, Cj = Carpal joint, Tj = Tarsal joint.

Cranial: It describes the part of the neck, trunk and tail positioned towards the head from any given point. It also describes aspects of the limb facing the head and above the carpal and tarsal joints.

Rostral: It describes the parts of the head positioned towards nares from any given point on the head, e.g., nostrils are rostral to the eyes.

Caudal: Parts of the head, neck and trunk positioned towards the tail from any given point, and aspects of the limbs facing tail but proximal to the carpal and tarsal joints.

Palmar: It replaces caudal in the forelimb from the carpal joint distally.

Plantar: It replaces caudal in the hindlimb from the tarsal joint distally.

Dorsal: It describes the following:
  i) Upper aspects of the head, neck, trunk and tail, also meaning towards the vertebrae or back.
  ii) Aspects of the limb from the carpal or the tarsal joint distally.

Ventral: Lower aspects of the head, neck, trunk and tail, also means towards lower aspects of the animal.

Proximal: Describes nearness to the point of origin of a structure e.g., the scapula is proximal to the humerus.

Distal: Describes a point farther away from the point of origin of a structure e.g., the radius is distal to the humerus.

The term volar is no longer used. Superior and inferior are still used to describe the upper and lower dental arches. In anatomical terms, anterior and posterior are still used for certain locations such as for parts of the eye, ear and brain.

The nomenclature committee also recommended the use of standard abbreviations which are listed in table 1-1.

In a view when only two directional terms are used, standard listing is done e.g., ventrodorsal, dorsopalmar etc. In ventrodorsal term, it is clear that central ray enters from the ventral side to exit from the dorsal side; and so on. For complex terms, as in case of oblique views, a hyphen is added to indicate the point of entrance and exit of the central ray, e.g., dorsolateral-palmaromedial oblique. When a special position, other than the routinely used, is employed appropriate term should be used for the same, e.g., a dorsopalmar view of the flexed carpal joint would be written as dorsopalmar (flexed).

TABLE 1-1 : Abbreviations of directional terms to describe radiographic views (recommended by the nomenclature committee of the American College of Veterinary Radiologists).

Directional term	Abbreviation	Directional term	Abbreviation
Left	Le	Medial	M
Right	Rt	Lateral	L
Dorsal	D	Proximal	Pr
Ventral	V	Distal	Di
Cranial	Cr	Palmar	Pa
Rostral	R	Plantar	Pl
Caudal	Cd	Oblique	O

For most oblique views, combinations of basic directional terms would be required. To standardise nomenclature for such views, the nomenclature committee recommended as follows:

i) 'Right and left' should precede other terms, e.g., right cranioventral.

ii) 'Medial' and 'lateral' should be subservient when used in combination with other terms, e.g., dorsomedial, ventrolateral etc.

iii) On the head, neck, trunk and tail, the terms 'rostral', 'cranial' and 'caudal' should take precedence when used in combination with other terms, e.g., rostromedial, craniodorsal etc.

iv) On the limbs, the terms 'dorsal', 'palmar', 'plantar', 'cranial' and 'caudal' should take precedence when used in combination with other terms e.g., dorsoproximal, palmarodistal etc.

REFERENCES

Christensen, E.E., Curry, T.S. and Dowdey, J.E. 1978. An Introduction to the Physics of Diagnostic Radiology. 2nd edn., Lea and Febiger, Philadelphia.

Emmerson, M.A. 1975. Anatomy in radiology. In: Sisson and Grossman's The Anatomy of the Domestic Animals. Ed., Getty, R. 5th edn, W.B. Saunders Co., Philadelphia.

Gillete, E.L., Thrall, D.E. and Lebel, J.L. 1977. Carlson's Veterinary Radiology. 3rd edn., Lea and Febiger, Philadelphia.

Gloyna, E.F and Lebetter, J.D. 1969. Principles of Radiological Health. Marcel Dikker Inc., New York.

Kealy, K. 1992. Veterinary radiology - A historical perspective. Vet. Rad. **33**, 5.

Smallwood, J.E., Shively, M.J., Rendano, V.T. and Habel, R.E. 1985. A standard nomenclature for radiographic projections. Vet. Rad. **26**, 2.

Webbon, P.M. 1987. Radiology in veterinary science. Brit. Vet. J. **137**: 349.

Willamson, H.D. 1978. The new photography - A short history of veterinary diagnostic radiology. Vet. Rec. 130: 84.

SELECTED QUESTIONS

1. Define the following terms:
   i) Veterinary radiology ii) X-rays iii) Roentgenogram iv) Rostral v) Palmar vi) Plantar.

2. List the possible uses of radiography in veterinary practice.

3. Name the persons credited with the following achievements:
   i) Discovery of X-rays ii) Production of first veterinary radiograph iii) Discovery of electron iv) First noble prize in physics v) Invention of moving grid.

4. Who is considered father of veterinary radiology and why?

5. What directions are required for continued development of veterinary radiology without acquiring very expensive equipments and by using conventional machines?

CHAPTER

2 GENERAL CONCEPTS OF RADIATION

KULDIP SINGH
ASHOK KUMAR

RADIATION OR RADIANT ENERGY

All the matter of the universe is made up of mass, electricity and energy. Einsteen, in his theory of relativity, predicted that matter and energy are interchangeable: $E = Mc^2$ where E represents energy (joules), M represents mass (Kg) and c represents velocity of light in vacuum (m/second). Both matter and energy can neither be created nor destroyed-the law of conservation of energy or matter. It is only the form which can be changed. Like matter, energy may exist in many forms, e.g. kinetic energy, electrical energy, potential energy, chemical energy, nuclear energy, heat energy, electromagnetic energy etc. Energy of one form can easily be converted into another form. For example, X-rays are produced in X-ray machine from electrical energy. **Energy emitted and transferred or propagated through the matter is called radiant energy or radiation.**

Any type of energy or matter-energy combination capable of removing one or more orbital electrons from the atom after interaction is known as ionising radiation (The process called ionisation) and forms the basis of biological effects. For example, some fast moving particles such as X-rays, alpha, beta or gamma rays, types of ionising radiation, can cause serious injury to the living beings, if used indiscriminately, X-rays constitute largest source of man made ionising radiation. The other most intense source of ionising radiation is natural environment, e.g., cosmic rays from terresterial sources, both external (principally from potassium[40] radium[228], carbon[14]) and internal (principally from uranium[238], thorium[232] and potassium[40]). All these contribute significantly to annual whole body radiation dose.

Types of radiation

Ionising radiation can be classified (Table 2-1) into two basic types:
  i) Particulate or corpuscular radiation
  ii) electromagnetic radiation

i) Particulate or corpuscular radiation: These radiations are composed of subatomic particles of matter and may either be electrically charged or be neutral. The energy carried by the particles depends on their mass and speed. Typical particulate radiations are alpha and beta particles, protons, electrons (cathode rays), neutrons, nuclear fragments etc.

ii) Electromagnetic radiation: Production of electromagnetic radiation (EMR) is another form of transporting energy through the space. The EMR consists of both electrical and magnetic fields set up by vibrating electrons. The changing magnetic field is perpendicular to the electrical field

TABLE 2-1 : Types of ionising radiation and their characteristics.

Types of ionising radiation	Atomic mass number	Charge	Energy (Mev)	Specific ionisation (ip/cm air)	Range in air	Range in soft tissue	Origin
Particulate radiation							
Alpha	4	+2	4-7	20,000-60,000	1-10 cm	Upto 0.1 mm	Nucleus
Beta	0	-1	0-3	100-400	0-1 m	0-2 cm	Nucleus
Electromagnetic radiation							
X-rays	0	0	0-10	Upto 500	0-100 m	0-30 cm	Electron cloud
Gamma rays	0	0	0-5	Upto 500	0-100 m	0-30 cm	Nucleus

[Adapted from Bushong S.C. 1975. Radiological Science for Technologists – physics, biology and protection. The C.V. Mosby Co St. Louis]

(Fig 2-1). The frequency of EMR ranges from 10^1 to 10^{24} Hertz (cycles/second) or perhaps higher and photon wavelength varies from 10^{-16} to 10^7 meters. Grouped together these radiations make up the EM-spectrum. The wavelength of various electromagnetic radiations are : heat waves = 10^{-3} cm, radio waves = 3×10^5 cm, light waves = 5×10^{-5} cm, infrared rays = 10^{-2} cm, UV range = 10^{-6} cm, X-rays = 10^{-8} cm, gamma rays = 10^{-10} cm and cosmic rays 10^{-11} cm. Each EMR has same general form and is associated with extremely tiny amount or bundle or quantum of energy, photon, that depends on the frequency of the wave. **Photon is a Greek word which means smallest amount of energy. The energy transported by a photon is measured in electron-volt which is the amount of energy that an electron gains as it is accelerated through potential difference of one volt.**

X-rays travel by a transverse wave phenomenon as actual disturbances are at right angle to the direction of propagation of disturbance. The waves have both electrical and magnetic force fields that are continuously changing in a sinusoidal fashion. This type of variation is called **sine-wave or sine-curve** (Fig.2-2) and has following features:

i) Wavelength - It measures quality of radiation and is the distance between the two successive corresponding crests or troughs on a wave.

ii) Velocity - It refers to the distance travelled by the waves in a second. It is measured as meters per second.

iii) Frequency - It refers to the number of crests or valleys on a wave that passes the point of observer per unit time or identified as oscillations (rise and fall) per second or cycles per second. Its unit is Hertz (Hz).

iv) Amplitude - It is a function of vigours of wave and is the distance from crest to valley.

The mathematical inter-relationship of these features is: Speed or Velocity = Frequency × Wavelength. In other words, frequency and wavelength are inversely proportional to each other. Higher penetrating X-ray photons have high frequency and shorter wavelength and vice versa.

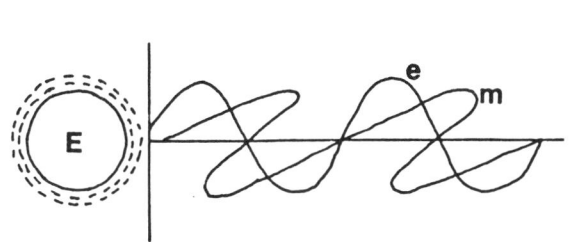

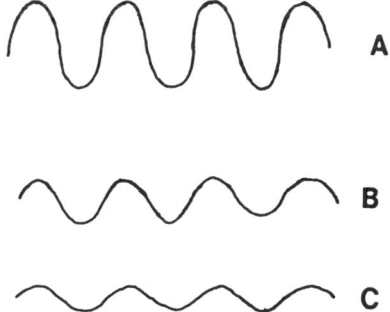

Fig. 2-1 : Electrical (e) and magnetic (m) components of electromagnetic radiation are at right angle to each other. E = Vibrating electron (Adapted from Selman, J. 1976. The Basic Physics of Radiation Therapy. 2nd edn., Springfield, Thomas).

Fig. 2-2 : Sinewave features of X-rays. Wave A has highest amplitude and wave C lowest.

X-RAYS

X-rays are electromagnetic radiations of high energy and low wavelength produced by the conversion of kinetic energy of electrons into electromagnetic radiations. These rays have sufficient energy to ionise the atom and the ability to penetrate solid matter, animate as well as inanimate, and so are used to examine the internal structures.

The physical properties of X-rays are listed below:

  i) These are electromagnetic radiations and behave as waves. Their short wavelength ranges from 0.1 to 0.5 A° with energy levels of 25-125 keV.

ii) These are high energy radiations with high frequency. Their energy is inversely related to the wavelength.

iii) Because of their high energy, X-rays are able to penetrate materials which readily absorb and reflect visible light.

iv) X-rays travel in straight line with the speed of light (3×10^{10} cm/second).

v) These rays cannot be focused by a lens.

vi) X-rays do not possess mass or charge and thus are unaffected by electrical or magnetic field.

vii) X-rays interact with matter, are absorbed or scattered and liberate minute heat on passing through it. Thicker and denser the material, more is the absorption and scattering.

viii) X-rays affect photographic film by a chemical reaction initiated in a way similar to that by visible light.

ix) X-rays produce phosphorescence/fluorescence in certain crystalline material (e.g. calcium tungstate, zinc cadmium sulphide).

x) These rays are able to produce indirect ionising effects on gases because of their ability to remove electrons from atoms. Air is made electrically conductive. The electrical properties of solids and gases are also influenced.

APPLICATION OF X-RAYS

A wide range of X-rays are produced for application in medicine, research and industry. These uses are listed in Table 2-2.

TABLE 2-2 : Some of the wide range of X-rays produced for application in medicine, research and industry.

Type of X-ray	Approximate energy	Application
Diffraction	Less than 10 kV$_p$	Research : Structural and molecular analysis
Grenz rays	10-18 kVp	Medicine : Dermatology
Superficial	50-100 kVp	Medicine : Therapy of superficial tissues
Diagnostic	30-150 kVp	Medicine : Imaging anatomical structures and tissues
Orthovoltage[a]	200-300 kVp	Medicine : Therapy of deep lying tissues
Supervoltage[a]	300-1,000 kVp	Medicine : Therapy of deep lying tissues
Megavoltage	Greater than 1 MV	Medicine : Therapy of deep lying tissues. Industry : Checking integrity of welded metals

a - These radiation therapy modalities are slowly being phased out.

(From Bushong, S.C. 1975. Radiological Science for Technlogists -physics, biology and protection. The C.V. Mosby Co St. Louis, Used with permission)

BASIC INTERACTION OF X-RAY WITH MATTER

For an X-ray examination, the part to be examined is kept between the X-ray source and an X-ray film (Fig.2-3). Thus the X-ray beam emitted by the machine traverses through the part to be examined to reach the film, carrying useful information which is recorded as an image on the film. While passing through the patient:

i) Some X-rays are differentially transmitted through the patient carrying useful information.

ii) Some photons are absorbed and cease to exist.

iii) Some are deflected from their course as scatter radiation which carries no useful information and rather decreases the quality of a radiograph by causing fog on the film.

The discussion that follows details these points and importance of interaction of X-rays with matter as far as diagnostic radiology is concerned. Ionisation of tissues and harmful effects of X-rays are discussed in chapter 8.

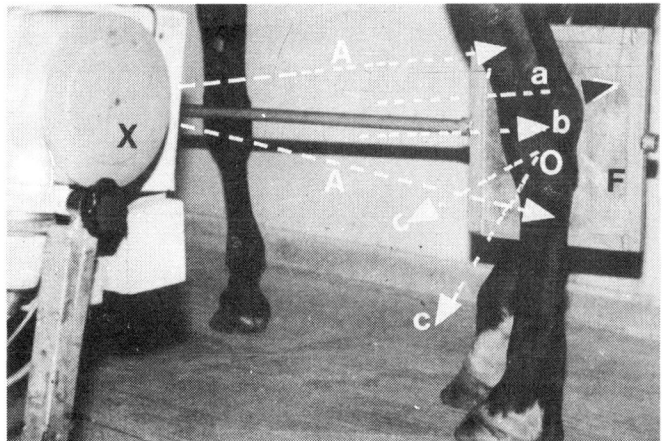

Fig. 2-3 : For an X-ray examination, object to be radiographed (O) is kept between an X-ray film in a cassette (F) and X-ray source (X). A part of the X-ray beam (A) passes through the object (a) carrying useful information for the film, some X-rays (b) are absorbed in the object and some (c) are deflected.

All elements have atoms. Each atom has a central nucleus, and electrons at the periphery arranged in shells (Fig.2-4). The nearest shell to the nucleus is designated as K shell and consecutive shells as L, M and so on. When an incident beam of X-ray strikes matter, there is interaction with the target atom, a single or all electrons may be involved. This interaction produces certain effects which are mostly related to the intensity of incident photons required to initiate an interaction and the extent of energy transfer that occurs in the process. The electrons present in each shell are specific in number, e.g. inner most K shell contains two electrons, if more electrons are present they must move to L shell which can contain eight electrons. Since nucleus is positively charged and electrons are negatively charged, there is a force of attraction called **binding energy**. The binding energy of K shell, i.e. nearest to the nucleus, is maximum. This binding energy decreases towards the periphery and the electrons of the outermost shell are loosely bound or almost free. To dislodge an electron from its orbit, an energy greater than its binding energy is required. Once this basic atomic physics is understood, it becomes easier to understand basic interaction of X-rays with matter.

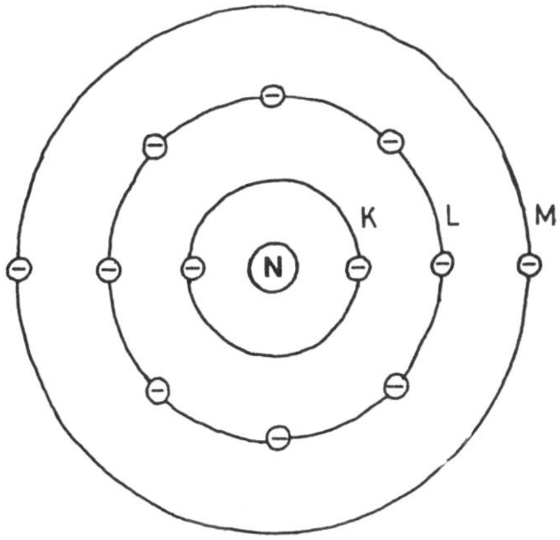

Fig. 2-4 : An atom with a nucleus (N) and electrons arranged in shells (K, L, M)

X-ray photons can interact with matter by five ways: coherent scattering (unmodified or classical scattering), photoelectric effect, compton effect, pair production and photodisintegration. Last two processes do not occur in photon energy range used in diagnostic radiology (40-150 keV) and hence are not of much interest to a radiologist. In coherent scattering, interaction causes only a change in the direction of photon without transfer of energy and thus no ionisation. In this effect, either a single or all electrons do vibrate to produce radiation to cause some fog on the film, but percentage of scatter radiation is so low (5%) that it is also not of much interest in diagnostic radiology. Therefore, only photoelectric and compton effects, of much importance in diagnostic radiology, are being discussed here.

Photoelectric effect

This effect is mostly produced when X-ray photons interact with inner shell (K L or M) electrons of an atom. **Since the effect is inversely proportional to the third power of the photon energy, relatively low energy photons are involved.** The effect is shown in fig 2-5. When an incident photon with an energy slightly greater than the binding energy of a K shell electron encounters the latter, the K shell electron is ejected from its shell. The photon disappears as most of its energy is utilised to overcome the binding energy of the K shell electron. The free electron flies off as a **photoelectron.** Another electron from an adjacent or outer shell of the same atom or from outer shell of another atom immediately falls into the void created by the ejected electron. As this electron drops into the created void, it gives off energy in the form of radiation called **characteristic radiation** (see Chapter 3) of which energy is equal to the difference of binding energies of the electrons of the shells involved. The shifting of electrons continues till stable energy stage is reached in the atom. If a K shell electron void is filled by an electron from outside the atom, characteristic photon of highest energy is produced. In elements with low atomic number, most interactions occur at K shell. If the atomic number is high, energy of incident photon is frequently insufficient to eject a K shell electron and thus most photoelectric effects occur at L or M shell levels.

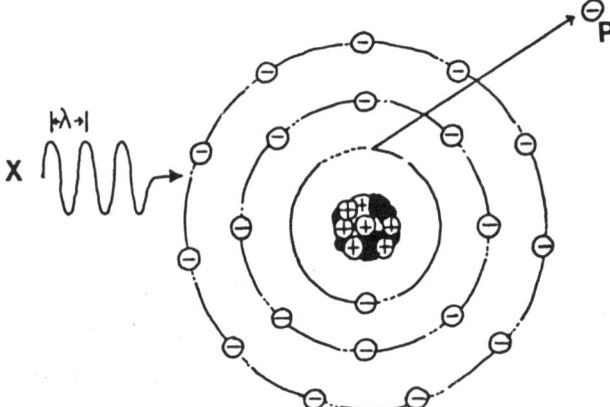

Fig. 2-5 : Photoelectric effect. When an incident photon of X-ray (X) encounters an inner shell electron, the electron is ejected from its orbit as a photoelectron (P). X-ray photon ceases to exist as it is absorbed (From Bushong, S.C. 1975. Radiological Science for Technologists. The C.V. Mosby Co Saint Louis. Used with permission).

Photoelectric effect thus produces: i) **a photoelectron, a negatively charged ion of low energy which is absorbed in the patient being exposed** and ii) **characteristic radiation.** Importance of photoelectric effect in diagnostic radiology is related to two factors :

i) The effect increases as the atomic number of the element (absorber) increases and is roughly proportional to the third power of the atomic number of the element. Since in a body different tissues are made up of different elements, contrast of the image produced on a film is enhanced. Moreover, there is no scatter radiation and thus radiograph of an excellent quality is obtained.

ii) Photoelectric effect increases the radiation dose of the patient as all the energy produced by the effect is absorbed by the patient.

The probability of photoelectric effect decreases as incident photon's energy increases and so patient dose can be decreased by using high kVp technique.

To sum up, photoelectric effect occurs more with low energy incident photons and high atomic number elements, provided that the photons have sufficient energy to overcome electron binding energy in the atom.

Compton effect

This effect is produced when an incident moderate or high energy photon encounters a free electron of the outer shell of the atom. The probability of this type of effect is nearly independent of the atomic number of the element but depends on the density of the absorber and energy of the incident X-ray beam. **Compton effect produces almost all the scatter radiation encountered in diagnostic radiology.**

The compton effect is shown in fig 2-6. **As an incident photon encounters a free electron of the outer shell of the atom, the photon is deflected to travel in a new direction as scatter radiation.** The electron is ejected from its orbit (recoil electron) at a specific angle. Since outer electrons are loosely attached, little energy is spent in ejecting them from their orbit and the photon retains most of its energy. The energy retained by the incident photon depends on two factors : i) its initial energy and ii) the angle of its deflection from the original direction. At narrow angle of deflection, photon retains most of its energy. Compton effect has following implications in diagnostic radiology:

i) Photon deflected at narrow angle is able to reach the X-ray film being exposed to cause fog.

ii) Compton scatter radiation is difficult to remove because it is too energetic for filters and its angle of deflection is too small for the grid to be effective. Because this type of scatter is difficult to remove, quality of radiograph is decreased.

iii) **Compton scatter radiation is a major radiation hazard especially during fluoroscopic examinations.**

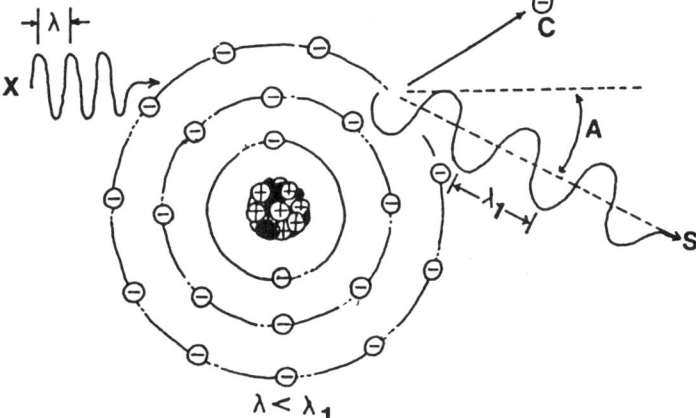

Fig. 2-6 : Compton effect. When moderate or high energy X-ray photon (X) encounters an electron of outer shell, the photon is deflected as scatter X-rays (S). C = Secondary or compton electron ejected from the orbit. The wavelength of scatter radiation is greater than that of incident X-ray. A = angle of deflection (From Bushong, S.C. 1975. Radiological Science for Technologists. The C.V. Mosby Co, Saint Louis. Used with permission).

If relative frequency of different interactions is considered, coherent interaction contributes only about 5%. Compton effect predominates except at very low photon energies, if water is the absorber. Because contrast agents have very high atomic number, photoelectric effect predominates exclusively. The bone is intermediate in density to water and contrast agents and frequency of compton and photoelectric effects is roughly equal. As the kVp increases, percentage of photoelectric interaction decreases while percentage of photon interaction and per cent transmission increases. For example, in exposures of soft tissues at 50 kVp photoelectric effect is 78.9 %, compton effect is 21% and per cent transmission is 0.002% while at 100 kVp these values are 31%, 63% and 6%, respectively.

ATTENUATION

Intensity of an X-ray beam is a product of number of photons (quantity) and energy of photons (quality). Attenuation refers to the reduction in the intensity of the X-ray beam by way of absorption or deflection. In a monochromatic or homogenous beam of radiation, the change occurs only in the quantity while in polychromatic radiation both quantity and quality changes. In simple words, **attenuation refers to the total reduction in the number of X-rays remaining in an X-ray beam after passing through the structure being exposed.** It means all photons do not reach the X-ray film and this has important bearing on the image formation. If all photons were to reach, the film will be totally black after processing. If all photons were attenuated, none will reach the film which will then appear as totally white. However, **the extent of attenuation by different tissues varies and thus a contrast is obtained which makes the image visible on the film.** It also explains the definition of a radiograph i.e. photographic record of the extent of penetrability of X-rays through given structures.

As has already been stated, attenuation of X-ray beam occurs by processes of absorption and deflection or scattering. In absorption process, energy of radiation beam is transferred to the absorbing tissues through ionisation. In scattering, there is no transfer of energy to the absorbing tissues rather photons are simply deflected from their original direction (Fig 2-7) Monochromatic radiation gets attenuated exponentially i.e. there is no fixed number of photons that are absorbed for each incremental thickness of the tissues that rays penetrate, rather for every 1 cm of thickness a constant percentage of the photons are removed. It means every succeeding thickness absorbs less number of photons than preceding one though percentage absorbed is same. For example, if radiation beam that passes through a given thickness of tissues has 100 photons and first centimeter of thickness of a given structure absorbs 50 photons (50% of 100), second centimeter will absorb 25 photons (50% of 50), third will do 12 photons (50% of 25) and so on. The exponential attenuation of the photons is explained by the fact that photons have no mass or charge and thus travel at a constant velocity. In contrast, alpha and beta particles have a fixed range, e.g., beta particles with energy of 2 MeV have a range of about 1 cm in soft tissues beyond which there is no substantial penetration of energy.

Attenuation of polychromatic or heterogenous X-ray beam is more complex as it changes both in quantity and quality. Polychromatic X-ray beam contains a whole spectrum of energy and the mean energy lies between one third and one half of its peak energy. On passing through an absorber, the number of photons decreases thus changing the quantity of the beam. Since low energy photons are readily absorbed than high energy photons, the mean energy increases. In other words, on passage through each successive layer of the absorber, X-rays become more hard and more nearly monochromatic and act like a monochromatic beam. Therefore, there is also a change in the quality of polychromatic X-ray beam.

Measurement of attenuation

In diagnostic radiology, attenuation can be measured in terms of two coefficients: linear attenuation coefficient and mass attenuation coefficient.

Linear attenuation coefficient: It is defined as the fractional decrease in the intensity of the beam per centimeter of attenuating material (absorber). This is for monochromatic radiation and is specific for certain energy of X-ray beam as well as for certain types of absorbers such as water, bone and muscle. When energy of X-ray beam increases, the number of X-rays that are attenuated decreases, and so does the linear attenuation coefficient.

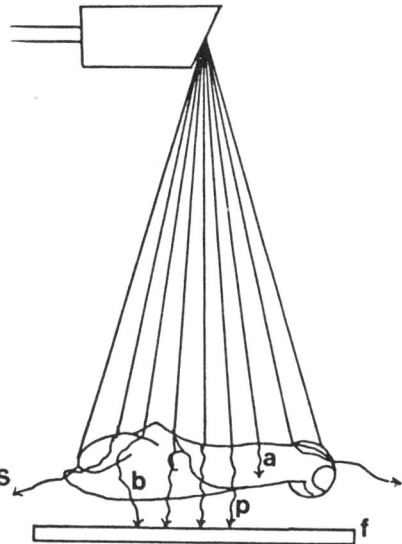

Fig. 2-7 : When an X-ray beam arives at the part being radiographed, some X-rays pass through without ineracting to carry useful information from the patient (p), some are absorbed (a), some are scattered away (s) from the film (f) and some are scattered in the patient (b) to reach the film. Scattering and absorption attenuate X-ray beam.

A heterogenous beam has a multitude of attenuation coefficients and hence use of linear coefficient is not practical. Alternate method of specification used in such a case is **half-value layer (HVL). The HVL refers to the thickness of the absorber or material required to reduce the intensity of original beam by its one half.** It is a common term used to express the quality of X-ray beam. A beam with high HVL will be more penetrating than a beam with low HVL. For determining HVL, aluminium filter is used in usual diagnostic range but for high radiant energies upto 3 MeV, copper may be used and for still higher energies, lead may be used.

A homogenous beam will have same HVL as original beam after passing through several centimeters of material. However, in case of a heterogenous beam (like X-rays), as linear attenuation coefficient decreases after attenuation of the beam, the HVL becomes greater. Thus a heterogenous beam becomes more monochromatic after passing through the absorber.

Since both HVL and linear attenuation coefficient may be used for describing characteristics of radiation, a relationship worked out is explained by the equation:

$$\text{HVL} = \frac{0.693}{\text{Linear attenuation coefficient}}$$

Therefore, if linear attenuation coefficient remains constant, so will be the HVL.

Mass attenuation coefficient: This attenuation coefficient quantitates the attenuation of materials independent of their physical state. Mass attenuation coefficient is equal to linear attenuation coefficient divided by density of the absorber. The linear attenuation coefficient of water, ice and water vapour, the three physical states of water, is 0.214, ,0.196 and 0.000128, respectively. The density of water, ice and water vapour is 1.0, 0.917 and 0.000598, respectively. When linear attenuation coefficient of each of these is divided by the density, mass attenuation coefficient in all the three comes out to be 0.214. This is of little practical importance in diagnostic radiology. Mass takes into consideration grams per square centimeter and thus a gram of water and a gram of vapour.

absorbs the same amount of radiation. However, in practice, patient's thickness for radiographic purposes is measured in centimeters and not in grams, and one centimeter of water will absorb much more X-rays than will do one centimeter of vapour.

Factors affecting attenuation of X-ray beam

The extent of attenuation depends on: energy of radiation or intensity of the incident beam, atomic number of the absorber, thickness of the absorber and density of the absorber.

Energy of radiation (intensity of the incident beam): In general, higher is the energy of the X-ray beam, more are the X-rays that transmit without interaction (i.e. lesser attenuation) except in case of absorbers with high atomic number. When the energy of incident photons increases, the photoelectric absorption decreases and probability of compton scatter or percentage of transmitted photons increases. At low energy radiation, most effects are photoelectric, regardless of the atomic number of the absorber and thus few photons are transmitted. **To image small differences in the soft tissues, one must use low kVp technique to get maximum differential absorption. High kVp techniques can be used for examination of the bone resulting in much less patient exposure.**

In case of high atomic number absorbers, X-ray transmission may actually decrease with increasing beam energy as there is abrupt change in the likelihood of photoelectric interaction as radiation energy reaches binding energy of the inner shell elelctrons. X-ray photons can not affect an electron unless these have more energy than electron binding energy. Thus a low energy photon has more chances of being transmitted.

Atomic number of absorber: Each element has a specific atomic number which refers to the number of positive charges in the nucleus of the atom. The difference in the atomic number of elements is of great importance in diagnostic radiology. At lower kV range, the probability of photoelectric absorption is proportionate to the third power of the atomic number of the absorber. That is why more X-rays are absorbed photoelectrically in the bone (atomic No. 13.8) than in soft tissues which have atomic No. 6 (Fig.2-8). In other words, the probability of photoelectric absorption is seven times greater in the bone than in the soft tissues. High atomic number of iodine (53) and barium (56) makes them good contrast materials and similarly lead (82) makes a good material to be used as a shield against radiation protection. At high kilovoltage range, the absorption capacity of elements is more directly related to the atomic number. Compton scattering of X-rays is independent of the atomic number of the absorbing material.

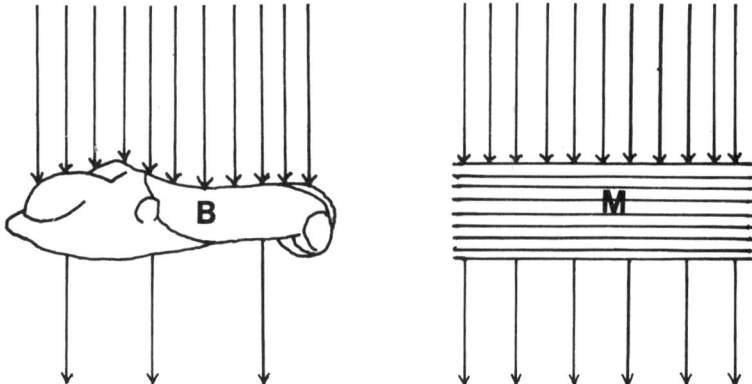

Fig. 2-8 : Higher atomic number of bone (B) is responsible for more absorption of X-rays than in soft tissue (M)

Thickness of the absorber: It has previously been explained in this chapter that for every centimeter of the thickness of the absorber, a constant percentage of photons are absorbed. Therefore, if an absorber is very thick, a very high percentage of X-ray photons will be absorbed. This is important in large animal radiography when very thick soft tissue areas are involved.

Density of the absorber: Tissue or absorber density is another important factor that affects attenuation. Density refers to the quantity of matter per unit volume (g/cm³) and denotes how tightly the atoms of a substance are packed. Differential absorption of X-rays in different tissues of the body is due to the density difference and this makes possible an X-ray image for viewing. Attenuation is directly proportional to the density of the absorber. Tissue density also influences the frequency of both compton scatter and photoelectric absorption. In general, density is directly related to the atomic number of elements (Table 2-3). However, certain exceptions exist, e.g. although lead has higher atomic number (82) than gold (79) yet its density (11.0) is lesser than that of gold (19.3). Due to high density of the bone, the quantity of X-rays absorbed and scattered is double than that done by soft tissues. Density of soft tissue is 773 times that of air for similar thickness and thus most X-rays are transmitted through air.

Table 2-3 : Effective atomic number and density of materials of interest in diagnostic radiology.

Material	Effective atomic number	Density
Body tissue		
Muscle	7.4	1.0
Fat	6.3	0.91
Bone	13.8	1.85
Lung	4.0	0.32
Contrast material		
Barium	56	3.5
Iodine	53	4.93
Air	7.6	0.001293
Others		
Lead	82	11.35
Concrete	17	2.35
Tungsten	74	19.3

(Adapted from Morgan J.P. and Silverman, S. 1984. Techniques of Veterinary Radiography. 4th edn., Iowa State University Press).

REFERENCES

Bushong, S.C. 1975. Radiologic Sciences for Technologists.The C.V Mosby Co. Saint Louis.

Curry, T.S., Dowdey, J.E. and Murry, R.C. 1984. Christensen's Introduction to the Physics of Diagnostic Radiology. 3rd edn., Lea and Febiger, Philadelphia.

Douglas, S.W. and Williamson, H.D. 1980. Principles of Veterinary Radiography. 3rd edn., Willilams and Wilkins Co., Baltimore.

Morgan, J.P. and Silverman, S. 1990. Techniques of Veterinary Radiography. 4th edn., Iowa State University Press, Ames.

Ridgway, A. and Thumm, W. 1968. The Physics of Medical Radiography. Addison-Wesley Publishing Co., London.

Selman, J. 1977. The Fundamentals of X-ray and Radium Physics. 6th edn., Charles, C.Thomas - Publisher, Illinois.

SELECTED QUESTIONS

1. What do you understand by the following ?
   i) Ionising radiation ii) Sine-wave iii) Photoelectric effect v) Compton effect v) Attenuation of X-ray beam vi) Half value layer vii) Radiant energy viii) Photon ix) Electron volt x) Photoelectron.

2. What are the implications of Compton effect in diagnostic radiology?
3. List the physical properties of X-rays.
4. Name the factors which affect attenuation of X-ray beam.
5. How atomic number of the absorber affects attenuation of the X-ray beam?
6. Name various types of ionising radiations with examples?
7. What is the difference in the origin of X-rays from the atom in comparison to alpha, beta and gamma rays?
8. Why lead is used as a shield against radiation protection?
9. What is the advantage of using high kVp exposures?
10. Why soft tissue imaging with X-rays is better using low kVp techniques?

```
-RADIATION-
HANDLE ME CAREFULLY
I CAN BE DANGEROUS
```

```
-BAD RADIOGRAPHIC IMAGE-
NOT GOOD FOR YOUR IMAGE
```

3 DIAGNOSTIC X-RAY MACHINE

HARPAL SINGH
KULDIP SINGH
T.K. GAHLOT

DIAGNOSTIC X-RAY TUBES

Primary function of an X-ray tube is to convert electrical energy into X-rays. An X-ray tube is the largest thermionic diode type electronic vacuum tube that consists essentially of glass insert or tube containing a cathode and a stationary or rotating anode placed 1-3 cm apart. The tube itself is fitted into an oil filled casing. Essential components of the tube include a tungsten filament cathode, a tungsten target anode, evacuated glass envelop and two circuits to heat the filament and to drive electrons to the anode. Stationary and rotating anode tube encasings are shown in Figs. 3-1 and 3-2.

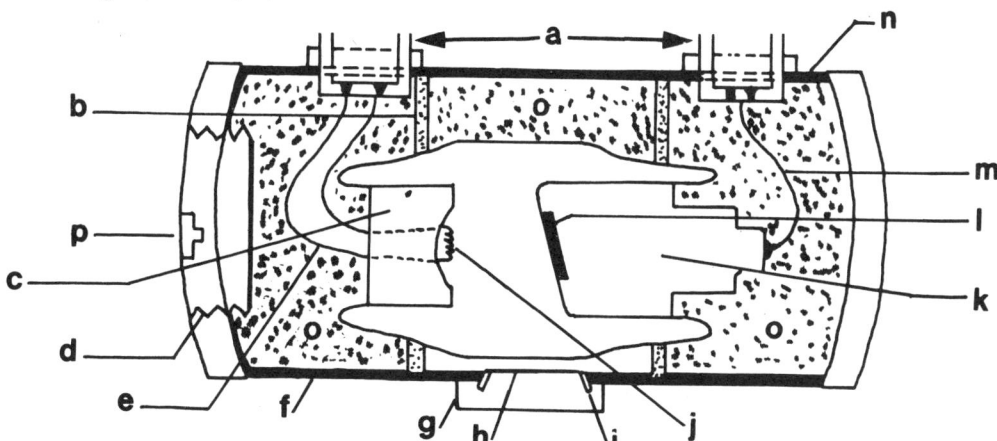

Fig. 3-1 : Stationary anode X-ray tube and casing. a = High tension cables, b = Support for insert, c = cathode block, d = Oil expansion bellows, e = High tension and filament connections, f = Lead lining, g = Attachment for cones etc, h = Radiolucent window, i = Fixed collimator, j = Filament, k = Copper anode, l = Target, m = High tension connection, n = Steel casing, o = oil, p = Thermal cut out switch (Figs. 3-1 and 3-2 are from Forster, E. 1985. Equipmet for Diagnostic Radiography. MTP Press Ltd, Kluwer Academic Publishers, Netherland. Used with permission).

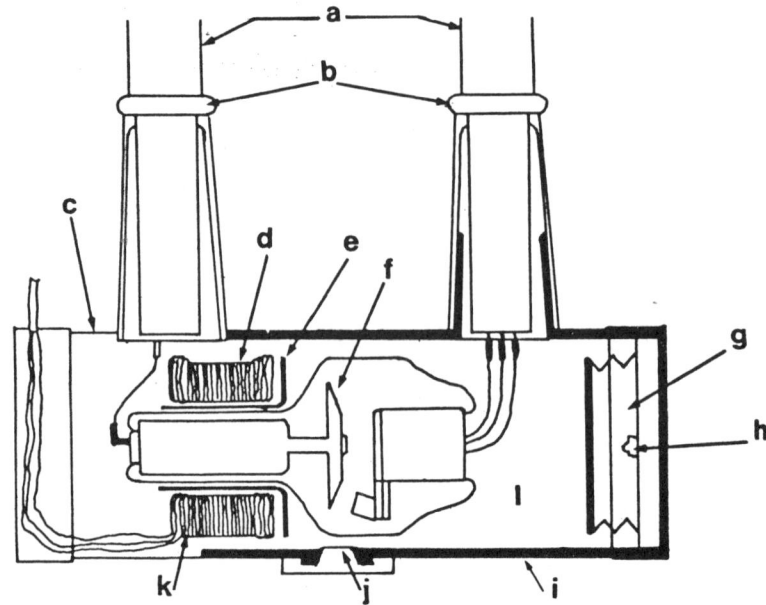

Fig. 3-2 : X-ray tube insert and casing with a rotating anode. a = Shock proof cables. b = Collar and screw caps, c = Lead lined steel casing, d = Iron core, e = Heat shield, f = Anode, g = Oil expansion bellows, h = Thermal cut out switch, i = Extra thickness of lead, j = Radiolucent window, k = Stator winding, l = Oil.

Glass envelop and tube housing

Anode and cathode of an X-ray tube are enclosed in a vacuum glass tube or envelop. **Vacuum allows unobstructed path for the electron stream (tube current) and prevents oxidation and burning out of the filament,** thus ensuring efficient X-ray production and more tube life due to low heat production. A radiographic tube is generally cylindrical in shape measuring 12-18 cm in length and 9 cm in diameter. The envelop is usually made of pyrex or borosilicate glass to enable it withstand extremely high temperatures. X-ray tube has a thin glass window which allows maximum emission of X-rays with minimal absorption in the glass envelop.

X-ray tube is housed in a metal housing which is lined with lead except at the window. The protective metal housing contains sealed oil to serve as an electrical insulator for high voltage cables that feed into the tube, and also helps in heat dissipation. Some protective housings have a fan to cool the tube. The tube housing also provides mechanical support to the X-ray tube, protects the tube from any external damage and reduces the level of leakage radiation while useful beam passes through the window (Fig 3-3).

Cathode

The negative side of the X-ray tube is called cathode. The cathode assembly consists of a filament and its supporting wires, and a focusing cup (Fig 3-4). **Terms cathode and filament are often used interchangeably.** The cathode serves as a source of electrons for the X-ray tube and directs their flow towards the anode. The X-ray tube current measured in **milliamperage (mA) refers to the number of electrons flowing per second from the filament to the target.**

i) **Filament:** Filament, a spiral coil, is usually made of tungsten wire. Tungsten is preferred because of its high melting point (3370°C), which allows the filament to withstand higher tube current. Tungsten has little tendency to vaporise and, therefore, prolongs the life of the X-ray tube.

Moreover, tungsten provides higher electron emission than other metals because of its high atomic number (74). However, the composition of most filaments in modern machines has been changed to a tungsten-rhenium alloy to enhance thermionic emission efficiency, and to prolong the tube life. The filament is supported by two stout wires which connect it to a proper electrical source. A low voltage filament current (10 volts, 3-5 amp) is sent through one wire to heat the filament. One of these wires is also connected to high voltage source which provides the high negative potential needed to drive electrons towards the anode. At low filament current, no tube current flows because sufficient heating of the filament for thermionic emission is not achieved. An increase in the filament current increases the temperature of the filament and rate of electron emission thereby causing a large increase in the tube current (mA). This relationship between filament current and tube current depends upon the tube voltage. The X-ray tube current is adjusted by controlling the filament current.

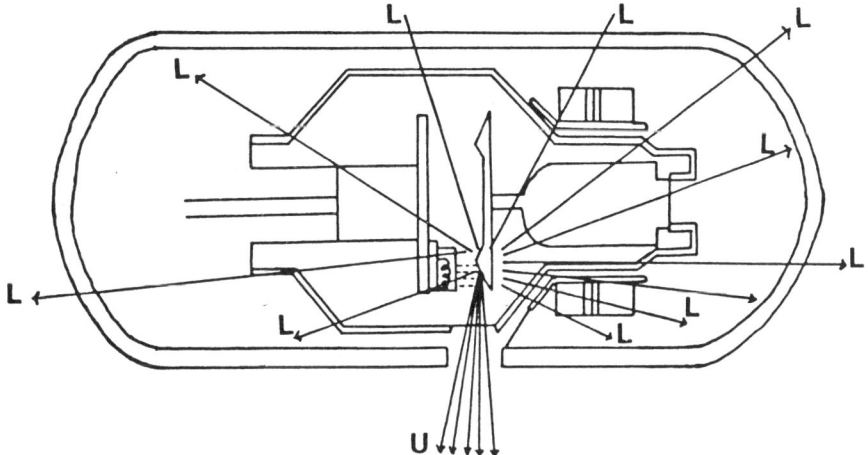

Fig. 3-3 : X-ray tube housing reduces the level of leakage radiation. L = Leakage radiation, U = Useful beam.

Most modern diagnostic X-ray tubes are provided with two filaments mounted side by side (Fig 3-5). The filaments usually differ in size, producing two focal spots of different sizes on the target. These X-ray tubes with two focal spots are called **dual focus** tubes. The selection of one or the other is generally made with the mA station selector on the control panel. The two filaments are supplied by three stout wires one of which is connected in common with both filaments.

ii) Focusing Cup: The filament is embedded in a concave metal shroud called focusing cup. It is usually made of nickle or molybdenum. Since the electrons accelerated from the cathode to the anode are negatively charged and tend to spread out, a negative electrical potential is maintained in the focusing cup so as to restrict the electrons to a small beam towards the anode target called **focal spot** or tube focus. (Fig 3-6). The effectiveness of the focusing cup is determined by its size and shape, its charge, the filament size and shape, and the position of the filament within the focusing cup.

Anode

Anode is the positive side of the X-ray tube. It is of two types; stationary and rotating. A stationary anode is used in dental X-ray machines, portable X-ray units, and special purpose units, where high tube current and power are not required. Tubes with rotating anodes are used in X-ray units of larger capacity capable of producing high intensity X ray beam in a short time. The anode serves two main functions in an X-ray tube : a) provides mechanical support for the target and b) acts as a good thermal conductor for heat dissipation.

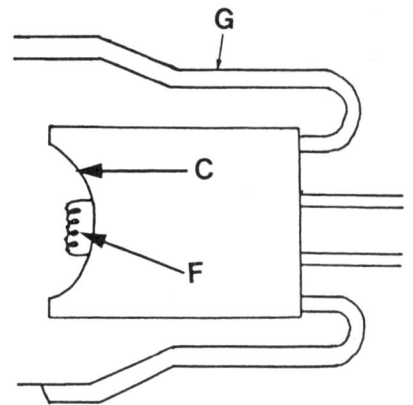

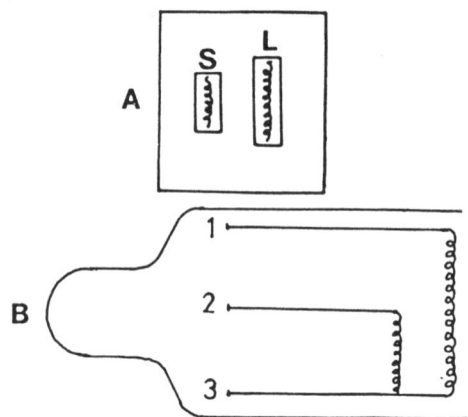

Fig. 3-4 : Focusing cup. F = Filament, C = Focusing cup, G = Glass envelop of X-ray tube.

Fig. 3-5 : A = Two filaments, small (S) and large (L) mounted together. B = Wire connection of two filaments, 1 and 2 low voltae conductor, 3 is high voltage conductor which is common to both filaments.

i) Stationary Anode: A stationary anode consists of a block of copper in which is embedded a small square or rectangular plate of tungsten - rhenium alloy metal, 2-3 mm thick, called target (Fig.3-7). Alloying of tungsten allows added mechanical strength, improved ability to dissipate heat and reduced roughening or crazing of the target surface. The life of the tube is thus prolonged.

Useful X-rays are actually produced from the target area of the anode. **When electrons strike the target almost 99% of the kinetic energy gets converted to heat and only about 1% is converted to X-rays.** The tungsten is selected as the target material due to following reasons:

a) It has high melting point (3370°C) and can thus withstand high temperature produced during exposure. b) It has high atomic number (74) and, therefore, has high X-ray production efficiency. c) It has reasonably good thermal conductivity and thus allows rapid dissipation of heat. d) It has low vapour pressure at high temperature which helps to preserve vacuum within the glass envelop.

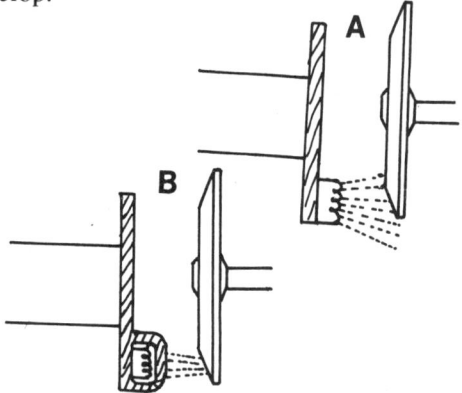

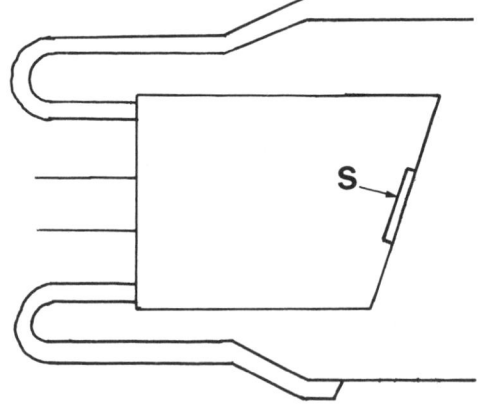

Fig. 3-6 : If focusing cup is not used (A) the electron beam spreads beyond the anode. Focusing cup restricts the beam to te anode (B).

Fig. 3-7 : Stationary anode (S).

Despite high melting point, tungsten target cannot withstand heat of repeated exposures and, therefore, is embedded in a copper block. Copper is a much better conductor of heat and so total thermal conductivity of the anode is enhanced. Heat loading capacity of an anode can further be increased by circulating air, water or oil around the anode to enable more exposures than would otherwise be possible

ii) Rotating anode: Since heat dissipation is a limiting factor in X-ray tubes with a stationary anode, rotating anodes (Fig 3-8) were introduced. Rotating anode X-ray tube provides several hundred times more larger target area for electron beam to interact and thus heat generated during an exposure is spread over a larger area of the anode. Increase in target area is achieved due to rotation of the anode. During rotation the anode constantly turns a new face to the electron beam so that heat does not concetrate at one point. **The use of rotating anode allows much higher exposure in a much shorter exposure time.** For instance, with a stationary anode, the upper limit for a 4 mm focus is 200 mA, whereas with rotating anode, the upper limit for a 2 mm focus is 500 mA, maximum exposure time being the same in both cases. Special tubes have been designed to tolerate current as large as 1000 mA for very short exposure time operating on three phase current. The usual speed of anode rotation varies between 3000 to 3600 revolutions per minute (rpm). However, anode of some high capacity X-ray tubes rotate at 8000 to 10,500 rpm.

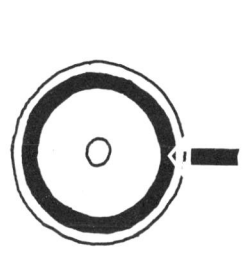

Fig. 3-8 : Rotating anode. Arrow indicates focal tract.

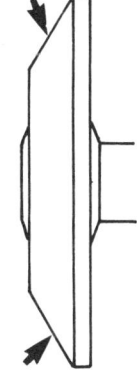

Fig. 3-9 : Rotating anode has a bevelled edge (arrows).

Rotating anode is made of a molybdenum disc coated with a strip of tungsten - rhenium alloy, and molybdenum stem. The tungsten strip has a bevelled edge (Fig 3-9). The degree of bevelling is called the target angle or anode angle. The anode angle determines not only the effective focal spot size but also the area covered by the beam at a given focal film distance. Typical disc diameter ranges from 6.5-10 cm, although tubes of high capacity may have an anode of 12.5 cm diameter.

Rotating anode is driven by an electromagnetic induction motor, which consists of two main parts, rotor and stator, separated from each other by the glass envelop. The part that rotates inside the glass envelop is called rotor (Fig 3-10). The non-rotating part outside the glass envelop is called stator which consists of a series of electromagnets equally spaced around the neck of the X-ray tube. Current flowing in the stator produces a magnetic field which induces a current in the rotor making it to rotate. Rotation of the anode at high speed requires excellent bearings with protection from heat generated during an exposure. Molybdenum stem meets this requirement because it has a high melting point and is a poor heat conductor. The molybdenum stem should be as short as possible because increasing the length of the stem increases the load on the bearing surfaces. Self lubricating bearings coated with metallic barium or silver reduce friction.

iii) Focal spot and target angles: The focal spot on the target surface of the anode is that area which is bombarded by the electrons from the cathode during an exposure. The size and shape of the focal spot is determined by the size of the filament, shape and size of the focusing cup and the positioning of the filament in the focusing cup. The size of the focal spot may be small, medium or large. Smaller the focal spot, sharper is the radiographic definition. However, as the focal spot size decreases, heating of the target concentrates at smaller area. So it is necessary to have large focal area for spread of heat over a larger area. These conflicting requirements are met to some extent by using **line focus principle.** As has already been stated, rotating anodes have a bevelled edge and the degree of this bevelling is called anode angle or target angle which may be 10°, 12°, 17°, or 20°. By utilising the angle of the target surface, the effective area of the target is made much smaller than the actual area of electron interaction (fig. 3-11). This principle influences the projected or effective focal spot size and is referred to as line focus principle. The effective focal spot size is the area projected on the patient and film and is the area viewed perpendicularly to the surface of the target. The actual focal spot is rectangular in shape while effective focal spot is almost square in shape. The size of the focal spot in X-ray machines vary from 0.3 to 3.0 mm. Lower the target angle, smaller is the effective focal spot size (Fig. 3-12). However, practically there is a limit to which target angle can be decreased due to heel effect discussed further in this chapter. For general diagnostic radiography, target angle is usually not less than 15°. The line focus principle provides the sharpness of image of a small focal spot and the heat accommodation of a large focal spot.

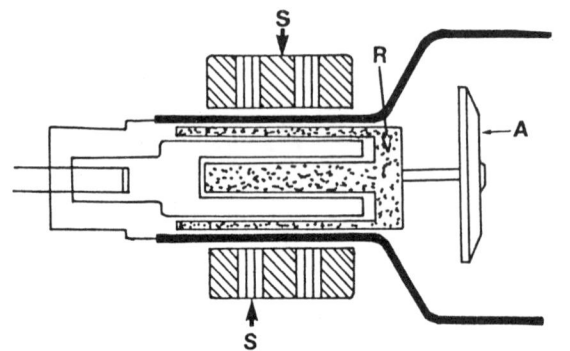

Fig. 3-10 : Main parts of electromagnetic induction motor of rotating anode. A = Anode, R = Rotor, S = Stator.

Fig. 3-11 : Line focus principle. Y = Actual area bombarded (actual focal spot), Z = Apparent focal area (effective focal spot).

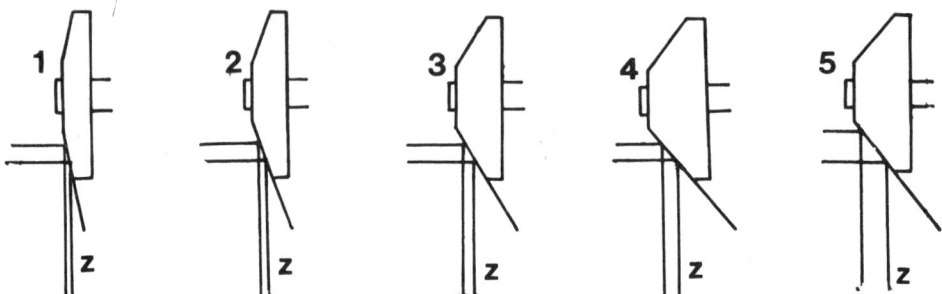

Fig. 3-12 : Effect of focal spot size and target angle on effective focal spot size. Target angle may vary from 1 to 4 and focal spot size may vary from 4 to 5. Z = Effective focal spot size, X = Focal spot size (Adapted from Morgan, J.P. and Silverman, S. 1990. Techniques of Veterinary Radiography. Iowa State University Press, Ames).

In stationary anode tubes, focal spot size must be larger (1.5, 2.0, 3.2, 4.2 and 4.5 cm) so as to accommodate the heat produced. Focal spot size always refers to the effective focal spot size. Rotating anodes generally have small effective focal spot size than that of stationary anodes and usually incorporate two filaments with two different focal spot sizes. Rotating anode tubes have many combinations of focal spot sizes i.e. 0.3 and 1.0; 0.3 and 1.2; 0.3 and 1.5; 0.3 and 2.0; 0.6 and 1.6; 0.6 and 2.0; 0.8 and 1.8; 1.0 and 2.0; 1.2 and 2.0; 1.5 and 1.5; 2.0 and 2.0 mm. However, most diagnostic rotating anode tubes have 1 and 2 mm focal spot sizes. A smaller focal spot is used for lower heat producing exposures to obtain sharper radiographic details. A larger focal spot is used in higher heat producing exposures required to penetrate a thicker part or to make repeated exposures in a shorter time interval.

iv) Heel effect: Radiation intensity on the cathode side of the X-ray beam is higher than on the anode side (Fig.3-13). This **difference in the intensity across the X-ray beam, which may be as high as 40%, is called heel effect.** Heel effect is a disadvantage of line focus principle. X-rays are emitted from the tungsten target in all directions and those emerging parallel to the angled target (from the heel of the target) are attenuated more by the target itself (Fig. 3-14). Thus intensity of the X-ray beam emerging from the anode side is low. Heel effect on a radiograph appears as variations in the density, film being more blacker on the cathode side because of relatively higher penetration by high intensity of X-ray photons on that side. **Therefore, while taking a radiograph of a part of unequal thickness, thicker or denser side should be positioned towards the cathode and thinner towards the anode.** Heel effect is more noticeable when film is of a larger size or FFD is less (Fig. 3-15). Heel effect also varies according to the angle of the anode, smaller the angle, more pronounced is the heel effect.

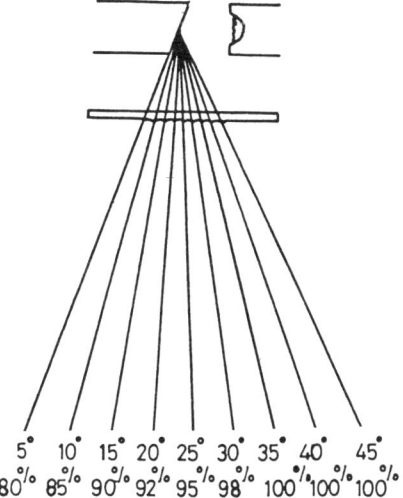

Fig. 3-13 : Distribution of X-ray intensity according to an angle measured from the plane of the target on the anode face. Per cent X-ray intensity on the cathode side is higher than on the anode side. This phenomenon of unequal beam intensity is called "heel effect" (Figs. 3-13 to 3-15 are from Morgan, J.P. and Silverman, S. 1990. Techniques of Veterinary Radiography. Iowa State University Press, Ames. Used with permission).

Causes of X-ray tube failure

The malfunctioning of an X-ray tube can be divided into two types i.e. early and late tube failure. Common causes of early tube failure are thermal overloading (overheating of the tube), damage to filament and glass envelop, and electrical and insulation failure. Among reasons for late failure are progressive pitting or crazing or roughening of the anode surface, anode cracking, anode failure, thinning of filament, tube housing problems, oil leaks, stator problems, burned out or frozen bearings or their malformation etc. Out of all these causes, filament evaporation and overheating of the X-ray tube are more common. By exercising control, one can significantly extend the useful life of the X-ray tube.

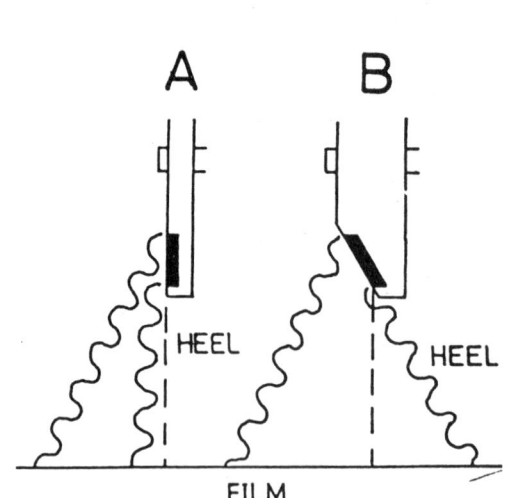

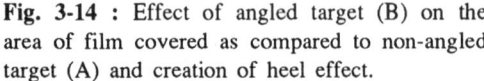

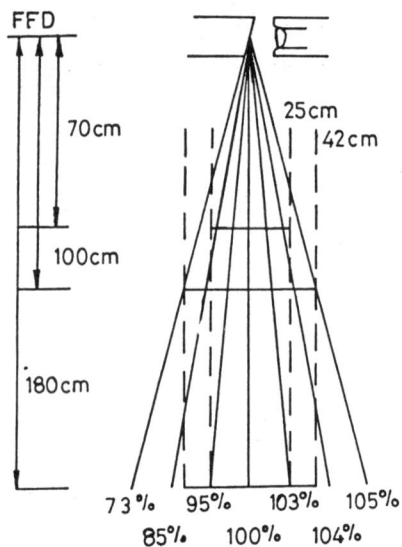

Fig. 3-14 : Effect of angled target (B) on the area of film covered as compared to non-angled target (A) and creation of heel effect.

Fig. 3-15 : Heel effect is more noticeable when FFD is less or film is of larger size.

Filament (cathode) failure

Prolonged heating of the filament by normal current or due to repeated exposures causes evaporation of filament metal causing progressive thinning of the tungsten filament. Even 10% thinning will cause the filament to break and thus to end tube life. Furthermore, as the filament evaporates, some of the vaporised tungsten gets deposited on the internal wall of the glass tube. When deposition occurs on the window of the tube, inherent filtration increases and so X-ray output decreases. Tungsten deposits may impair insulation and tube may get punctured or cracked by electrical spark-over.

As the tube ages, a progressively lower filament current setting is required for a desired tube current (milliamperage) because thinner filament has higher electrical resistance (R) while original filament circuit is designed to maintain constant current (I). Thus I^2R increases which raises filament temperature with resultant increase in tube current (mA). In order to correct this, the filament current is adjusted downward just enough to obtain a proper electron emission for required tube current. **To extend the filament life, radiographer must ensure proper patient positioning and correctness of exposure factors before activating the exposure switch. Failure of needle movement on the mA meter during an expected exposure is a strong indication of filament failure.**

Anode failure

If tungsten target on the anode is overheated due to any reason, it may melt resulting in roughening of its surface. This will result in abnormal absorption of X-rays by the metal (Fig.3-16) and production of a beam of non-uniform intensity at each successive exposure, especially with a rotating anode disc. Excessive heating of the anode may be the result of a single exposure that exceeds permissible settings or may be due to repeated exposures in a short time. Damage to rotating anode may include its pitting or melting of anode stem. Vaporised tungsten coats the glass envelop and these deposits act as secondary anode targets to attract electrons from the filament.

The metal deposits also act as an additional filter layer which reduces the X-ray output of the tube. This effect of internal metal deposits can be minimised by metal lining in the X-ray tube.

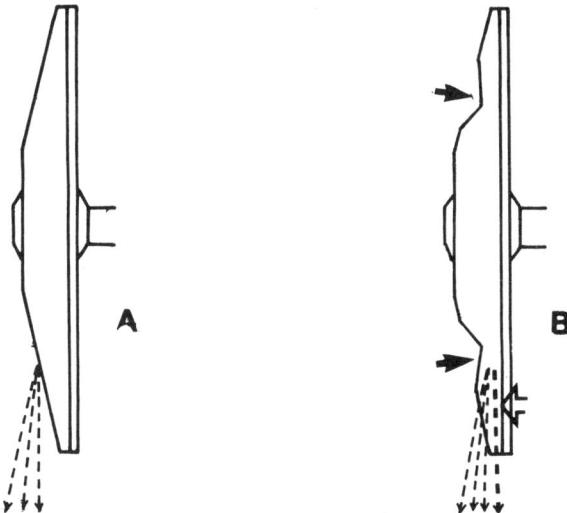

Fig. 3-16 : Roughening (closed arrows) of the anode results in abnormal absorption of X-rays by metal (open arrow) in B as compared to a normal anode in A.

Damage to anode bearings causes imbalance and vibrations of the anode with the possibility of anode stem fracture. **Increase in noise intensity made by the rotating anode indicates cracking or bending of the anode assembly.** Bearings may expand beyond acceptable limits and may either decrease rotation or may completely seize rotation. Slow anode rotation due to damaged bearings also leads to overheating and damage of the target on anode disc surface which finally damages the target. Failure of anode bearings usually occurs due to frequent thermal overloading of the tube and housing. Overheating of anode occurs when the switch is unnecessarily held in the pre-exposure position. **Exposures should be made at high kVp and low mAs settings so that heat production is less. Maximum radiographic exposure factors should never be applied to a cold anode.** The anode should first be warmed by pre-exposure rotation. Radiographic tube rating chart and anode cooling chart provided by the manufacturer for each type of X-ray tube should be consulted because such a chart indicates maximum "safe" exposure time for any selected combination of kV and mA for a single exposure with a relatively cool tube. One must allow enough time to elapse for cooling of the anode.

Glass envelop and tube housing failure

The glass envelop may crack due to secondary arcing from the filament to the metal deposits on the glass wall as a result of tungsten evaporation. This arcing is noticed when the tube is used at higher kVp ranges. As discussed earlier, a metal lining can prevent this arcing. Oil leaks due to a broken seal may also cause tube failure. Breakage of glass housing and breakdown of oil circulation during exposure are other factors which contribute to housing failure.

Steps to extend tube life

Following precautions help to extend tube life:
  i) Anode should be warmed up before actual exposure is made. Use warm up time suggested by the manufacturer. Prewarming of the anode using two stage switch prevents cracking of the anode.

ii) Do not switch on mA or kVp settings while rotor is engaged because this causes torque force on the anode and bearing assembly and increased filament evaporation.

iii) Use high kVp and low mA settings to minimise overheating of the anode.

iv) Consult correct tube rating chart to avoid overheating of the anode.

v) Do not run anode unnecessarily because this shortens the life of bearings.

vi) Adequate cooling of the tube housing must be ensured to avoid excessive heating of oil in the tube housing.

vii) Do not allow overheating of the filament by repeated exposures in a short time.

SPECIAL PURPOSE X-RAY TUBES

Grid controlled X-ray tube

In addition to the electrodes cathode and anode, a grid controlled X-ray tube contains a third electrode called control grid. In such machines, focusing cup surrounding the filament acts as the third electrode. The focusing cup can be made electrically negative relative to the filament. The electrons flow through the tube from cathode to anode as long as the grid does not carry electrical potential. However, if the grid is made negative, there is no flow of electrons across the tube. In this manner, the grid serves as a switch to control the production of X-rays. This type of tube is of value when short repeated X-ray exposures are necessary in certain situations e.g., during cine studies after cardiac catheterisation.

Microfocus tubes

Microfocus tube is designed to utilise a microfocal spot in magnification radiography. The microfocus tube utilises an electron lens to focus the cathode beam to strike a small diameter spot on the anode. The focal spot is thus kept roughly one fourth the size of the conventional tube. This allows fine image detail and eliminates film fogging from off focus radiation. The tube has stationary anode and cooling is accomplished by a closed loop liquid cooling system. The power requirement of such a tube is very low i.e. 115 volts. The microfocus tube has low tube rating which limits mA to a maximum of 2 to 5 and the kVp to a range of 30 to 80.

X-ray therapy tubes

X-ray therapy tubes operate at a low tube current (5 to 20 mA) but at high kVp settings and for relatively long period of time. These tubes are operated to provide X-rays of four main kVp ranges: i) low voltage (50 to 120 kV) for treatment of skin lesions, ii) intermediate voltage (130 to 150 kV) for treatment of lesions located a few centimeter beneath the skin, iii) orthovoltage (160 to 300 kV) for treatment of deep lying lesions, and iv) mega/supervoltage (300 to 1000 kV) also for deep lying lesions.

The therapy tube has a stationary anode, single large filament and a much larger focal spot size (5 to 8 mm). This permits use of a larger tungsten target and a massive copper anode as compared to a diagnostic X-ray tube. The target angle is greater (30°) in the therapy tube than in the diagnostic tube. Therefore, X-rays expose a rather larger field. The therapy tube is used at much closer focal-object distance.

The effective heat dissipation is brought about by use of much heavier cooling equipment and also by more volume of oil around the therapy tubes as compared to the diagnostic tube. In addition, a flow of water through a coil within the shield is necessary to provide adequate cooling for the tube. Apart from this, a heat exchange unit is often placed in an adjacent room to remove the heat from the circulating water.

GENERATION OF X-RAYS

X-ray production

The evacuated X-ray tube or coolidge tube or hot filament X-ray tube is a device for producing free electrons from tungsten filament (cathode), the process called **thermionic emission,** by heating the filament (2300-2500°C). **Mostly outer orbital electrons are separated to form a cloud nearby, called space charge or thermions, in vacuum around filament.** The electrons remain in a constant agitated motion and their number increases as the temperature of the filament is increased. Under the influence of high potential difference (kVp) applied across the X-ray tube, the filament or the cathode is given a very high negative charge (negative electrical potential) and the target or anode is given an equally high positive charge or positive electrical potential. The resulting strong electrical field causes the cloud of electrons near the filament to rush up at high speed through the vacuum tube towards the anode. **This stream of electrons is called tube current or cathode rays.** The speed of these electrons approaches nearly one half the speed of light. **The electron stream in the tube is confined and concentrated on a small spot on the anode face called focus or focal spot** by a negatively charged molybdenum focusing cup that surrounds the filament. At 100 mA, approximately 6×10^{15} electrons per second travel from the cathode to the anode. If a machine is operated at 70 kVp, each electron arrives at the target with a maximum kinetic energy of 70 kev or 1.2×10^{-7} ergs. On interaction of these projectile electrons or cathode rays with the target of X-ray tube operated in diagnostic kVp range (40-150 kVp), more than 99% of the kinetic energy of electrons is transformed into thermal energy as a result of collision with outer shell electrons of tungsten (target). This results into excitation of these electrons and dropping back to their normal energy state with emission of infrared radiations responsible for heat generation in the anode of the X-ray tube. Approximately less than 1% of the remaining kinetic energy is irradiated as X-rays. Tungsten target has the ability to generate energies in diagnostically useful range. The energy developed or rate of energy dissipation in the anode of the X-ray tube is given by $P = VI$, where P is the power (joules/sec or watt), V is the potential difference in volts across the circuit and I is the current (amperes) flowing through the circuit. High current flow (amperage) due to exposures that exceed capabilities of an X-ray tube is not permitted. Excess heating of the target results in melting and vaporisation of the tungsten filament with resultant decrease in the X-ray output.

Tube efficiency for X-ray production

It refers to the percentage of kinetic energy of projectile electrons that is converted to X-rays. The efficiency for X-ray production is directly proportional to the atomic number of the target, applied potential difference (kVp) and a large tube current as expressed in the equation: $Px = kZIV^2$ Where Px is the X-ray power output in watts (ampere × volts), Z is the atomic number of the target material, I is the RMS value (0.707 of peak value of the current or voltage in a sinusoidal A.C waveform, the average of these quantities being 0.636 of the peak value in a full wave rectification) of tube current in amperes, V is the RMS value of the tube voltage (kVp) and k is numerical coefficient (1.4×10^{-9}/volt) which is affected by the nature of the applied voltage (i.e. single phase or three phase power). Total energy output (joules) = Power output (watts) × time of operation (seconds). As the voltage is increased, the X-ray power output is increased as the square of voltage. The tube percentage efficiency is given by the equation:

$$\frac{\text{X-ray power output}}{\text{Cathode ray power (watt)}} \times 100$$

where cathode ray power (watts) equals VI (voltage and current being expressed in RMS volts and amperes, respectively. In other words:

$$\text{Percentage efficiency (E)} = \frac{kZIV^2}{IV} \times 100 = kZV \times 100$$

Where k = 1.4 x 10^{-7} Volt^{-1}. To cite an example, the percentage efficiency of a tungsten target in an X-ray tube operating at a setting of 100 kVp i.e. 0.707 × 10^5 volts is calculated as : EZ = (kZV)% = (1.4 × 10^{-7}/volts × 74 × 0.707 × 10^5 volts) % = 0.734%, where 74 is the atomic number of tungsten, V (voltage) = 0.707 Vp (RMS value of volts), the peak value of 100 kVp = 0.707 × 10^5 volts.

It means that the efficiency is less than 1% i.e. 0.73% of total kinetic energy of the electron stream appears as a X-rays and the remaining energy i.e. over 99% of the input energy is converted to heat at the anode. At 60 KVp only 0.5% of the kinetic energy of electrons is converted to X-rays.

Electron target interaction

An accelerated electron must undergo several interactions with the target atom before losing all its energy. The energy loss occurs through the processes of collisional interaction and radiative interaction.

i) Collisional interaction: It involves the outer electrons of the target atoms and produces heat. This interaction is of two types. In the first type, an incoming electron may excite an outer orbital electron of the target atom by passing near to its orbit and thus transferring energy. The outer electron thus moves away from the nucleus and its return to its original orbit occurs with release of some energy which is irradiated as heat. In second type, the incoming electron passes on sufficient energy to remove an outer orbital electron from a target atom i.e. causes ionisation. The ejected electron and incoming electron further undergo additional interactions with target atoms till they lose all their energies. The total energy loss is given off as heat.

ii) Radiative interaction: This interaction is responsible for generation of X-radiation. With this type of interaction between the electrons and target atoms, two types of radiation are given off: a) characteristic or line radiations and b) Bremstrahlung or braking or general or white radiations.

a) Characteristic or line radiation: It is produced when the projectile electron interacts with the electron of inner-shell (K-shell) of the target atom rather than with electron of the outer shell. As has already been discussed in chapter 2, this interaction results is ejection of the electron from the inner shell provided that energy of the projectile electron exceeds the binding energy of the ejected electron. Both projectile and ejected electrons leave the atom in which interaction takes place. When the K-shell electron is ejected, a temporary electron vacancy is produced in the K-shell into which an electron from an outer shell or from another atom falls. This shifting of electrons continues until a stable energy state is reached. Each transition of an electron from an outer shell to an inner shell results in the emission of an X-ray photon (Fig 3-17). The energy of this X-ray photon emitted is equal to the differences in the binding energies of the orbital electrons involved. For example, in the tungsten atom, the binding energy in K-shell is about 70 keV and in the L-shell is about 11 keV. Thus the characteristics X-rays emitted will have energy of (70-11 keV) 59 keV. The energy of X-ray photon will always be the same for tungsten regardless of the energy of the electron that ejected the electron of the K-shell. **Therefore, the X-ray photon energy is a characteristic of the K-shell of a tungsten atom or of any other element used as the target.** When the vacancy in the K-shell is filled up by the electrons of the L-shell, the vacancy created in the L-shell may be fillled up from M-shell resulting in another X-ray photon. However, the energy of the L-shell characteristic radiation is much less than that of the K-shell because the tungsten L-shell binding energy is 11 keV and that of M-shell isabout 2 keV. Thus L-shell characteristic X-ray photons have an energy of 9 keV. Similarly M, N and even O characteristic X-rays can be produced in a tungsten target. Though many characteristic X-rays can be produced but except for K-shell X-rays **most other characteristic radiations produced in other shells are of low energy to be useful in diagnostic range.**

b) General or Bremstrahlung radiation (white radiation): A projectile electron that completely avoids the orbital electrons on passing through a tungsten atom of the target of the X-ray tube may come sufficiently near the nucleus of the atom to come under its influence. Because the nucleus is positively charged and electron is negatively charged, the electron is attracted towards the nucleus. The electron thus gets slowed down and is deflected from its original course. In this process the electron may lose its kinetic energy. This loss in kinetic energy by the electron is emitted directly in the form of an X-ray photon (Fig.3-18). The X-ray produced by this process is called general radiation or Bremstrahlung. Brems or Bremstrahlung radiation is also called as Bremsen radiation resulting from the breaking of incoming electron by the nucleus (in German bremsen means brake). Usually the electron gives up only a part of its energy in the form of radiation each time it is braked so that a continuous spectrum of X-ray energy is produced. Occassionally the electron will colloid headon with the nucleus. In this thype of interaction, all the energy of the electron is converted into a single X-ray photon. It means that an electron can lose all, none or an intermediate level of its kinetic energy in Bremstrahlung interaction. **Brems radiation is heterogenous (polyenergetic) in nature because the amount of braking or deceleration varies among electrons.**

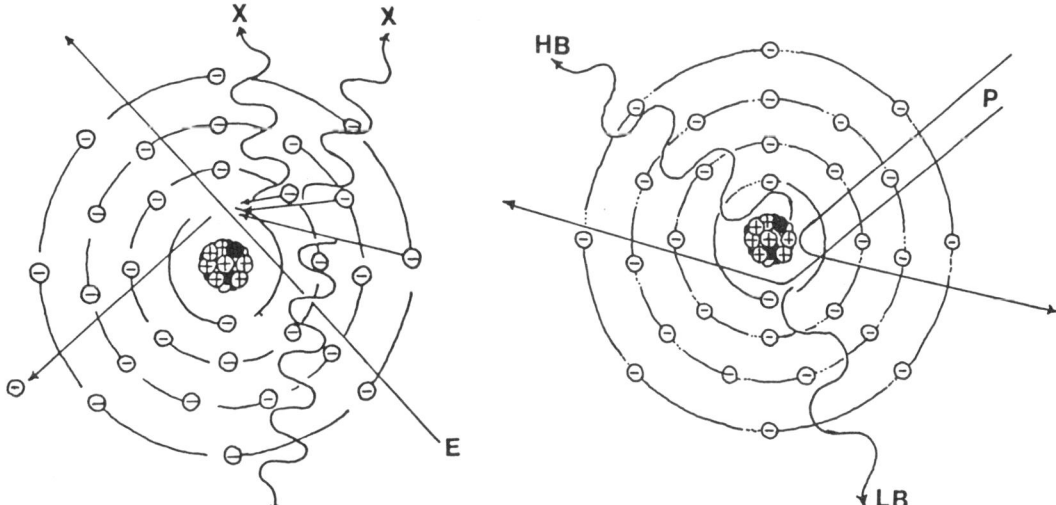

Fig. **3-17** : Production of characteristic X-rays. After ionisation of K-shell electron and its ejection, vacancy is filled up by outer shell electron. Shifting of electron produces these X-rays. E = Projectile electron, X = Characteristic X-rays (Figs 3-17 and 3-18 are from Bushong, S.C. 1975. Radiological Science for Technologists-physics, biology and protection. The C.V. Mosby Co., Saint Louis, Used with permission).

Fig. **3-18** : Bremstrahlung X-rays are produced from an interaction between a projectile electron and a target nucleus. Slowing down of the electron and its deflection produces X-rays. P = Projectile electron. LB = Low energy Brems radiation, HB = High energy Brems radiation.

X-ray emission (X-ray quantity and quality): An X-ray beam or X-ray spectrum may be described on the basis of its two main properties - its quantity and quality.

A. X-ray quantity: Quantity of an X-ray beam refers to the output intensity of the X-ray machine and is measured in roentgen (R) or milliroentgen (mR). In other words, X-ray quantity is defined as intensity of radiation flowing per second at a given point through a unit area of a surface perpendicular to the direction of the break. **The term radiation exposure is also used for X-ray quantity or X-ray intensity.** The units of measurement of X-ray radiation used in radiology are discussed in chapter 8. Following factors affect X-ray quantity:

i) Tube current (milliamperage or mA): Since tube current controls the number of electrons released from the cathode, a change in the X-ray tube current (mA) results in a proportional change in the amplitude of X-ray emission at all energy levels. Thus increasing the mA increases the X-ray output.

ii) Tube potential (kVp): Voltage applied between the cathode and anode controls the speed of the electrons towards the anode. X-ray quantity varies rapidly with change in kV. If kVp is doubled, the X-ray quantity would increase by a factor of four due to higher mean energy of its X-ray beam. Mathematically the change can be expressed as follows:

$$\frac{I_1}{I_2} = \frac{(kVp_1)^2}{(kVp_2)^2}$$

Where I_1 and I_2 are X-ray intensities at kVp_1 and kVp_2, respectively.

In order to maintain a fixed exposure of film or to maintain constant film density, an increase in kVp by 15% would require a reduction in mAs by one-half.

iii) Focal film distance (FFD): The X-ray quantity varies inversely with the square of the distance from the target. This relationship is known as **inverse square law.** As the distance from the target increases, the number of rays in each square centimeter of the beam decreases, and thus the X-ray quantity decreases. This has been discussed further in section C of chapter-4.

iv) Filtration: Aluminium filters positioned in the path of an X-ray beam absorb low energy X-rays which have no use in diagnostic radiology. Thicker the filter more is the effect on the quantity. However, main effect of filtration is on the quality of X-rays as will be discussed later.

B. Quality of X-rays: The quality of an X-ray beam refers to its ability to penetrate matter. As the effective energy of an X-ray beam increases, its penetrability also increases. **Highly penetrating X-rays are called high quality or hard X-rays while those with low penetrability are called low quality or soft X-rays.** Quality of X-rays is characterised by half-value layer (HVL) which is defined as that thickness of a specified absorbing material which reduces the X-ray intensity (quantity) to one half of its initial value. HVL is affected by Vp and filtration of the X-ray beam. HVL has already been discussed in Chapter-2.

Kilovoltage and filtration are the two main factors which influence the quality of X-rays:

i) Kilovoltage: An increase in kVp causes an increase in the effective energy of the beam thus making it more penetrable.

ii) Filtration: As has already been discussed, use of aluminium filters allow removal of low energy photons through photoelectric effect. As low energy photons are removed, the beam is left only with hard X-rays and thus quality of X-rays is changed. Filtration has been discussed in detail in chapter-4.

Heterogenous nature of X-rays

An X-ray beam is heterogenous in nature as it consists of photons that vary in their energy because of variations in wavelength. This polyenergetic or heterogenous nature is due to the following reasons:

i) It has already been discussed in this chapter that during X-ray production two types of radiations are produced i.e. characteristic and Brems and that brems are polyenergetic in nature. Thus a variety of energies in the primary beam makes it heterogenous.

ii) The energy of X-ray beam is dependent upon kilovoltage. Fluctuations in kilovoltage cause variations in the energy of photons.

iii) A number of encounters of the electrons with the target atom before being finally stopped produce X-ray photons of various energies.

X-RAY MACHINE CIRCUITS

A number of auxillary devices are necessary for the proper operation of an X-ray machine. These can be described under two main heads:

i) Sources of electricity

ii) Main X-ray circuits

Sources of electricity

Most X-ray units work from a power source of 220 volts, 60 hertz though some units can operate on 110 volts or 440 volts. An X-ray tube requires electrical energy for two purposes: to evict electrons from the filament and to accelerate these electrons from the cathode to the anode. There are two main lines of electrical supply to the X-ray tube:

i) One to produce high voltage (AC) which is applied across the tube's cathode and anode. This AC through rectification is converted to DC. AC is necessary for operation of X-ray transformer as only AC can be stepped up by the transformer. The timer and exposure switch control the high tension supply and thus the time of exposure. A voltage compensator alongwith its own voltmeter is used to compensate for changes in main voltage as fluctuating voltage affects the output of X-ray machine.

ii) Second line heats the X-ray tube filament and is provided by the voltage compensated autotransformer which supplies a stable voltage. The filament temperature controls the tube current (mA) and its control has an attached mA selector.

A simplified but complete X-ray circuit is shown in Fig. 3-19. The circuit of X-ray generator has two main aspects :

    i) Primary circuit (low voltage)

    ii) Secondary circuit (high voltage).

i) Primary or low tension circuit: This circuit is at low tension i.e. 240 or 415 volts and a part of the circuit is connected to primary coil of transformer. It includes main switch, fuses, circuit breakers, autotransformer, main voltage compensator, kilovoltage control, prereading of kV meter, primary winding of high tension transformer, timer circuit, filament heating circuit and various compensating circuits.

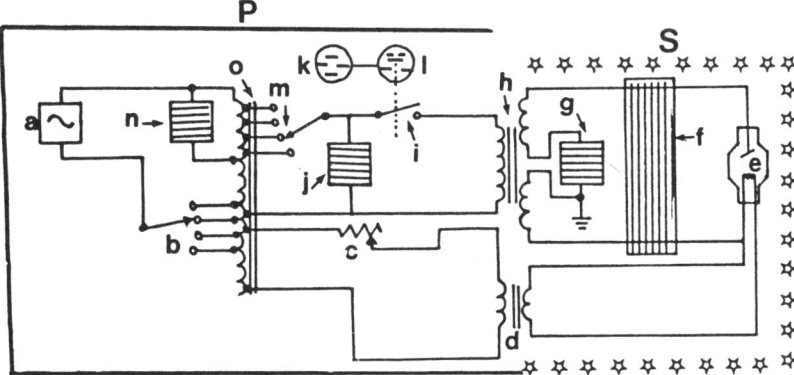

Fig. 3-19 : A simplified X-ray machine circuit. P = Primary circuit, S = Secondary circuit, a = Mains voltage, B = Line voltage compensator, C = mA selector, d = Filament transformer, e = X-ray tube, f = Rectifier, g - mA meter, h = High tension transformer, i = Exposure switch, j = kV meter, k = Hand control for exposure, l = Timer, m = kV selector, n = Voltage meter, O = Autotransformer (Adapted from Ridgway, A. and Thumm, W. 1968. The Physics of Medical Radiography. Addison-Wesley Publishing Co, Reading).

ii) Secondary circuit or high tension circuit: It consists of winding of high tension transformer, high tension rectifiers, X-ray tube and scondary winding of the filament heating transformer. Its part is connected to secondary coil of transformer.

The primary and secondary coil transformers are electrically insulated from each other. Some parts of the equipment are connected to both low and high tension sides.

There are two main components of an X-ray generator :

    i) Control panel

    ii) Transformer assembly

Control panel

It is a separate unit connected electrically to the X-ray machine (Fig 3-20). It contains meters and switches to select kVp, mA and exposure time. The control panel vary with the type of X-ray machine but most often following components, or some of them exist.

On-off switch: It is a main switch to turn the unit 'on'. The switch permits flow of current to the tube at 'on' position and prevents the same at 'off' position. For the safety of the X-ray tube and also to avoid an accidental exposure, the switch should remain in 'off' position when machine is not being used.

Voltmeter and voltage compensator control : Most X-ray machines are designed to operate on a 220 voltage power source. A voltmeter measures the voltage of electric current and voltage compensator allows adjustment of voltage. In most machines these days such a system is automatic.

Kilovoltage selector: It allows precise selection of desired kV. In some machines this control is automatically linked to a certain milliampere (mA) value. In such a case, a high kVp is available at a relatively low mA and vice versa.

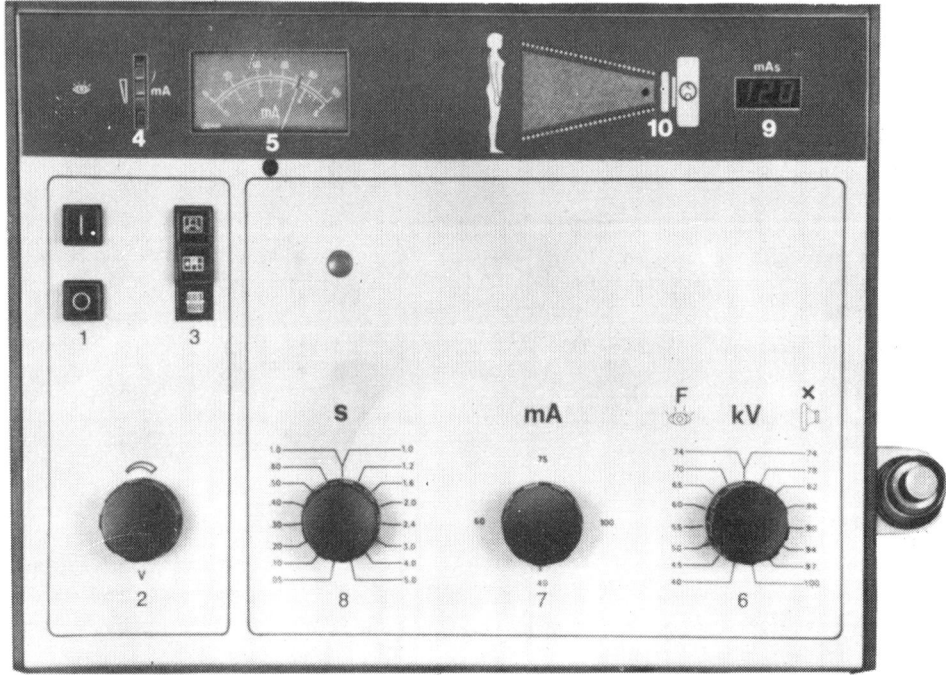

Fig. 3-20 : Control panel. 1 = Generator "ON-OFF" button, 2 = Power line voltage compensation, 3 = Selection of fluoroscopy/radiography/bucky radiography functions, 4 = Fluroscopic current setting, 5 = Tube current indication for radiography and fluoroscopy as well as indication of power line compensation in connection with No. 2, 6 = Setting of fluoroscopic and radiographic kV, 7 = mA selection, 8 = Exposure time selector, 9 = LED display of mAs, 10 = Pilot light indicator for fluoroscopy and radiography, 11 = Exposure release switch, F = Fluoroscopy symbol, X = X-ray symbol (Courtesy Siemens India, Pvt Ltd).

Milliammeter and milliamperage control: A milliammeter is placed within the high tension circuit. It indicates the current passing through the tube during an actual X-ray exposure. Its needle deflects during exposure to indicate flow of current across the tube and X-ray production. In modern units, a milliampere-second meter is added to indicate the product of mA and time i.e.

mAs. Milliamperage control is a push button or a knob to activate the proper resistor to obtain correct filament current for desired mA. It also simultaneously selects the focal spot size.

Timer and exposure button : In any given radiographic examination, the quantity of X-rays reaching the film is directly related to the X-ray tube current and the time for which the tube is energised i.e. the exposure time. The range of exposure time in available machines is large with minimum setting being as short as 0.001 second. An exposure device mostly consists of a two-stage exposure button of which first half depression rotates the anode and a complete depression, after a short pause, causes actual radiographic exposure. Various mechanical or electronic timing devices are available to 'make' and 'break' high voltage across the X-ray tube. Most machines these days have an electronic timer.

Timers can be of following types :

i) Mechanical timers: These are simplest of the timers and are used in low power portable and dental X-ray units. These function much like a spring driven clock.

ii) Synchronous timers: These are similar to mechanical timers except that movement is caused by a synchronous motor and recycling time is fairly long. These can not be used for rapid serial exposures.

iii) Electronic timers : These are most accurate and most widely used. These have resistor and capicitor circuits and allow selection of wide range of time intervals. The accuracy may be as short as one millisecond and can thus control almost 60 exposures per second.

iv) mAs timers: An mAs timer monitors the product of mA and time, exposure is terminated when desired mA is attained. Such timers provide maximum safe tube current for shortest possible exposure time for any mAs selected.

Fluoroscopy control: Most diagnostic X-ray machines are supplied with a special radiographic - fluoroscopic change-over switch so that machine can be used either for conventional radiography or for fluoroscopy. When switched to fluoroscopy, the timer is bypassed and equipment can be operated for any desired time by exercising control through a foot switch or exposure button. The part of control panel to be used for fluoroscopy is marked by an eye while part to be used for conventional radiography is marked by cone (Fig 3-20).

Transformer assembly

This assembly is enclosed in a metal box filled with a special type of oil which serves as an insulator. The assembly consists of low voltage transformer for filament circuits and rectifiers for high voltage circuit. A transformer is an electromagnetic device used for increasing or decreasing the voltage (potential difference) of incoming electrical energy to an appropriate level without appreciable loss of energy. It also transfers electrical energy from one circuit to another by mutual electromagnetic induction.

Transformer consists of two highly insulated coils of wire lying side by side and wound around the opposite sides of an iron ring or iron core. These coils are called primary and secondary coils. The primary coil is an input side and is supplied with AC and the secondary coil develops AC by mutual induction and is an output side. AC causes marked intensification of magnetic flux within the coil because of magnetisation of the core. There are two main types of transformers: open core and closed core. In open core, there is loss of magnetic flux at ends of cores. In closed core, commonly used in X-ray equipment, heavily insulated coils are wound around a square or circular and laminated core made of layers of silicon steel metal plates. Lamination of the core hinders formation of eddy currents in the core produced by electromagnetic induction during transformer operation.

Flow of current through the primary coil produces a magnetic field within the iron core and this field induces a current in the secondary coil. However, this induction of current in the secondary coil occurs only when magnetic field is changing and not if it is in a steady state. Due to this reason, direct steady current, like that from a battery, can not be used to induce current in the secondary coil. AC is used because it changes continuously in magnitude and periodically in polarity to produce a changing magnetic field. AC produced in the secondary coil induces electromotive force (EMF) in the priamry coil opposite in direction to the primary current. Since EMF is produced in each coil by changing magnetic flux in the other, it is called mutual induction.

Laws of transformer

There are two laws of transformer. First law states that the voltage in two circuits is proportional to the number of turns in two coils:

$$\frac{Np}{Ns} = \frac{Vp}{Vs}$$

Where Np = number of turns in the primary coil, Ns = Number of turns in the secondary coil, Vp = EMF or voltage in the primary coil, Vs = EMF or voltage in the secondary coil. It means that if number of turns in the secondary coil is twice to the number of turns in the primary coil then voltage in the secondary coil will be twice the voltage in the primary coil. **Such a transformer which increases the voltage in the secondary coil is called step-up transformer. Reverse is true for a step-down transformer.** EMF would be same if both the primary and secondary coils have equal turns.

Second law is about energy conservation i.e. transformer cannot creat energy or power (voltage × current). It means power output cannot be greater than power input (power here indicates kV × amp or kilowatt i.e. energy per unit time). An increase in voltage must be accompanied by a decrease in current, the product of voltage and current in two circuits thus must remain equal : Vp Ip ; = VsIs. Where Vp = Voltage in primary coil, Ip = Current in primary coil, Vs = Voltage in secondary coil, Is = Current in secondary coil. In other words, **if voltage is increased as in a step-up transformer, the amperage is decreased. Reverse is true for a step-down transformer.**

In an X-ray machine, step-up transformer receives 120 or 240 volts and multiplies this voltage to 30 to 150 kV necessary to drive the electrons with sufficient speed, and at the same time it decreases the current to mAmp (1/1000 of amp)

Autotransformer

It is a special transformer which controls the voltage supplied to the primary coil of transformer thereby providing voltage of varying magnitude to several different circuits of the X-ray tube, mainly to the filament circuit and high tension circuit. It is used between the source of AC and primary side of the high voltage transformer (step-up) and is actually a kVp selector located in the control panel. It has a single core of insulated wire wound around a large iron core. The single winding has a number of connections or electrical taps located along its length. A single coil serves as both the primary and secondary coils, the number of turns being adjustable. Fig 3-21 illustrates the set up. Two of the primary taps A, A' conduct input power to the transformer. Some of the secondary taps, such as C, are located closer to one end of the winding then the primary taps. This allows the autotransformer to increase or decrease the voltage. At regular intervals along the core, insulation is interrupted and bare points connected or tapped off to metal buttons. A movable contractor varies the number of turns included in the secondary circuit thereby varying its output voltage.

The law of autotransformer is:

$$\frac{Vs}{Vp} = \frac{Ts}{Tp}$$

Where Vs = Secondary voltage of autotransformer, Vp = Primary voltage of autotransformer, Ts = Number of windings enclosed by secondary taps, Tp = Number of windings enclosed by primary taps. Autotransformer can be used only when there is relatively small difference between its input and output voltage. A step-up transformer alters voltage pre-selected by autotransformer from volt to kV as per equation:

$$\frac{Vp}{Vs} = \frac{Np}{Ns}$$

Where Vp = Voltage on primary side, Vs = Voltage on secondary side, Np = Number of turns on primary side Ns = Number of turns on secondary side.

Line voltage compensator (mains)

A constant value of AC voltage from the mains is must to get a desired X-ray output from an X-ray machine. Therefore, line voltage compensator is used to compensate for voltage fluctuations and thus to maintain output voltage. The voltmeter associated with compensator indicates voltage across a fixed number of turns.

kV meter compensator

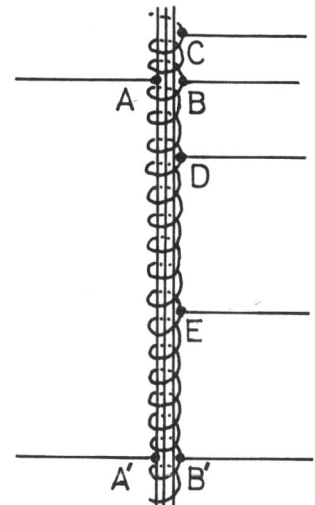

Fig. 3-21 : A simplified form of autotransformer. See text for details.

It is used for voltage compensation of the meter before exposure. At the time of exposure the voltage output of autotransformer drops to a lower value, and this voltage drop across a source depends on the current drawn from the source i.e. on the tube current or mA. For a large mA, the voltage drop will be greater. So voltage compensation of meter before the exposure is required and is accompanied by kV meter compensator. A kV meter compensator is connected mechanically to mA selector so that for large mA selections, a slightly larger voltage is applied to compensate the kV meter.

Filament supply circuit (low tension circuit)

The purpose of this circuit is to regulate current flow through the filament of the X-ray tube. The typical values of voltage and current in the tube filament are 6-12 volts and 3-5 amp. The current is reduced by a rheostat or variable resistor. Ten volts produce a current of 3-5 amp through the filament. The primary side of the filament transformer is supplied from the autotransformer and the value of current is around 0.5 amp. In the primary side of the circuit, there is a variable choke which constitutes a greater or smaller reactance to AC in this circuit and thus controls the amount of total current flowing to the filament transformer. In modern machines, mA conltrol is done by a type of rheostat wherein there are a number of resistors in series with a variable tapped connection. It produces the step-wise control of mA. The controls for mA circuit are arranged in the primary side of the filament supply circuit.

Stabilisation of filament heating

Every effort is made to maintain a constant predetermined value of current in the filament heating circuit as even a 5% variation in the filament supply may cause sufficient change in the filament heating current to cause mA change (tube current) of about 30%. It is thus important to supply the primary winding of the filament a constant AC voltage or, in other words, a constant predetermined value of current to the filament heating circuit. Slowly occuring changes in the mains voltage, ultimately affecting filament heating, are compensated by the mains voltage compensator. Since this mains voltage compensator provides the autotransformer with a properly adjusted voltage per turn, the filament transformer which is supplied from the autotransformer will also be protected against slow changes in mains voltage.

For rapid changes in the mains voltage occuring during an actual exposure, a voltage stabiliser is required to stabilise voltage to the filament transformer. There is also a frequency compensator to compensate for voltage change across the primary side of the filament transformer due to frequency dependent impedence change in the primary circuit and is discussed further.

Input winding of the voltage stabiliser is supplied from autotransformer and the current in it sets up a magnetic field throughout its central core piece. This stabiliser is frequency dependent and requires frequency compensator. By using static voltage stabiliser, rapid occuring variations from the mains and autotransformer which takes place during an exposure are stabilised before reaching the filament transformer.

Frequency compensator

If frequency of the supply changes, the output voltage of the stabiliser will also change. For instance, if frequency increases by 1 cycle/second, the voltage output of the stabiliser may increase by 3 or 4 volts. So a frequency compensator is required. Input voltage comes from voltage stabiliser and voltage output from the compensator forms the input of the X-ray tube filament transformer. It consists of an inductor and a capacitor connected in parallel which is then connected in series with the supply to the filament transformer.

Space charge compensator

Space charge around the X-ray tube filament results in an increase in mA as kV is increased. It is an undesirable feature as independent variation of mA and kV is prevented. So compensation is brought about by a space charge compensator which applies a small opposing voltage to the supply voltage of the filament transformer. The variable control of space charge compensation is gauged to the kV selector of the X-ray machine (Fig 3-22). As kV is increased so is the reverse voltage from the space charge compensator so that the resulting decrease of emitted electrons from the filament, now at a lower temperature, just balances the increase in number of electrons crossing the tube from the space charge. In other words, increase in kV results in an automatic reduction in the filament heating current.

Secondary coil of X-ray transformer (High tension circuit)

It consists of many turns of electrically insulated wire that is thinner than the wire in the primary coil because of smaller current in the secondary circuit. The transformer steps up the primary voltage to provide the high voltage required to operate the X-ray tube; the step-up ratio being about 500:1.

Milliammeter is connected in series in high voltage circuit to measure the tube current in mA in the X-ray tube. It is grounded together with mid-point of the secondary coil of the X-ray transformer and is at zero potential and so safely mounted in the control panel without hazard to personnel. The filament current is measured by an ammeter placed in the low voltage filament circuit. The milliammeter measures average values and gives no indication of the peak values of the tube current.

Rectifier

Except in self rectified units, a system of rectification is included to change AC supplied by the transformer to DC. This helps in increasing the heat tolerance capacity of the X-ray tube and thus safely permitting larger exposures. For keeping electron flow from cathode to anode direction, the secondary voltage of the high voltage transformer must be rectified. **The process of converting AC into pulsating multi-directional DC is called rectification.** As AC periodically reverses direction and varies in magnitude, DC is used for operating the X-ray tube more efficiently. A rectifier is thus used for rectification and is incorporated in series between the secondary coil of the transformer and X-ray tube. Due to rectification, the tube can withstand large energy loading than if current is not rectified. This is so because if electron flow is reversed, as will happen if AC is used, the cathode assembly can not withstand tremendous heat generated by such an operation.

Rectifiers are valve tubes, thermionic diode tubes resembling X-ray tube, which allow flow of current in one direction only by suppressing inverse voltage of AC. Now a days valve tubes are being replaced by solid state rectifiers especially in high voltage X-ray circuits. Solid state rectifiers have longer life and do not require heating current, examples include copper oxide, germanium and silicon rectifiers.

Rectification is of two types:
   i) Half-wave rectification
   ii) Full-wave rectification

i) Half-wave rectification : In this type of rectification, the tube emits X-rays only half of the time that it is on (Fig 3-23). The inverse voltage of AC (non-conducting) is removed from the supply to the tube by rectification. Such rectification is not widely used in diagnostic X-ray machines. Some X-ray units, e.g. dental units, are self rectifying and X-ray tube itself acts as the rectifying diode. In such a case, half-wave rectification occurs (Fig.3-24).

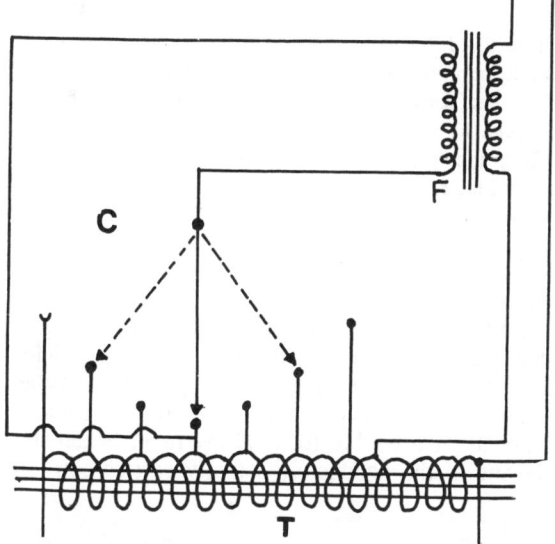

Fig. 3-22 : Space charge compensation circuit. C = kV control, T = Autotransformer, F = Space charge compensator transformer (From Forster, E, 1985. Equipment for Diagnostic Radiography. MTP Press Ltd, Kluwer Academic Publishers, Netherland. Used with permission).

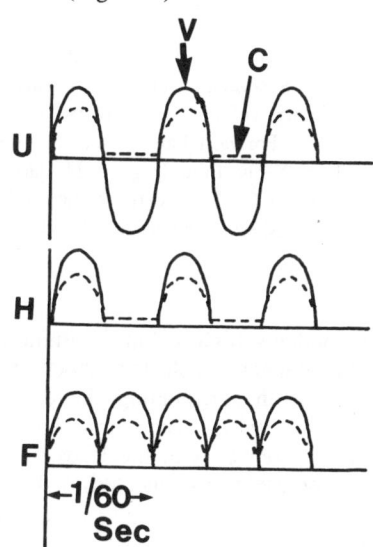

Fig. 3-23 : Three types of wave form of voltage. U = Unrectified, H = Half-wave rectification, F = Full-wave rectification, V = Voltage across X-ray tube, C = Current through X-ray tube (From Bushong, S.C. 1975. Radiologic Science for Technologists-physics, biology and protection. The C.V. Mosby Co, Saint Louis. Used with permission).

ii) Full-wave rectification : All modern X-ray generators supply full-wave rectified circuits (Fig 3-23). It means a DC pulsating current flows through the X-ray tube, accomplished by four valve tubes as shown in Fig 3-25. The negative half cycle corresponding to the inverse voltage is reversed so that positive voltage is always directed across the X-ray tube. In other words, both halves of AC voltage are used to produce X-rays and thus X-ray output per unit time is twice as large as it is with half-wave rectification. In half-wave rectification, output is 60 cycles/second whereas in full-wave, it is 120. Full-wave rectification has following advantages:

i) Exposure time is cut to half.

ii) Considerable increase in the tube rating or heat loading capacity is permitted.

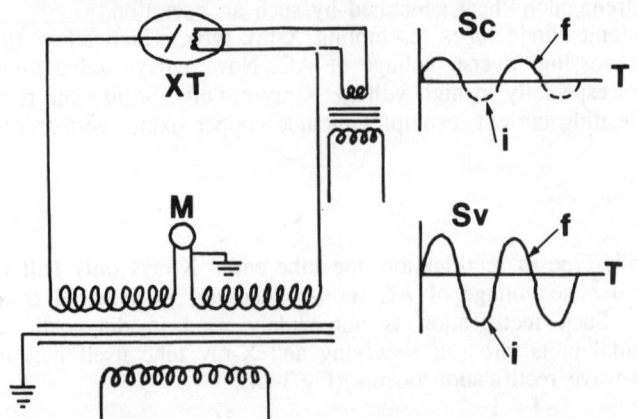

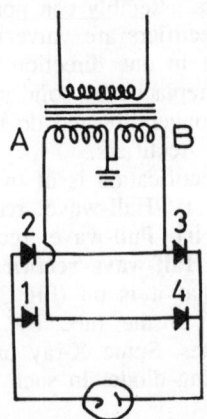

Fig. 3-24 : Voltage and current waveforms of a self rectified X-ray circuit. Sc = Secondary current with inverse half cycle suppressed, Sv = Secondary voltage with inverse half cycle larger than forward half cycle, f = Forward half cycle, i = inverse half cycle, M = mA meter, XT = X-ray tube (Figs 3-24 and 3-25 are from Forster, E, 1985. Equipment for Diagnostic Radiography. MTP Press Ltd, Kluwer Academic Publishers, Netherland. Used with permission).

Fig. 3-25 : Full-wave rectified (2 pulse) circuit-1 to 4 are valve tube rectifiers. Only two rectifiers are in circuit at any given time.

Cable

Cable conducts high voltage current from the rectifier to the X-ray tube. This is designed to eliminate the danger of electric shock, provided that the insulation remains intact. The grounded woven wire sheath surrounding the cable prevents danger of spark over to the patient or other objects near the cable. The cable has three conductors which when connected to the cathode filament makes contact with the leads of the two filaments of a double focus X-ray tube. When this three conductor cable is connected to the anode, only one of the conductors actually carries current.

X-ray tube

X-ray tube is in itself a part of the secondary circuit and has already been discussed earlier in detail in this chapter.

Safety devices

Certain safety devices are incorporated in an X-ray equipment to overcome electrical hazards and are discussed below briefly:

i) Switches: Three switches mainly operated by the technician are: the main switch, on-off switch and exposure switch. By operating the main switch, one can isolate whole X-ray machine from the electrical supply. The on-off switch, provided in the control panel, energises the autotransformer and many auxillary circuits such as of meters, filament etc. The exposure switch needs constant pressure in order to help it on 'ON' position.

ii) Fuses and circuit breakers: A fuse is a thin wire encased in a glass tube with sealed metal ends which are in contact with the electrical circuit. The thin wire of the fuse melts if a current higher than that required for a part flows through. Fuse is included either in the casing of main switch or in X-ray circuit immediately before the autotransformer. A blow fuse should never be replaced with a fuse of a higher rating than that required for a circuit. Electromagnetic circuits are often used in place of fuses. In case higher current than required flows through, circuit breaker opens to switch off the machine.

iii) Grounding : The control panel, high tension transformer tank, the tube housing and outer metal casing of high tension cables are grounded with a copper wire leading deep into the earth. This results in zero potential or ground potential in these parts to make them safe for touch. In the event of a short circuit, large part of the electrical current is grounded and thus limiting the danger of a shock to personnel.

iv) Shock proofing: It refers to insulation and grounding of the electrical components. For example, high tension cables are adequately insulated and also grounded to avoid electrical hazards.

v) Interlocking control circuit: The variables mA, kV and timer are interconnected with an electromagnetic relay. All three switches must close before an exposure is made. Timer switch will not close if kV and mA are not set below the values critical to the X-ray machine.

THREE PHASE X-RAY GENERATOR

Three phase power refers to generation of three simultaneous voltage wave forms out of step with one another. All these three are of equal frequency, wave from and magnitude but differ in phase and each lags behind the preceeding voltage by one-third of the cycle (Fig 3-26) . Since the demand for very short exposure time has increased, three phase high tension X-ray generators are now commonly used in X-ray machines. With three phase equipment, three transformers are needed for kV selection for each phase. Fig.3-26 shows that with three phase power, multiple voltage wave forms are superimposed on one another which results in an effective wave form that maintains a nearly constant high voltage showing six pulses per 1/60 second compared to two pulse characteristics of single phase power.

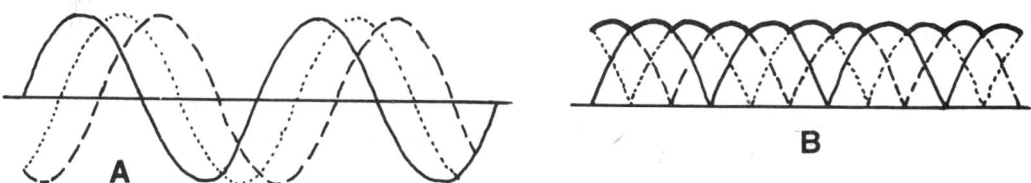

Fig. 3-26 : Three phase power voltage wave form. A = Three phase, B = Three phase six pulse (Adapted from Bushong, 1975).

Three phase generators have following advantages and disadvantages:

Advantages

i)   More X-rays of shorter wavelength for a given mA/kVp are produced from a three phase unit. This reduces exposure time.
ii) The variation between maximum and minimum voltage is less than 15% as compared to that of 100% with single phase power units.
iii) With three phase power, the life of X-ray tube is prolonged due to even thermal loading.
iv) Less soft radiation is produced resulting in lower patient radiation dose.

Disadvantages

i)   Three phase units are more expensive, but operating cost is less.
ii) Separate technique charts are required than those developed for single phase units as X-ray output and penetrability are higher.

TUBE STAND

A tube stand supports the tube to hold it in a stationary position during an exposure. In table tube stand combination, the tube stand should move the entire length of the table. The stand should:

a) permit the tube to move vertically; b) allow the tube to be rotated for horizontal exposure, and c) permit angulation of the tube. The table stand combination meets most of the requirements for small animal radiography. However, large animal X-ray machines require elaborate attachments and a number of alternatives are available. A ceiling mounted stand is ideal as there is maximum degree of tube movement within a limited space. It also permits a radiographer to make exposures from a protected area.

TYPES OF X-RAY MACHINES

X-ray machines can be grouped into three main categories: portable, mobile and fixed.

Portable X-ray machines

These machines are commonly used in veterinary practice because of their convenient transportation. Such machines can be suspended from a rack or a strap or can be positioned on a wooden block. Portable machines are equipped with small sized low weight transformer located within the tube head. Small control panel is attached to the tube stand or supported on a separate stand. These machines have stationary anode with both single and double focal spot. The maximum output usually varies from 70-110 kV and 15-35 mA.

Advantages

    i) Relatively cheap and require little maintenance.
    ii) Being light in weight, are easy to transport.

Disadvantages

    i) Because of low electrical output, these machines are of limited value in radiographic examination of anatomical structures above the carpus and tarsus of large animals.
    ii) Low mA range of such machines necessitate longer exposure times.

Mobile X-ray machines

These machines have higher output than portable machines by virtue of their larger transformers and are mounted on wheels (Fig.3-27). Some models can also be used for radiography of the digits in a standing animal. Usually these machines have rotating anode with output of 90-125 kV and 40 to 300 mA. Most machines are movable on smooth surface within the radiology section.

Fixed X-ray machines

These units are usually installed in a room specially constructed for the purpose. Their large transformers are capable of greater output and connected to the X-ray tube with high tension cables. Output of these machines vary from 120-200 kV and 300-1000 mA, but are expensive. These are suitable for almost all types of radiographic examination of large animals. A ceiling mounted telescoping tube support facilitates movement of the tube across the room and almost to the level of ground, apart from both horizontal and vertical X-ray beams (Figs. 3-28 and 3-29).

Advantages

    i) Suitable for all types of radiographic examinations in case of small and large animals.
    ii) Because of high kV and mA output, exposure time is less.

Disadvantages

    i) More scatter radiation occurs because of higher kV output.

ii) Expensive to purchase, and three phase electricity supply is required for installation of the machine.

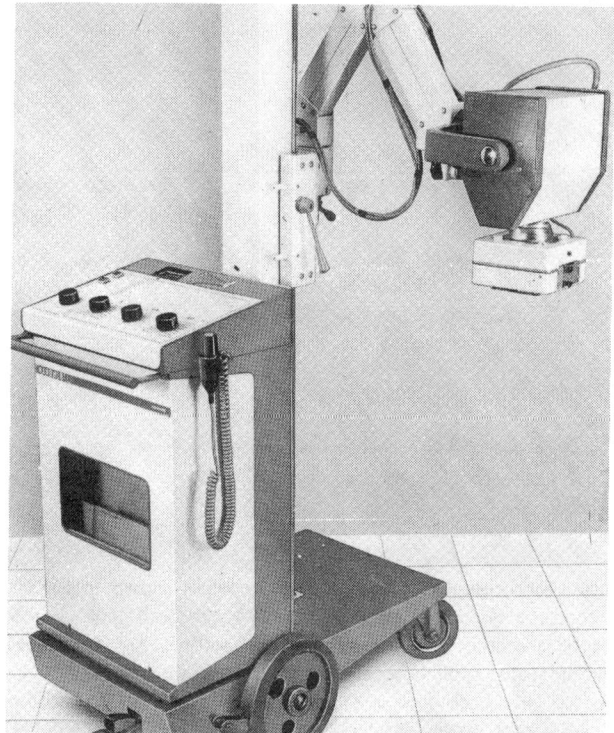

Fig. 3-27 : Mobile X-ray machine (Courtesy Siemens India Pvt Ltd).

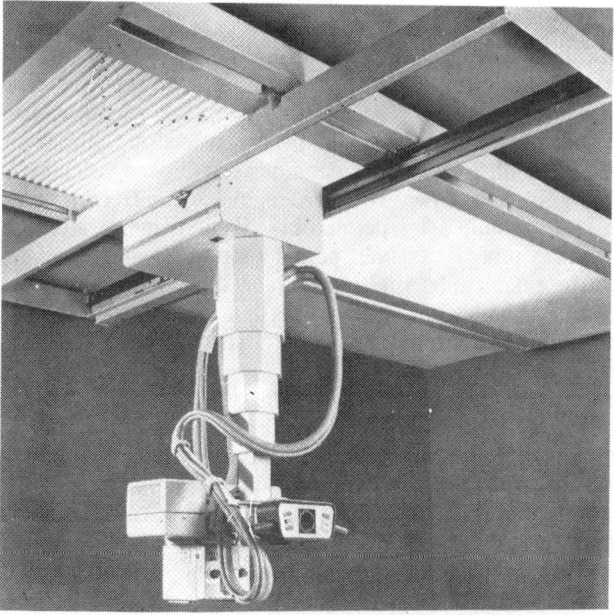

Fig. 3-28 : Ceiling suspension X-ray machine (Courtesy Siemens India Pvt Ltd).

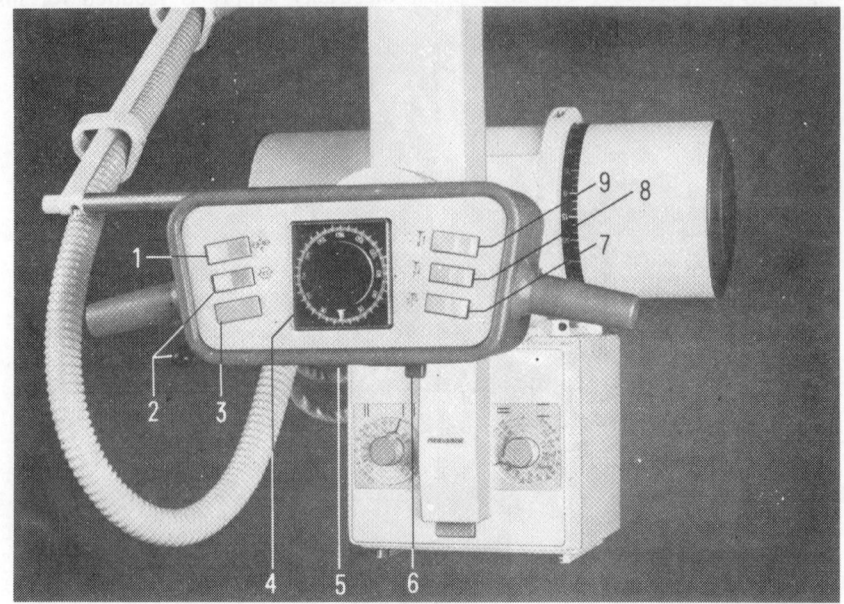

Fig. 3-29 : Controls of ceiling suspension X-ray machine shown in Fig. 3-28. 1 = Switch for fluoroscopy control device in conjunction with bucky wall unit, 2 = Switch to release the locking for tube assembly oblique or horizontal setting, 3 = Spare switch, 4 = Angle indicator for oblique setting of tube assembly, 5 = Light marker for centring the tube assembly on the catapult bucky, 6 = Tape measure for FFD, 7 = Switch for releasing transverse and longitudinal movement locking system, 8 and 9 = Switches for downwards and upwards motions of tube assembly (Courtesy Siemens India Pvt Ltd).

X-RAY TUBE RATING CHARTS

There is a maximum permissible limit of exposure factors within which an X-ray machine should operate to avoid any damage and it is always lower than its full capacity. The limited ability of the anode disc to accumulate and dissipate heat restricts the maximum allowable exposure. Moreover, kVp limit is set by insulation considerations. At higher voltages than permitted, there is danger of insulation being broken down between regions of different potential in the tube. The maximum tube current is also limited by permissible filament temperature. Additionally, maximum heat dissipation ability of the tube in a given period of time is an important factor. Knowledge of three charts is important to extend tube life and include tube rating chart, anode heat cooling chart and housing cooling chart. Although an interlock system operates in the X-ray machine to avoid damage to the tube yet intelligent use of the machine and these charts is necessary. The selection of a correct chart depends on the following conditions:

i)   Type of power supply: Tube rating is different for single phase and three phase power supply and also for short or long exposures.

ii) Type of voltage rectification: Maximum operation limit of the X-ray machine is less with half-wave rectification than with full-wave rectification.

iii) Application: A special chart is used for rapid sequence radiography as in serial angiography and tomography.

iv) Size of tube focus: The maximum operation limt of the machine is less with a small focus than with a large focus.

v)   Tube design: The design of the tube affects the maximum tube rating and so the chart supplied by the manufacturer should be consulted.

vi) Cold vs hot tube: The charts apply to a relatively cold tube that has not been subjected to heavy loading just prior to use.

Radiographic tube rating chart

Fig. 3-30 shows a typical but hypothetical tube rating chart. Of the three charts, it is most important as it indicates maximum safe exposure time for any selected combination of kVp and mA for a single exposure with a relatively cool tube. A series of radiographic rating charts are available for each machine which cover various modes of operation possible with that particular tube.

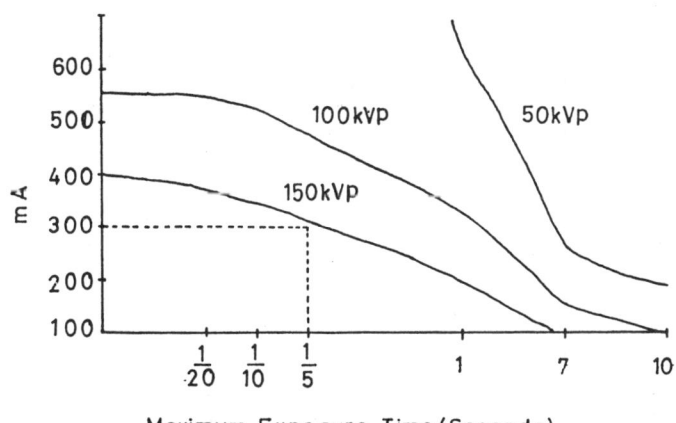

Maximum Exposure Time (Seconds)

Fig. 3-30 : A hypothetical tube rating chart. Dotted line example shows that if an exposure is to be made at 300 mA for 0.2 seconds, the maximum safe limit for kVp would be 150. Each X-ray machine has its own tube rating chart supplied by the manufacturer.

Anode heat cooling chart or curve

X-ray tubes are also rated for anode heating capacity measured in heat units. Heat units {one heat unit = 0.785 watt-sec (joule) or 0.188 calories} are the product of kVp × mA × exposure time in seconds. The unit is used to express heat built up during radiographic exposures in a single phase X-ray generator. In case of a three phase generator it is expressed as kilowatt rating i.e. 1/1000 of maximum kVp × mA at 0.1 second exposure for a particular tube. Since more heat is generated with a three phase generator a modified formula, 1.35 × kVp × mA × seconds, is used to calculate heat units.

Thermal capacity of an anode and its heat dissipation characteristics are contained in anode cooling chart or curve (Fig.3-31) which is not dependent on filament size or speed of rotation. It is used to determine the length of time required by the anode for complete cooling after any level of heat units accumulation. It means the curve or chart indicates maximum heat units that may be safely stored in the anode and also the time required for anode cooling between the exposures. Manufacturers now supply the information regarding the maximum number of exposures per second that are possible depending on the maximum load exposure. This information is of critical importance in procedures like serial angiography.

Tube housing cooling chart

The ability of the metal tube housing to store and dissipate heat contributes to tube life. In order to prevent tube housing damage and to permit adequate cooling of insert, the temperature of housing must be kept below 90°C. The housing cooling chart is almost similar to the anode cooling chart. Tube housing generally has maximum heat capacity in the range of 1-1.5 million heat units and complete cooling after maximum heat capacity requires 1-2 hours if an air blower has been provided in the X-ray unit.

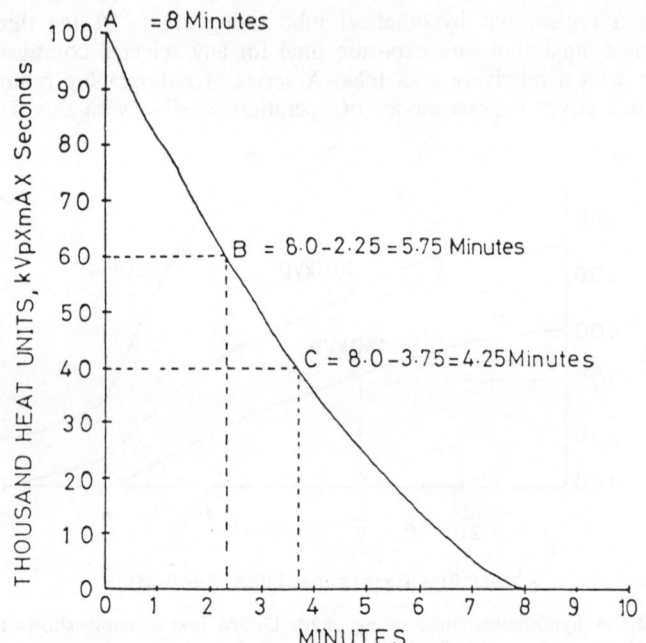

Fig. 3-31 : A hypothetical anode cooling curve. In this curve, anode takes 8 minutes to cool down from maximum permissible safe limit of 100 thousand heat units (A). As the heat units decrease, cooling time decreases, at 60 thousand heat units (B), cooling time is 5.75 minutes and at 40 thousand units, it is 4.25 minutes. Each X-ray unit has its own anode cooling curve or chart supplied by the manufacturer.

REFERENCES

Bargai, U., Pharr, J.W and Morgan, J.P. 1987. Bovine Radiology. Iowa State University Press, Ames.

Busk, R.L. and Ackerman, N. 1986. Small Animal Radiology. Churchill Livingstone, New York.

Bushong, S.C. 1975. Radiologic Science for Technologists. The C.V. Mosby Co., Saint Louis.

Curry, T.S. Dowdey, J.E. and Murry, R.C. 1984. Christensen's Introduction to the Physics of Diagnostic Radiology. 3rd edn., Lea and Febiger, Philadelphia.

Douglas, S.W. and Williamson, H.D. 1980. Principles of Veterinary Radiography, 3rd edn., Williams and Wilkins Co., Baltimore.

Ticer, J.W. 1984. Radiographic Technique in Veterinary Practice. 2nd edn., W.B. Saunders Co., Philadelphia.

Forster, E. 1985. Equipment for Diagnostic Radiography. MTP Press Ltd., Kluwer Academic Publishers, Netherland.

Gillette, E.L., Thrall, D.E. and Lebel, J.L. 1977. Carlson's Veterinary Radiology. 3rd edn., Lea and Febiger, Philadelphia.

Johns, H.E. and Cunningham, J.R. 1974. The Physics of Radiology. 3rd edn. Charls C Thomas, Publisher, Springfield.

Morgan, J.P. and Silverman, S. 1990. Techniques of Veterinary Radiography, 4th edn., Iowa State University Press, Ames.

Phillips, D.F. 1987. Radiology in your Practice: Choosing the right equipment. Vet. Med.**82**, 587.

Ridgway, A. and Thumm, W. 1968. The Physics of Medical Radiography. Addison Wesley Publishing Co., London.

Selman, J. 1977. The Fundamentals of X-ray and Radium Physics, 6th edn., Charles C. Thomas Publisher, Illinois.

Trigg, C.N. 1979. X-ray tube failures and preventions: A preliminary report. Rad. Tech. **50**, 430.

Van Der Plaats, G.J. 1980. Medical X-ray Technique in Diagnostic Radiology. 4th edn., Martinus Nijhoff Publishers, London.

SELECTED QUESTIONS

1. Why vacuum is created in the X-ray glass tube?
2. Name essential components of an X-ray tube.
3. Why tungsten is preferred for filament and anode in the X-ray tube?
4. What do you understand by the term "dual focus tube"?
5. Why a rotating anode is preferred over a stationary anode?
6. What is 'line focus principle'? What do you understand by effective focal spot?
7. What is heel effect?
8. List causes of X-ray tube failure.
9. List steps which can extend X-ray tube life.
10. What is a grid controlled X-ray tube?
11. What are characteristic and Brems radiations? Explain in brief.
12. What do you understand by:
    i) Inverse square·law ii) Tube current iii) Heterogenous nature of X-rays.
13. Name components of primary and secondary circuits of X-ray machine.
14. What are the advantages of full-wave rectification?
15. Name safety devices incorporated in the X-ray machine.
16. What are the advantages and disadvantages of fixed X-ray machines?
17. What do you understand by radiographic tube rating chart?

```
-X-RAY MACHINE-
    MUCH HEAT
IS GENERATED HERE
```

Malling IV, 1992, Relationship within Reaction Observed Ice Cone ... Quantum Vol. 76, pp 85-89.
...............1994, Samm IHC, 1990, Soul Medical .... W.B. Saunders Company, Ist Reprint.

Saw, AE, 1972, The Fundamentals of X-Rays and Radium Physics. 5th ed., Thomas C.C., Illinois Publisher, Illinois.

Tugg, Carl JCD, Xrays and Images .... A paper presented at third Symposium ...
Von Der Plaats, GJ, 1980, Medical X-ray Techniques in diagnostic Radiology. 4th edn., Martinus Nijhoff Publishers, London.

4 X-RAY IMAGE FORMATION AND RECORDING

SECTION-A

GEOMETRY OF IMAGE FORMATION

KULDIP SINGH

The purpose of radiography is to obtain as accurate an image as possible of the structure being radiographed. Two attributes that contribute to this accuracy are sharpness and the size and shape of the image of the object being radiographed. Following geometric factors affect the radiographic quality:

i) Image magnification
ii) Image distortion
iii) Image unsharpness

IMAGE MAGNIFICATION

Geometric rules applicable to ordinary shadow or a photographic and radiographic image are similar. If an object is placed in the path of a source of light, a shadow is formed. As the object is brought nearer to the source of the light, the shadow enlarges and vice versa (Fig. 4-1). This is related to the divergence of the light rays in the beam. In other words, shorter the distance between the object and light source, greater will be the magnification of the image. **Magnification of some degree is usually present in every clinical radiograph because the image formed is a two dimensional representation of a three dimensional structure. However, magnification has to be kept as low as possible except in vascular radiography.** The magnitude of image magnification is expressed either by magnification factor (MF) or by percentage magnification of image:

$$\text{Magnification Factor (MF)} = \frac{\text{Focus film distance (FFD)}}{\text{Focus object distance (FOD)}}$$

$$\% \text{ magnification} = \frac{\text{Image width - object width}}{\text{Object width}} \times 100$$

The extent of magnification will depend upon the distance of the object being radiographed from the film. Magnification is higher if the object is kept away from the film and is least if it is closer to the film (Fig.4-2). In other words, **to reduce magnification, part to be examined should be as close to the film as possible.** The distance between focal spot of the tube and the film also influences the magnification of the image. At higher FFD (> 75 cm), only parallel rays around the central ray form the image as the divergent or peripheral rays are cutout by the cone or diaphragm thereby reducing the magnification effects (Fig.4-3). For routine radiography in veterinary practice, FFD is kept at 90 cm in which case magnification factor is around 1.10. At 180 cm FFD, the magnification factor is 1.05. However, higher exposure factors are required with increasing FFD. Therefore, **a FFD between 90-100 cm represents a compromise between the magnification effect and increased exposure factors.**

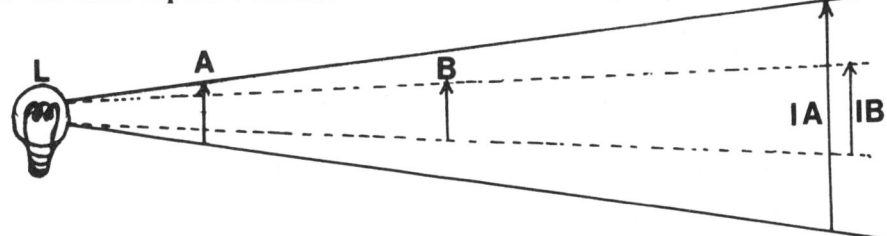

Fig. 4-1 : Effect of distance of object from light source on image magnification. When distance is short (A), image is magnified (IA).

IMAGE DISTORTION

A radiograph is not an exact picture of the structure being examined and differences of varying degrees are present in the shape and size and such misrepresentation of shape and size is called image distortion. In other words, **distortion means unequal magnification of different portions of the same object.** Image distortion can not be eliminated completely simply because the object to be radiographed is solid and three dimensional whereas the radiographic image lies in a single plane and is two dimensional. Distortion of the image depends upon two factors:

A) Shape of the object and B) position of the object.

A) **Shape of the Object :** The shape of the projected image depends on the shape of the object being radiographed as is evident from the folllowing examples:

a) The radiographic image of an oval opaque object is circular as object is not parallel to the film (Fig.4-4). b) The image of a rectangular opaque object is almost square (Fig.4-4). c) The image of disc will be circular regardless of the central position of the disc from the central X-ray beam (Fig.4-5). d) The image of sphere will be circular only along the projection of the central axis. e) The image of thick objects is distorted more than that of thin objects (Fig.4-6) because object to film distance with thicker objects changes measurably across the object. The part of thicker objects that is parallel to the film will be least distorted.

B. **Position of the object:** Image is not distorted if the object is positioned parallel to the film. The image of an inclined object will be smaller than the object itself and such a situation is called **foreshortening of the image.** The extent of foreshortening increases as the angle of inclination of the object increases (Fig.4-7). The degree of distortion is also affected if the central ray is not in line with the object being radiographed i.e. when the object is placed in a lateral position from the central X-ray beam (Fig.4-8). Spatial distortion can occur when objects of the same size are positioned at different distances from the film (Fig.4-9). This occurs due to unequal magnification of the object placed at different distances from the film and thus one of the images appears larger than the other one. Such an effect is minimal if objects lie along the central X-ray beam.

IMAGE UNSHARPNESS

A) **Penumbra or edge gradient:** Penumbra is a Latin word meaning almost (pen) shadow (umbra). **It is often called geometric unsharpness which refers to the region of partial illumination that surrounds the complete umbra (true shadow) causing undesirable blurred region on the radiograph.** In an X-ray machine, the focal spot is rectangular in shape (0.3×2 mm) and such a shape of the target leads to the formation of penumbra. An X-ray tube of 1 mm focal spot has a smaller effective spot on the anode side and a larger effective focal spot on the cathode side as has already been discussed in chapter 3. Fig 4-10 shows the zone of penumbra formed from an angled rotating anode. The width of the penumbra· is less on the anode side than on the cathode side of X-ray beam. In other words, **maximum sharpness of image is achieved by placing the object of great interest towards the anode side of the X-ray tube.** This assumes significance especially when X-ray tubes with low target angles are used at short focus film distance. Following factors affect formation of penumbra on a radiograph:

    i)   Large effective focal spot size
    ii) Short focus-film distance
    iii) Long object-film distance

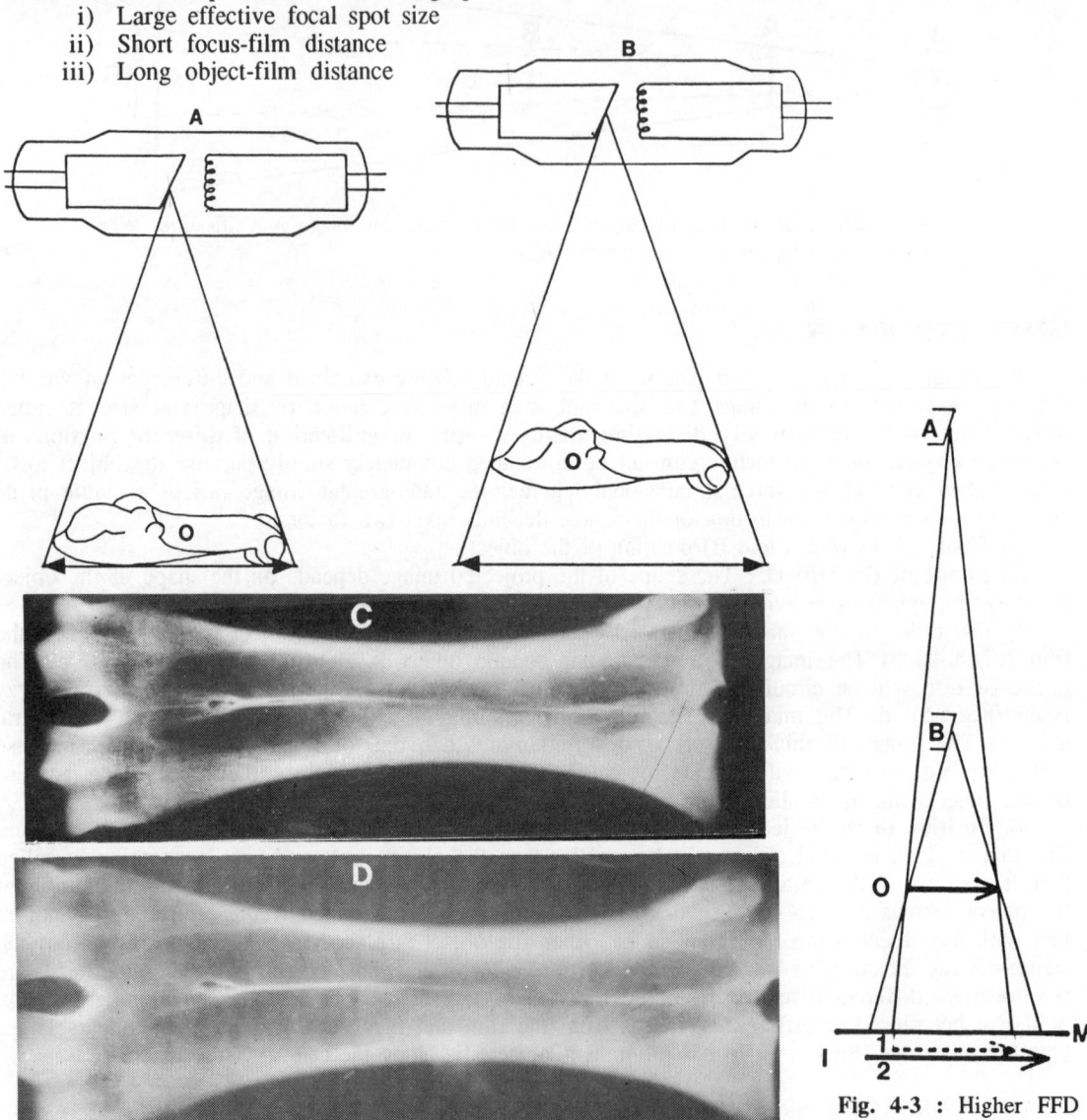

Fig. 4-2 : Effect of part-film distance. When the distance between object being radiographed (O) is kept at a distance from the film as in B, the image is magnified as compared to when object is near the film as in A. Same effect is shown in radiographs (C, D). C is normal with object as near to film as possible and in D₊ distance is six inches.

Fig. 4-3 : Higher FFD (A) results in less image magnification than lower FFD (B). O = Object being radiographed, M = Film, I = Image (Adapted from Bushong, 1975).

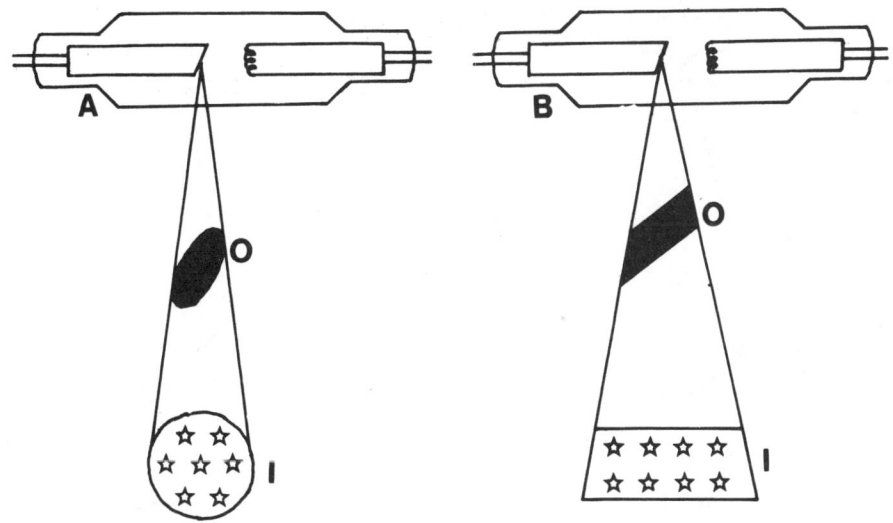

Fig. 4-4 : Effect of shape of the object. In A, image (I) of an oval object (O) is circular as object is not parallel to the film. In a similar situation, image of a rectangular object in B is almost square.

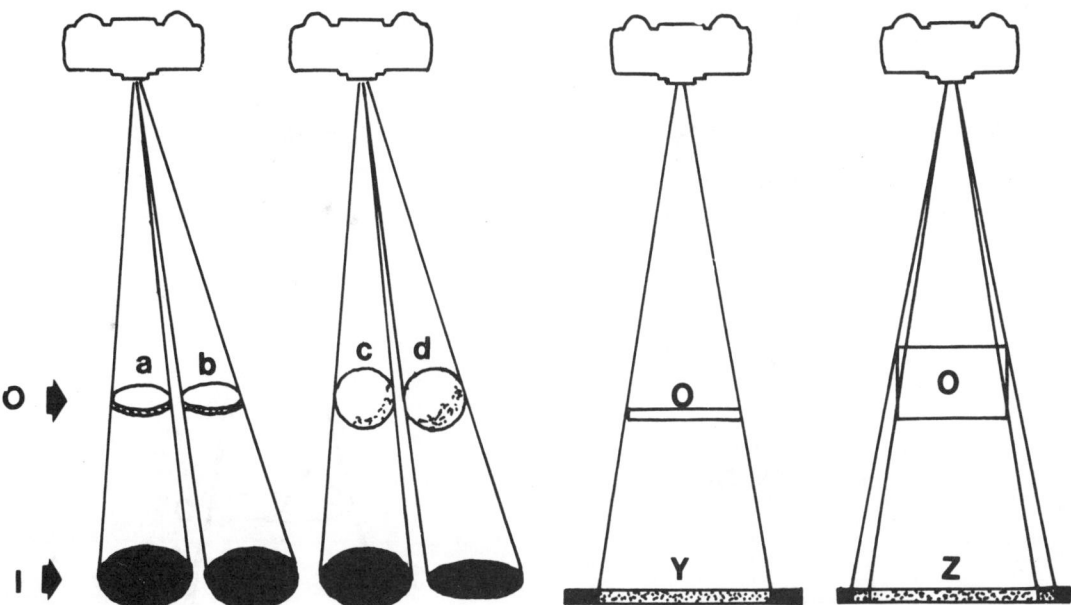

Fig. 4-5 : Effect of shape of the object. Image of a disc is always a circle (a, b) while a sphere image will be elliptical (d) when not on central axis. O = Object plane, I = Image plane (From Bushong, S.C. 1975. Radiological Science for Technologists-physics, biology and protection. The C.V. Mosby Co, Saint Louis. Used with permission).

Fig. 4-6 : The effect of thickness of the object. Thick object image (Z) is distorted more than that of thin object (Y) as object (O) to film distances changes more with thick objects (From Bushong, S.C. 1975. Radiological Science for Technologists-physics, biology and protection. The C.V. Mosby Co, Saint Louis. Used with permission.)

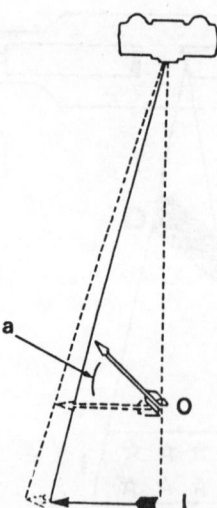

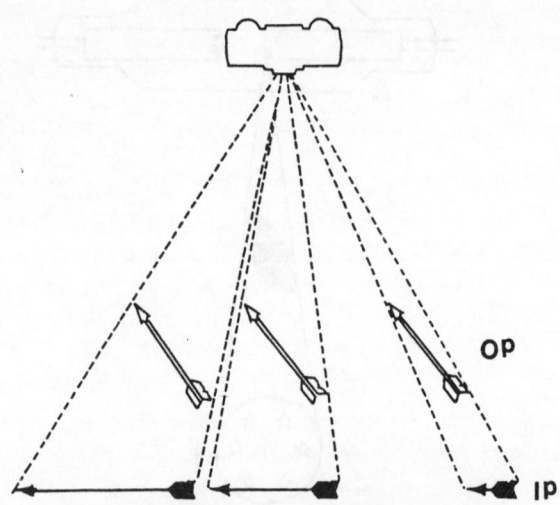

Fig. 4-7 : Angle of inclination (a) of an object (O) increases fore-shortening of the image (I). (Figs 4-7 to 4-9 are from Bushong, S.C. 1975. Radiological Science for Technologists-physics, biology and protection. The C.V. Mosby Co, Saint Louis. Used with permission).

Fig. 4-8 : Image distortion is more when an inclined object is lateral to the central X-ray beam. Op = Object plane, Ip = Image plane.

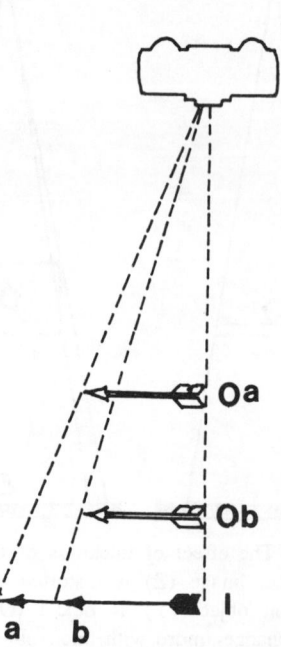

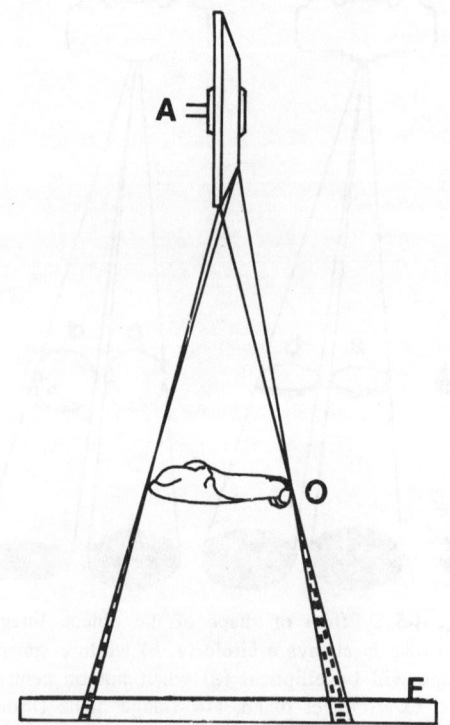

Fig. 4-9 : Spatial distortion of the image (I) occurs when objects of almost same size are positioned at different distances from the film.

Fig. 4-10 : Zone of formation of penumbra (arrows) on the film (F) from an angled rotating anode (A). O = Object.

Besides, factors governing magnification of the image also influence penumbra. The change in the geometry of the target (focus), object and image produces greater magnification and increased penumbra. **The object to be radiographed should be placed as near to the film as possible so that size of penumbra formed is smaller than the size of the effective focal spot.** As the distance between the film and object increases, penumbra size also increases and ultimately penumbra size becomes greater than the size of the effective focal spot.

$$\text{Penumbra} = (\text{Effective focal spot}) \; \frac{\text{Object-film distance}}{\text{Focus-object distance}}$$

Since generally a fixed FFD is used, penumbra is minimum if :-

    i) Object-film distance is minimum
    ii) Focus-object distance is large
    iii) Focal spot size is small

B) Motion unsharpness: It refers to the loss of radiographic quality due to the movement of either the patient or the X-ray tube or the film during X-ray exposure (Fig.4-11). Elimination of movement of the tube and film does not pose problem. However, movement of the patient in veterinary practice is not easily controlled. Resultant unsharpness due to this factor can be reduced by using shortest possible exposure time and appropriate restraining measures.

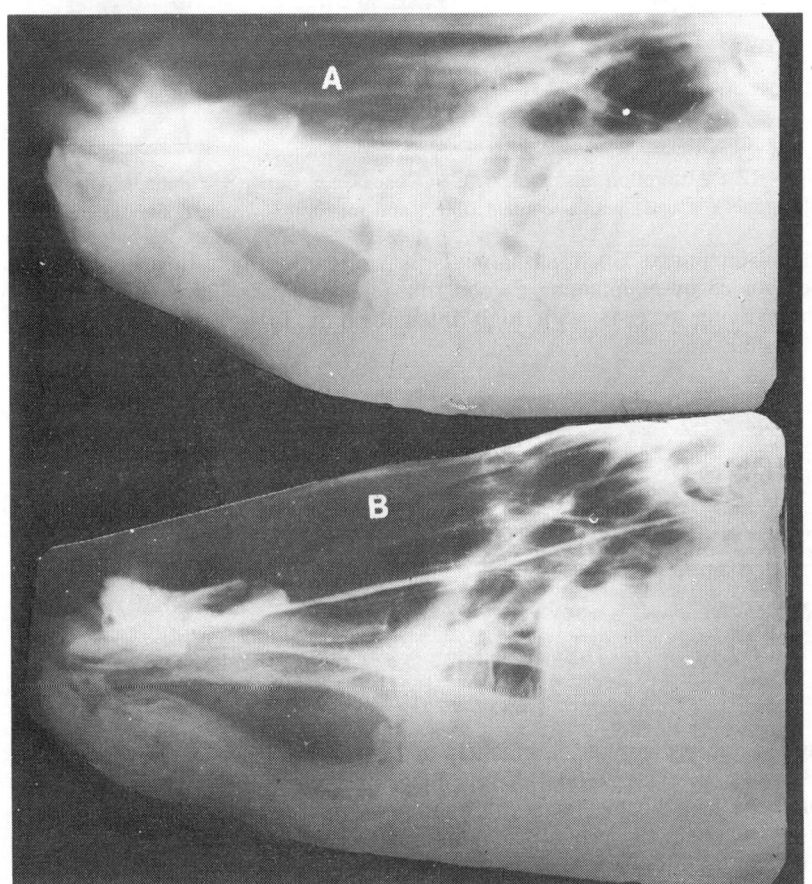

Fig. 4-11 : Motion unsharpness (A) in comparison to a normal radiograph without motion (B).

C) Absorption unsharpness: This type of unsharpness arises from gradual change in X-ray absorption across the boundary or periphery or centre of the object. This leads to unsharp edges.

The peculiar shape of the object or the part being radiographed influences this type of unsharpness. Fig 4-12 shows cone, cube and sphere of the same thickness and assumed to be made up of the same material. In case of a sphere, absorption unsharpness occurs across the entire image with maximum absorption of X-rays only in the centre of sphere. In case of a cube, fewer X-rays get absorbed along the sides and more get absorbed in the region of lower corners. In case of a cone, there is little absorption unsharpness as its edges are parallel to the diverging X-ray beam and are thus sharply defined on the film. **Since absorption unsharpness is related to the shape of the object, little can be done to control it.**

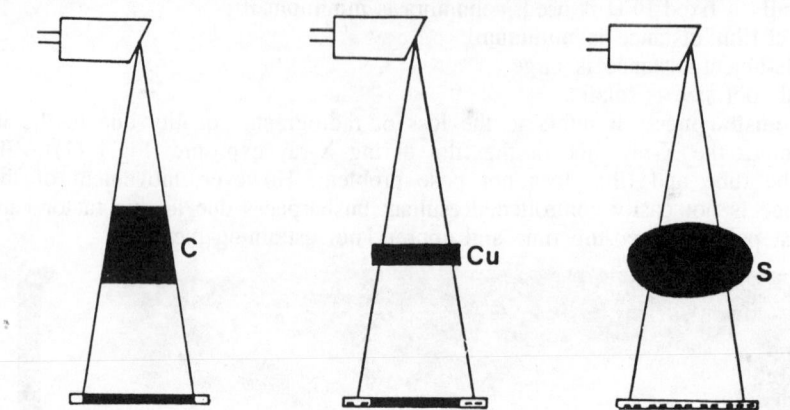

Fig. 4-12 : Absorption unsharpness. C = Cone, Cu = Cube, S = Sphere. Absorption unsharpness is least with a conical object and maximum with a spherical object.

D) Screen unsharpness: Such unsharpness is caused by light diffusion in screen phosphor layer. It can be avoided by maintaining a good film-screen contact and by using a fine grain screen. **The use of intensifying screens with high intensification factor reduces exposure factors but causes image unsharpness.**

SECTION - B

RECORDING OF IMAGE

GAJRAJ SINGH
B.A. MOULVI

X-ray image is the pattern of information that an X-ray beam acquires as it passes through different body tissues after interaction. X-ray image is present in the space but can not be seen unless it is converted to a visible image by appropriate means. A photographic film is one of the most widely used materials to decode useful information from an X-ray image. It also provides a permanent record of the information.

X-RAY FILM COMPOSITION

The composition of X-ray film (Fig.4-13) is similar to that of a photographic film. Photographically active or radiation sensitive emulsion is coated on both sides of a transparent base (double emulsion film).   A thin layer of adhesive is used to achieve firm attachment between the emulsion and base. The emulsion is protected from scratches, pressure or contamination during use by a thin layer of gelatin called supercoating.   Thickness of a radiographic film is about 0.25 mm.

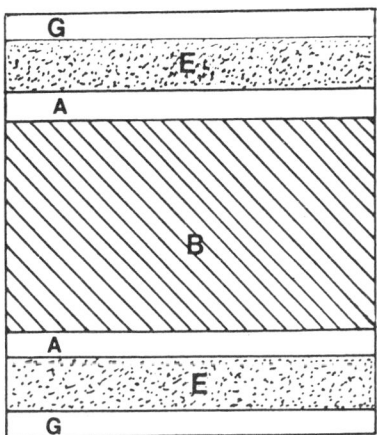

Fig. 4-13 : Diagrammatic representation of composition of X-ray film. G = Gelatin supercoating, A = Adhesive, E = Emulsion, B = Base.

Film base

Firm film base provides support to fragile photographic emulsion. An ideal film base should have following characteristics:

   i)   It should be flexible, inert, light in weight and easy to handle.
   ii) It should be relatively transparent.
   iii) Undesirable visible patterns should not be produced on it.
   iv) It should not absorb too much light when radiograph is being viewed.
   v)   The shape and size of the base must not change during developing process or during storage life.

The present day X-ray films have either polyester or cellulose triacetate base. Most bases are blue tinted because blue colour is more pleasing to the eye. Polyester base has the advantage of improved dimensional stability even under conditions of varying humidity and is stronger and thinner than the cellulose triacetate base. Polyester base is about 0.007 inch thick while cellulose triacetate base is about 0.008 inch thick. A thin layer of adhesive, applied to the base, maintains perfect union between the base and emulsion.

Film emulsion

Emulsion is composed of a homogenous mixture of gelatin and silver halide crystals. The gelatin is made from bone and offers following advantages:

   i)   It keeps silver halide crystals well dispersed and prevent their clumping.
   ii) Processing solutions can penetrate gelatin without affecting its strength or performance.

The variation in speed, contrast and resolution among different kinds of films are determined by the process by which silver halide crystals are manufactured and then mixed into gelatin. Silver halide is in the form of small crystals of silver iodobromide (silver bromide   90-99% and silver iodide 1-10%).   In a crystal, silver, bromine and iodine are present as ions arranged in a cubic lattice. The diameter of each crystal ranges from 1-1.5 microns. About 6.3×10^9 crystals are present per cubic centimeter of emulsion. Size, number and quality of the crystals determine the quality of the film emulsion. When a patient is exposed to X-rays, a latent image is formed on the film which however is not visible. Excessive pressure or heat may also form latent image. Altered crystals on exposed film are reduced to small grains of metallic silver after developing and unreduced crystals are removed by fixer. Remaining metallic silver appears black and a permanent negative is formed. If the crystals are not exposed to X-rays or light at all these are removed by fixer leaving only a transparent base. Thus variations in the extent of X-rays reaching the film, after traversing through a part being exposed form an image for viewing.

CHARACTERISTIC CURVE OF FILM

When a part to be radiographed is exposed to X-rays, certain exposure factors are used to produce a quality radiograph. It is important to obtain a correct density and a proper scale of contrast for the radiograph to be diagnostic. **The relationship between the exposure factors used and radiographic density obtained can be plotted as a curve called characteristic curve.** It can be done by exposing a film with an aluminium stepwedge kept on the cassette (to obtain different densities). The film is then processed and densities obtained are analysed using a densitometer. The resulting densities are then plotted against the known exposure. It is not necessary to undertake this procedure in each X-ray unit as manufacturers provide characteristic curve of each type of film supplied.

Understanding of characteristic curve offers following advantages:
i) It allows identification of film speed (Sensitivity) and latitude, and inherent contrast of film emulsion.
ii) It allows prediction of amount of change necessary to correct an exposure error.
iii) Curves of different types of film emulsions can be readily compared.

Typical characteristic curve of a screen type film with use of intensifying screens is shown in Fig.4-14. The exposure is expressed as relative exposure i.e. different areas of a film are exposed with constant kVp and mA but time of exposure varies. The log values of relative exposure and density are used in the curve so as to allow a very wide range of exposures to be expressed in a compact graph. The curve can be divided into three regions for clarity purposes: lower part (toe), upper part (shoulder) and in between a straight line. 'Toe' and 'shoulder' regions indicate that large variations in case of low and high exposure ranges cause small variation in density. At intermediate range of exposures, curve is nearly a straight line and thus small variations in exposure cause significant changes in density. Another point to remember is that even at zero exposure, density is not zero because base of the film and action of developer on the film produce some fog called **basic fog (basic density).** It is indicated in the curve upto from where the toe starts.

Characteristic curve can be used to judge the photographic characteristics of the film as follows:
i) Minimum the basic fog, better it is. In any case it should not exceed 0.2 density.
ii) In the toe region, indicated exposure begins to produce densities, above the basic fog, which can just be appreciated by the eye under normal viewing conditions.
iii) If the angle of the straight line is nearly vertical, film is of high contrast and narrow latitude (latitude is discussed further in this section).
iv) If the angle of the straight line is nearly horizontal, film is of low contrast and wide latitude.
v) **If a film requires comparatively less exposure to produce a given density, it is a fast film.** The characteristic curve of a faster film is more towards left than that of a slower speed film.

PHOTOGRAPHIC DENSITY OF FILM

Photographic density of the film measures degree of blackness of the film after processing. It is expressed by the equation:

$$D = \log \frac{I_o}{I_i}$$

Where D is density, I_o is the intensity of light incident on the film and I_i is the light transmitted by the film. The degree of blackness of the exposed film after processing is directly related to the intensity of radiation reaching the film. Even an unexposed film when processed produces a density of about 0.11 to 0.2 and, as has already been discussed, is due to basic fog. Diagnostically useful range of densities is 0.5 to 2.0.

FILM CONTRAST

Film contrast is inherent in the film and can be influenced by following factors:
  i)   Characteristic curve of the film
  ii) Photographic density of the film
  iii) Use of intensifying screens
  iv) Film processing

Importance of characteristic curve and film density has already been discussed. If a given exposure puts the developed densities on the 'toe' of the curve, the film is underexposed while reverse is true for densities in the "shoulder" region. Contrast refers to differences in various densities. If densities produced on a film are outside diagnostically useful range, contrast is lost. **Considerably more exposure is required to produce a given density if a film is exposed without using intensifying screens** because intensification factor of screens may range from about 15 to 50 or even more. **Contrast is always lower for a film exposed directly to X-rays without using intensifying screens (non-screen film).**

The degree of development of the film greatly influences the radiographic contrast because of the effect on film density and film fog. Following factors affect the degree of development of the film:
  i)   Development time
  ii) Temperature of the developer
  iii) Film agitation during developing
  iv) Composition of the developing solution

Time and temperature are considered more important because these factors can easily be controlled. Increasing the time of development increases the fog, and film contrast increases upto a point and then decreases because of increased fog. Therefore, manufacturer's recommendations regarding film development time and temperature should be followed for optimum results.

FILM SPEED

Speed of an X-ray film refers to the relative sensitivity to a given amount of radiation. Unlike photographic camera films, no numbers are given to indicate speed of X-ray films. Basically films are designated as standard, fast or ultrafast.

Fast films have following characteristics:
  i)   Larger silver halide crystals
  ii) Require less exposure (half of that required for standard film)
  iii) Produce grainy image that lacks definition
  iv) Possess narrow latitude.

A standard or medium speed film has wide latitude and small sized halide crystals. Ultrafast films require very less exposure (half to that for fast) but possess very narrow latitude and their storage life is short.

FILM LATITUDE

It refers to the range of exposure that produces diagnostically useful range of densities (Fig. 4-15). As has already been stated, film contrast is inversely proportional to the film latitude. **Films with wide latitude allow a longer grey scale to enable viewing of small changes in periosteal reactions, periarticular changes and other soft tissue abnormalitis.** So such films provide more diagnostic information and can withstand higher exposure error. Wide latitude films are especially recommended for radiography of extremities while using portable X-ray machines.

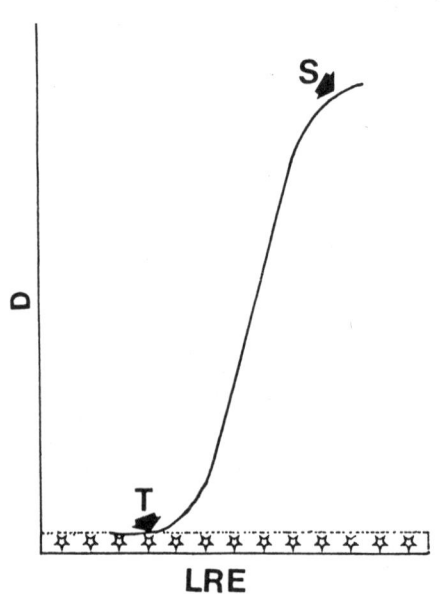

Fig. 4-14 : A typical characteristic curve of a screen type film. RLE = Relative log exposure, D = Density, S = Shoulder of curve, T = Toe of curve. Stars area indicates basic or fog density.

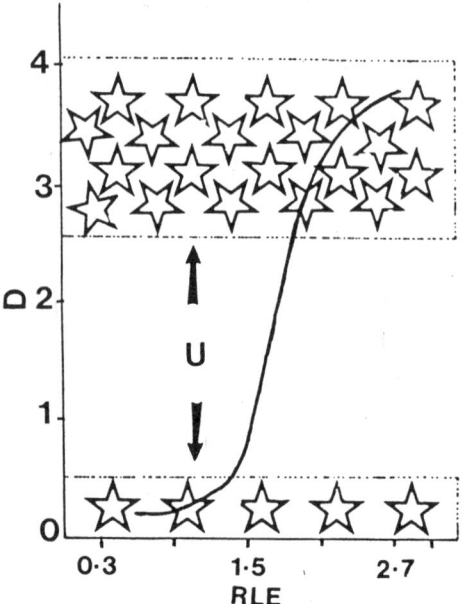

Fig. 4-15 : Film latitude. U = Range of radiographically useful densities. In the upper shoulder and lower toe regions (stars)contrast is lost. RLE = Relative log exposure, D = Density (After Bushong, 1975).

CROSSOVER EXPOSURE

In routine practice, a double emulsion coated film sandwiched between two intensifying screens is used in a cassette for radiographic work. Crossover exposure is additional exposure of one side of the emulsion to the light emitted by the screen on the opposite side of the emulsion (Fig.4-16) which results in additional film darkening. The main cause of this crossover exposure is partial absorption of light by adjacent emulsion which allows unabsorbed light to pass through the base to reach to the emulsion of the other side. The crossover light spreads by diffusion, scattering and reflection in the film base and interfaces between the emulsion and film base. Attempts are being made to design films with anticrossover coating to avoid crossover exposure and some success has been achieved.

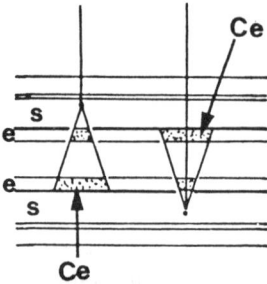

Fig. 4-16 : Crossover exposure. S = screen, e = film emulsion. Ce = Crossover exposure (After Morgan and Silverman, 1990).

TYPES OF X-RAY FILMS

Two types of X-ray films are used in diagnostic radiology, screen and non-screen types.

Screen films

These types of films are primarily more sensitive to ultraviolet and blue light range which originates from calcium tungstate crystals of the intensifying screens. Now a days newer films are available that are sensitive to green light (orthochromatic films). A screen film requires less exposure to produce an image as compared to non-screen film because of intensification factor of intensifying screens. The speed of screen films has already been discussed.

Non-screen films

These films are used for direct X-ray exposures i.e. without use of intensifying screens. Emulsion's thickness of non-screen films is more than that of screen films in order to absorb as much X-rays as possible and thus a prolonged processing time is required. Most non-screen films are more sensitive to direct ionising radiation and considerably less sensitive to blue light in comparison to screen films. These films require higher exposure because of absence of intensifying screens and more thickness of film emulsion. Such films are useful in radiographic examination of extremities to detect hair line fractures or slightly bony changes and in dental radiography to obtain better detail.

FILM SIZES

Following sizes of screen X-ray films are available in India:

Film size in inches	cm equivalent
5 × 7	12.7 × 17.74
6.5 × 8.5	16.5 × 21.6
6 × 12	15.3 × 30.5
6 × 15	15.3 × 38.1
8 × 10	20.3 × 25.4
10 × 12	25.4 × 30.5
11 × 14	27.9 × 35.6
12 × 12	30.5 × 30.5
12 × 15	30.5 × 38.1
14 × 14	35.6 × 35.6
14 × 17	35.6 × 43.2

SPECIAL TYPES OF FILMS

Automatic processor films:

These X-ray films have several characteristics so that rapid processing is possible in an automatic processor. Their increased hardness enables the films to be transported by a roller system. Such types of films are less expensive than ordinary X-ray films. Solutions used for processing these films are also different.

Occlusal films

These films are used for intra-oral radiography in human patients and can be used in small animals with or without specially designed intra-oral cassette.

HANDLING AND STORAGE OF UNEXPOSED FILMS

Following precautions should be taken while handling and storing X-ray films:

i)   Film boxes should be transferred to the place of their storage immediately after these are received.

ii) Film storage room should be cool (10-20°C) with low humidity (40-60 %).

iii) Film boxes should be kept vertically without any pressure on them.

iv) Films should not be stored near a source of heat, irradiation or water.

v)   Film should not come in contact with gases or vapours from chemical substances as fogging may occur.

vi) Films should be loaded into or unloaded from a cassette on a dry and clean bench inside the dark room under a proper safe light.

vii) Films should be handled delicately and any accidental splashing of processing solutions should be avoided.

viii) Shelf life of films is generally less than a year.

INTENSIFYING SCREENS

Intensifying screens interact with X-ray beam that has penetrated the patient and reached to the cassette. **Screens convert most of the radiant energy (95%) into visible light that has almost same information as the original X-ray beam.** The visible light thus produced and remaining X-rays (5%) interact with the film. **The use of screens intensifies the effect of X-ray beam on the film as it is far more sensitive to visible light than to X-rays. The screens thus allow reduction in the exposure factors required to obtain a diagnostic radiograph.**

Construction of screens:

An intensifying screen consists of four distinct layers (Fig.4-17) with a total thickness of about 0.4 mm. These layers are:

i) base, ii) reflecting layer, iii) phosphor layer in a binder and iv) protective layer.

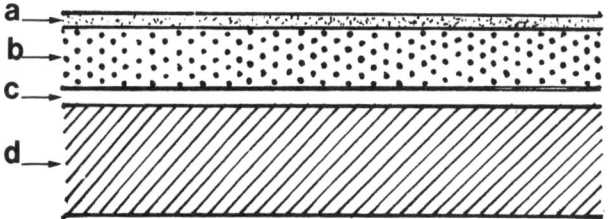

Fig. 4-17 : Diagrammatic representation of composition of intensifying screen.
a = Protective layer, b = Phosphor layer, C = reflecting layer, d = Base.

Base: Base provides mechanical support to active phosphor layer and is made of either high grade card board or polyester. It is farthest from the film placement in the cassette. Base should be chemically inert and moisture resistant. It should not suffer damage due to radiation and should not discolour with age.

Reflecting layer: Thin reflecting layer is spread between the base and phosphor layer. It is made of shiny white substance such as titanium dioxide or magnesium oxide. The light emitted by the interaction of X-rays and phosphor layer is directed in all directions including towards the base. The reflecting layer reflects back the light directed towards the base to the film. The effect of reflecting layer is shown in fig. 4-18A.

Phosphor layer: It is the active layer of the intensifying screen and its main function is to convert X-ray energy into visible light. The material used in the phosphor layer should possess following qualities:

  i) There should be no phosphorescence (phosphor afterglow).
 ii) Should be of high atomic number for higher X-ray interaction.
iii) Must efficiently convert X-ray energy into visible light.
 iv) Light emitted must be of proper colour to match sensitivity of the X-ray film.

The materials mostly used as phosphor in fluorescent screens are calcium tungstate, barium-lead sulphate and zinc-cadmium sulphide. Being physically strong with relatively high asborption coefficient (25%), calcium tungstate is most commonly used. It provides good detail but has poor X-ray to light conversion efficiency (3-5%).

Protective layer: It is a transparent layer placed close to the film. It consists of a cellulose compound and serves following purposes:

  i) Physical protection to the phosphor layer.
 ii) Prevents static electricity
iii) Provides a surface that can be cleaned without affecting the phosphor layer.

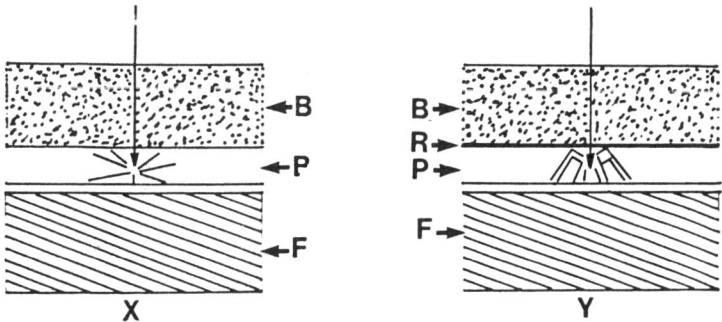

Fig. 4-18A : The effect of reflecting layer of intensifying screens. Screens with reflecting layer (Y) are more efficient as light emitted by interaction of X-rays and phosphor layer is directed towards film and thus more photons reach the film. B = Base, P = Phosphor layer, R = Reflecting layer, F = Film (After Bushong, 1975).

Intensifying action of screens

Basic function of an intensifying screen is to convert a part of the absorbed X-ray into visible light. This function depends on the absorption efficiency and conversion efficiency of the phosphor material used in the screen. The absorption efficiency is influenced by the thickness of the phosphor layer, e.g. the phosphor layer in fast speed screen is thicker than that of a medium speed screen. However, such **an increase in the phosphor layer thickness to increase the absorption efficiency of screens results in loss of radiographic detail.** To increase the absorption efficiency of rare earth intensifying screens, absorption characteristics of the phosphor layer are improved without increasing the thickness. The absorption efficiency of a pair of par speed tungstate screen is 20% and of fast speed is 40%. **Conversion efficiency (intrinsic efficiency) of the screen is the efficiency of the phosphor layer to convert X-rays to visible light.** The conversion efficiency of a calcium tungstate screen is about 5% while that of a rare earth screen is about 20%.

The intensifying action of a screen is measured by the intensification factor which is the ratio of X-ray exposure required to produce a given film density with and without screens:

$$\text{Intensification factor} = \frac{\text{Exposure required without screens}}{\text{Exposure required with screens.}}$$

Advantage of this intensification is that much less X-ray photons are required to obtain a diagnostic image than if no screens are used i.e. exposure factors are decreased. For example, if screens are used and 500 X-ray photons strike the cassette after penetration through the patient, 200 photons (40%) will be absorbed by the screens. Due to intensification, 170,000 light photons will be generated. Out of these, 85,000 will reach the film to form 850 latent image centres. If same number of photons i.e. 500 reach the cassette without screens, 25 will be absorbed by the film to make 25 latent image centres. In this example, the use of intensifying screens increases the photographic effect 34 times. Thus if kVp is kept constant, 34 times mAs will be required if film is exposed without screens. This example also illustrates that when screens are used, the number of latent image centres formed depends upon the light photons reaching the film after escaping from the screen. **The ability of the light photons, emitted by the phosphor, to reach to the film after escaping from the screen is called screen efficiency.** In the previous example, this efficiency is 50%.

Screen speed also influences screen efficiency as it quantitates relatively the conversion of X-ray photons into visible light. Available intensifying screens according to speed are : ultra fast, fast, medium or par speed, detail or low speed and ultra detail.

Some factors affect intensification factor but are inherent in the screens while others affecting the screen speed can be controlled.

Factors affecting intensification factor are:

i) Phosphor composition : The quality of calcium tungstate crystals used in the phosphor layer by the manufacturer affects the intensification factor. Phosphor layer used in rare earth screens has a high intensification factor.

ii) Thickness of phosphor layer : If thickness of the calcium tungstate phosphor layer is increased, absorption efficiency is increased. More is the absorption efficiency, more will be light photons generated.

iii) Crystal size and concentration : If the crystal size of the phosphor layer is bigger or their concentration is high, more light will be emitted. The crystal size of fast speed screens is bigger than that of par speed screens.

Following factors affect the screen speed:

i) Increasing the kVp will increase the intensification factor of calcium tungstate screens. This occurs because high kVp X-rays or heavily filtered X-rays are absorbed more in the tungstate screens by photoelectric process, thus increasing intensification factor.

ii) The screen speed is optimum at low temperatures. At extremes of temperature intensification decreases which in turn calls for higher exposures.

Rare earth intensifying screens:

There is increasing trend to use rare earth intensifying screens in veterinary practice because of advantages offered by these screens. X-ray absorption efficiency of a pair of rare earth screens is 60% i.e. higher than that of a pair of tungstate screens (20% to 40%). The conversion efficiency (20%) is much greater than the tungstate screens (5%). Rare earth screen-film combination has 12 times more fast speed than par speed tungstate screen-film combination and thus X-ray exposure is reduced by 15-50%. **The increased absorption efficiency of rare earth screens is due to improved absorption efficiency and not due to its increased thickness of phosphor layer.** Particular types of crystals used in the phosphor layer of rare earth screens are difficult to isolate from the earth or source and hence the term rare earth. Basic types of these phosphors are: terbium - activated gadolinium oxysulphide; terbium-activated lanthanum oxysulphide; terbium activated Yttrium oxysulphide and thulium-activated lanthanum oxybromide. These crystals continue to emit small amount of afterglow and so films loaded in rare earth screen cassettes should not be kept for a longer period as otherwise black spots will apear on the film after processing.

Advantages of rare earth screens:

i) Exposure time is reduced and thus less motion unsharpness.

ii) Patient radiation dose is decreased.

iii) Scatter radiation is decreased.

  iv) Use of smaller focal spot is possible.

  v) Increased tube life due to use of low mA.

Disadvantages of rare earth screens:

  i) Initial cost of screens is high.

  ii) May require green sensitive X-ray films (Terbium-activated gadolinium oxysulphide phosphor emits green light and thus requires orthochromatic films. Others emit blue green light and can be used with conventional blue light sensitive films).

  iii) Exposure chart is more complex as screen response is kVp dependent.

Care of intensifying screens:

Intensifying screens must be handled with care as rough handling leads to creation of radiographic artifacts on the processed film. Routine inspection and cleaning of screens should be done atleast once in a month. Well circumscribed white artifacts on a finished radiograph are generally produced by screen contaminants.

Following points be noted for the care of screens:

  i) These should be kept free of marks, dust and stains.

  ii) Do not allow abrasive liquids to come in contact with screens.

  iii) Do not allow processing soulltions to splash over screens.

  iv) Do not touch protective surface except at the time of cleaning.

  v) Use cleaners specified for the job as otherwise fogging effect may occur on the film.

  vi) Maintain good screen-film contact.

Screen-film contact:

Good screen-film contact is essential to obtain a quality radiograph. Proper contact can be checked by exposing a flat sheet of wire mesh or few paper clips placed over the cassette. Lack of sharpness of wire or clips images indicates poor screen-film contact while sharpness indicates good contact (Fig.4-18B).

Following are the causes of poor film-screen contact:

  i) Worn out contact felt.

  ii) Defective or broken hinges or latches of the cassette.

  iii) Defects in cassette frame.

  iv) Foreign material under the scrfeen.

The cause should be removed to obtain good quality radiographs.

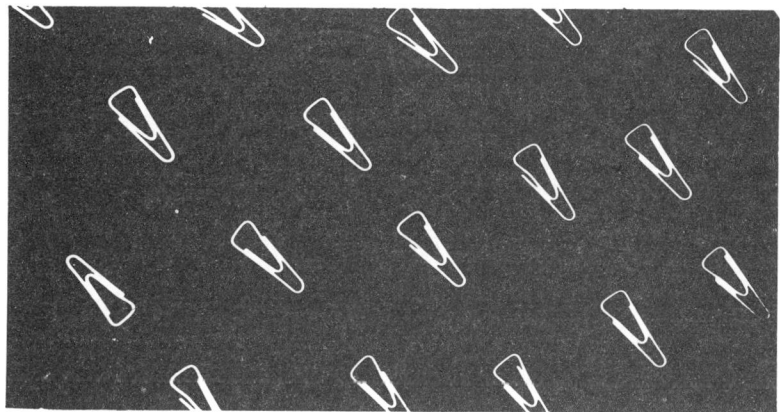

Fig. 4-18B : Test for screen-film contact. Sharp edges of paper clips indicate good contact. See details in text.

CASSETTES (FILM HOLDERS)

Cassettes are basically light proof boxes designed to hold film and screens for making an X-ray exposure. The use of cassettes serves following purposes:

   i) To protect the film from getting exposed to light.
   ii) To protect screens from physical damage.
   iii) To maintain good film-screen contact.

Non-screen film is contained in a light proof paper, cardboard or plastic holder. Provision of a lead sheet on back side of non-screen film holders minimises back scatter radiation and increases radiographic detail. Non-screen films are also available in ready to use light tight holders with no requirement to reload in the film holder.

A cassette (Fig.4-19 A and B) for screen type film is a light-tight metal or plastic holder designed to contain a pair of intensifying screens and a film. Such cassettes are expensive but require less exposure factors as compared to non-screen film holders. The cassette front (facing the X-ray tube) is radiolucent but strong enough to protect the film. It is made of plastic or low atomic number metal such as aluminimum or magnesium. Phenolic compounds, polycarbonate and carbon fiber are also used, the last being better because of its strength, rigidity, low weight, lesser deformability, less X-ray attenuation, reduced scatter radiation and reduced patient exposure. The back side of the cassette is made of heavy metal or lined with lead to absorb some X-rays transmitted through the film-screen combination. Two screens are sadwiched between the front and back of the cassette. A pad of oil free felt, glass fiber or isocyanite foam is placed between the cassette edges to maintain close film-screen contact. The back and front of the cassette are held tightly together by spring clips or by pivoted metal bars on the backside of the cassette which can slide into grooves in the frame. A leather tag is often present on the back side to facilitate opening of the cassette. Cassette sizes usually correspond to available film sizes. A cross section of a loaded screen cassette is shown in fig. 4-20.

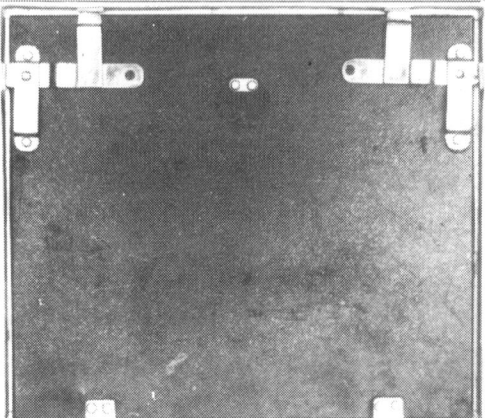

Fig. 4-19A : A cassette for screen type films.   **Fig. 4-19B :** Interior of a cassette for screen type films.

Care of cassettes

General care of cassettes is aimed at avoiding rough handling. It may be helpful to mark each cassette by numbering so that defects noted on a radiograph can be traced easily to the particular cassette. Hinges and clips of cassettes are subject to stress, and their proper functioning is of utmost importance to avoid accidental exposure of the film to light. So cassettes must be regularly inspected to check such faults. Care should be taken to avoid spilling of abrasive or processing solutions over them.

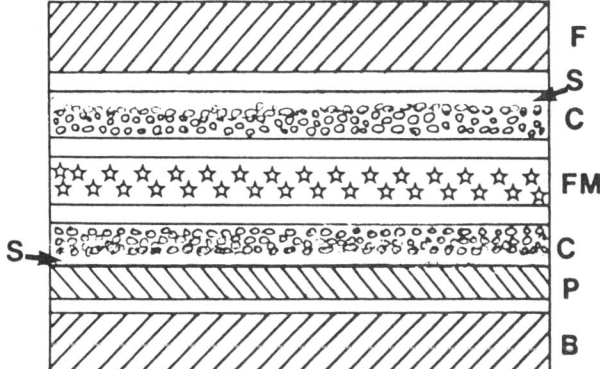

Fig. 4-20 : Cross-section of a loaded cassette. F = Front of plastic or aluminium, S = Screen support, C = Luminescent crystals, FM = Film, P = Felt padding, B = Lead back.

SECTION - C

RADIOGRAPHIC QUALITY

A.P. SINGH
RISHI TAYAL

Radiographic quality refers to the accuracy with which anatomical structures of the part being radiographed are represented on a radiograph. In other words, if visibility and sharpness of the structures is good, radiograph is of good quality. Many factors affect general quality of a radiograph. It is important to understand these factors so that every effort can be made to obtain a good quality radiograph. Three main visual requirements of a good diagnostic radiograph are:
  i) Excellent detail
  ii) Correct density
  iii) Proper scale of contrast
This section deals with detail, density and contrast, and the factors affecting these requirements.

DETAIL

Detail means degree of sharpness or definition of an object on a radiograph. Detail is good if structural and contour lines of the object are sharp. **Following factors influence detail :**
  i) Geometric factors
  ii) Intensifying screens
  iii) Motion of the patient, film and X-ray tube
  iv) Differential absorption of X-rays
  v) Double emulsion of the film

    vi) Radiographic mottle
   vii) Exposure factors used
  viii) Film processing
    ix) Scatter radiation and fog

Factors listed at i) to iv) have already been discussed under geometry of image formation in section-A of this chapter.

Double emulsion of the film

Since films are coated with film emulsion on both sides, an image is formed on both sides separated by the base of the film. This causes parallel unsharpness of the image which, however, is negligible.

Radiographic mottle

Appearance of non-uniform densities on a radiograph causing loss of detail is called radiographic mottle. It may either be due to intensifying screens or be due to film graininess. Screen mottle has two components: quantum and structural, former contributing principally to motttle. Quantum mottle occurs due to fluctuations in the absorption of X-ray photons exposing the film. Since use of fast speed films and intensifying screens decreases the radiation exposure, quantum mottle is increased to cause some loss of detail. Structural mottle is the result of unevenness in the thickness of phosphor layer of the screen and imperfections in crystal size. It is, however, usually negligible as manufacturers observe strict quality control measures.

Film graininess is due to random distribution of silver grains in the film emulsion and is visible only when radiographs are examined under magnification.

Exposure factors used

Each X-ray unit must workout its own exposure technique chart for particular type of machine being used (see chapter 6). Overexposure or underexposure produces too dark or two light radiograph with loss of detail (Fig.4-21)

Film processing

Correct processing of the film is absolutely necessary to obtain a good quality radiograph. Processing errors cause loss of radiographic detail, therefore, standardised norms should be followed for film processing (see chapter 5).

Scatter radiation and fog

Scatter radiation, discussed further in this chapter, and film fog reduce radiographic detail.

RADIOGRAPHIC DENSITY

Radiographic density is the measure of the degree of blackness on a processed film and is directly related to the number of X-rays reaching the film. More the number of X-rays that reach the film, blacker it is i.e. higher is the radiographic density. Film areas exposed lesser appear whiter after processing. Radiographic density is primarily influenced by subject density and its thickness. **Subject density is the weight per given volume of the subject.** Radiographic density is inversely proportional to the subject density as denser the object more it absorbs X-rays so that less photons reach the film. Those objects which allow X-rays to readily pass through them appear blacker on the film and are radiolucent while those which inhibit most X-rays appear whiter and are radiopaque.

Main densities which can be appreciated on a radiograph are : i) metal, mineral and bone; ii) fluid (soft tissue), iii) fat and iv) gas (Fig. 4-22). Mineral content of the bone does not allow X-rays to pass through readily while almost all X-rays can pass through air. Fluid allows more X-rays to pass through than bone but inhibits more than air does.

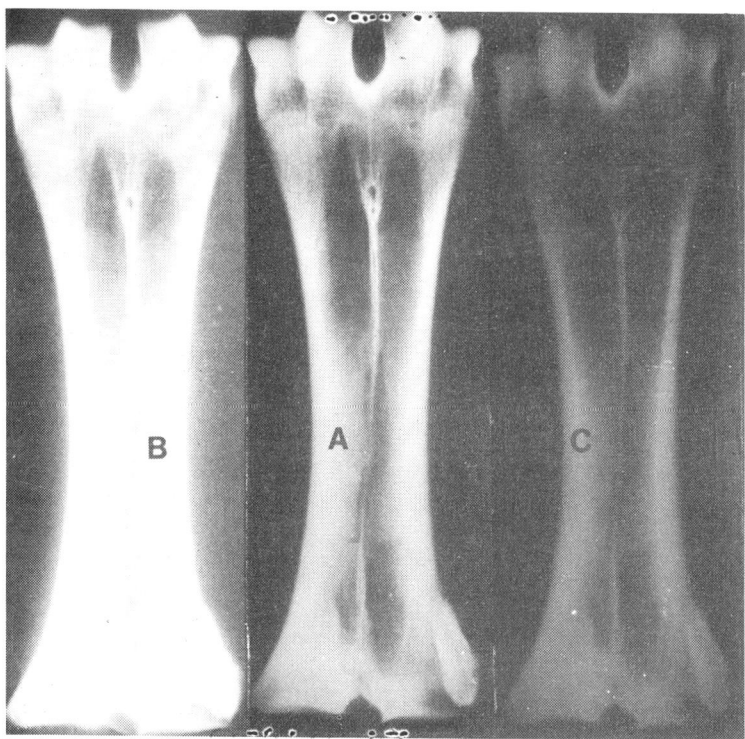

Fig. 4-21 : Underexposed (B) and overexposed (C) radiographs in comparison to normal (A).

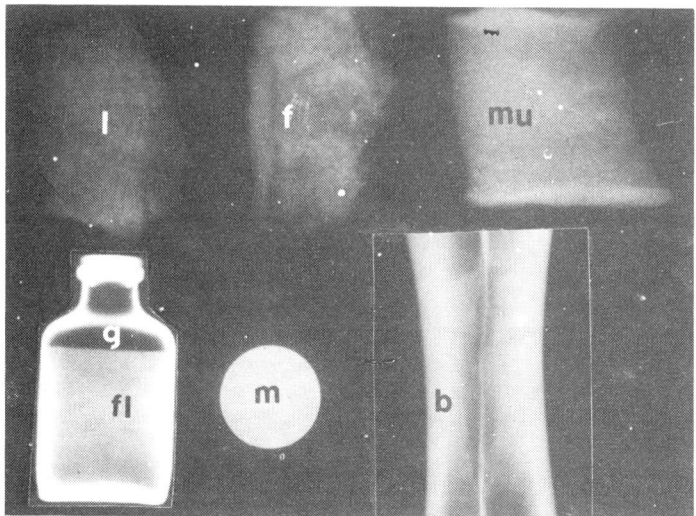

Fig. 4-22 : Radiographic densities of various types. l = A piece of lung (density is low than would be in a living subject), f = Fat, mu = Muscle, fl = Fluid, g = Gas, b = Bone, m = Metal.

Following factors affect radiographic density :
  i) Milliamperage
 ii) Exposure time
iii) Kilovoltage
 iv) Focal-film distance
  v) Speed of the film and intensifying screens
 vi) Developing time and temperature
vii) Grid ratio

Milliamperage : Main factor which influences radiographic density is mA as it controls radiation quality. As mA increases, more X-rays are able to reach the film. Thus **radiographic density is directly proportional to milliamperage provided that all other factors remain constant.**

Exposure time: As in case of mA, **radiographic density is directly proportional to exposure time if all other factors remain constant as total production of X-rays is increased.**

Killovoltage: kVp controls quality of X-rays, higher kVP producing high energy X-rays with high penetrability. **Increase in kVp increases radiographic density.** This effect, however, is more pronounced in low kVp range as small variations cause more change in X-ray quality. At higher ranges of kVp small variations have little effect. For example, **if kVp is increased from 30 to 45, it will cause a greater change in the radiographic density than if the same change is affected at a higher range i.e. say from 80 to 95.**

Focal-film distance : **Due to FFD change, variations in density are inversely proportional to the square of distances.** Thus if FFD is reduced to half of the original, radiographic density will increase by a factor of four. For example, if original FFD is 100 cm and it is reduced to 50 cm, density change will be:

$$\frac{(100)^2}{(50)^2} = \frac{10,000}{2,500} = 4 \text{ times increase.}$$ Fig. 4-23 shows the effect of change in FFD.

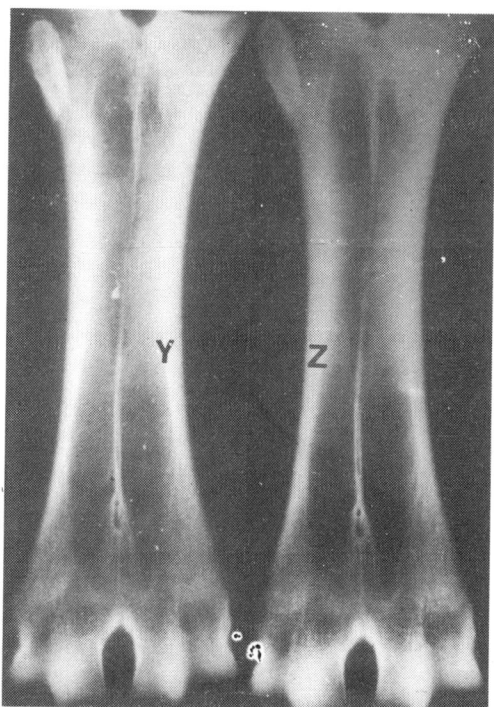

Fig. 4-23 : Effect of focal film distance change. Y = FFD 90 cm, Z = FFD 60 cm. A reduction in FFD increases radiographic density

Speed of the film and intensifying screens: As has already been discussed in section B of this chapter, speed of both film and screen affects radiographic density.

Developing time and temperature: Overdeveloping of the films, as a result of increase in time and/or temperature of the developing, leads to increase in radiographic density.

Grid ratio: Higher grid ratio allows more absorption of scatter radiation to affect radiographic density.

RADIOGRAPHIC CONTRAST

The difference in various densities of adjacent areas on a radiograph is radiographic contrast. It appears as a difference between blackness, greyness and whiteness due to variations in subject densities (Fig. 4-24) and enables the radiologist to interpret diagnostic information contained on a radiograph. An image with many densities i.e. black, dark grey, light grey, white and so on is said to have **low or long scale contrast.** If an image has only two densities i.e. black and white then contrast is high **(short scale).** **A good radiograph is expected to have a long scale contrast so that details of different structures are well visualised.**

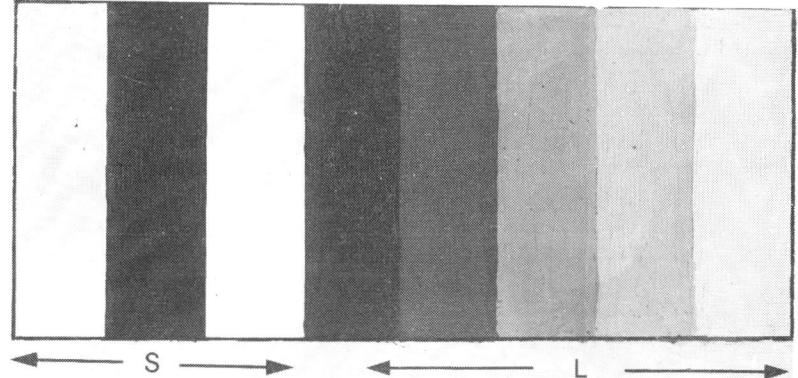

Fig. 4-24 : Variations in density create contrast. S = Short scale of contrast, L = Long scale of contrast.

Radiographic contrast is the product of film contrast and subject contast. Film contrast has already been discussed in section B of this chapter. So only subject contrast is being discussed here.

Subject contrast is the difference in the relative densities of various anatomical structures of a particular subject. When an X-ray beam strikes the structure being radiographed, there is differential absorption of X-ray photons across the structure. For example, bone will absorb more X-rays than muscle of equal thickness. Due to this difference in beam attenuation, X-ray photons reaching to the different areas of the film vary in number causing difference in densities and thus contrast is produced. **Following factors affect subject contrast:**

    i)   Thickness of the part
    ii) Density of the part
    iii) Effective atomic number of tissue
    iv) Killovoltage
    v)   Fog and scatter radiation.

Thickness of the part : The intensity of the transmitted beam is inversely proportional to the thickness of a given part. If thickness is increased, transmitted X-ray photons are reduced. For example, a 6 cm thick layer of muscle will attenuate X-ray beam more than a 4 cm thick muscle layer and thus more X-ray photons will be transmitted through the latter.

Density of the part : Greater the density (mass per unit volume) of a tissue, greater is the ability of the tissue to attenuate X-ray beam. For example, a bone will attenuate more X-rays than muscle of the same thickness as subject density of the bone is higher. Thus **radiographic density is inversely proportional to the subject density.**

Effective atomic nimber : Attenuation of the X-ray beam by photoelectric absorption is increased in substances with high effective atomic number, as has already been discussed in chapter 2, and is more true when low kVp is used. **The difference in effective atomic number of tissues is an important factor influencing contrast.** Since the difference is high between bone and muscles, a contrast is produced. For the same reason, contrast medium of high atomic number are used to alter densities of certain structures under examination so as to influence subject contrast and radiographic contrast.

Kilovoltage : The ability of X-rays to penetrate through a thickness of tissues depends upon the energy of X-rays which in turn is dependent upon kilovoltage applied. **Higher kVp generates X-rays of shorter wavelength with higher penetration power. As a result there is little variation in the penetration of thick and thin tissues. This results in low or long scale of contrast.** At lower kVp, lower energy X-rays are generated which pass through thin tissues but fail to penetrate thicker tissues. Thus high or short scale of contrast is obtained. Radiographic contrast thus varies inversely with kVp.

Fog and scatter radiation : Film fog due to light leakage in the cassette, or due to processing errors reduces contrast. Similarity scatter radiation reaching the film produces undesirable densities which cause loss of radiographic contrast.

SECTION- D

SCATTER RADIATION AND ITS CONTROL

RISHI TAYAL
JIT SINGH

Scatter radiation refers to radiation which deviates from the primary beam both in direction and wavelength after interacting with a medium or a patient being exposed to X-rays. Scatter radiation may follow different directions, if angle of scattering from the primary beam is less than 90°, it is **forward scatter** and if angle of deflection is more than 90°, it is called **back scatter.** At low kVp most scatter is backwards (Fig.4-25) while at high kVp most is forward (Fig 4-26). Some amount of scatter radiation is always able to reach the film to cause fogging, resulting into loss of contrast. **If no device is used to check this scatter radiation, 50% blackening of a processed film may be entirely due to scatter radiation.** Apart from affecting the quality of a radiograph, scatter radiation is also hazardous to personnel working in radiology section. It is thus important to understand factors influencing scatter radiation and also the methods that can be used to reduce its quantity. It has already been discussed earlier in chapter 2 that most scatter radiation in diagnostic radiology is the result of compton effect.

FACTORS AFFECTING SCATTER RADIATION

The relative intensity of scatter radiation is directly proportional to the following factors :
   i)   Kilovoltage
  ii)   Body part thickness
 iii)   Field size
Kilovoltage : **As the kVp increases, scatter radiation due to compton effect, and per cent transmission of X-rays through the patient increases but patient dose due to photoelectric**

effect decreases. Low kVp results in less scatter radiation and high radiographic contrast alongwith increased patient dose. High kVp techniques are basically used to control patient dose of radiation. In large animal radiography, high kVp has to be used because of thicker body parts. **At low kVp, high mAs can be used for sufficient X-ray's penetration but it may increase the patient dose to unacceptable level.**

Body part thickness : The quantity of scatter radiation increases with increased body part thickness (Fig.4-27) and control can not be exercised over this factor. **Increased body thickness results in multiple scattering** e.g, at 70 kVp, scattering in a 3 cm thick part is about 45% while at thickness of 30 cm, it is almost 100%. This is the reason why even high powered X-ray machines can not be used for obtaining abdominal radiographs in large animals. Certain devices can be used to control scatter radiation to some extent and are discussed further in this section.

Field size : Larger the field size, more is the scatter radiation reaching the film. **Radiographic quality can be increased by using a small field size as scatter radiation gets a large angle to escape away from the film** (Fig.4-28)

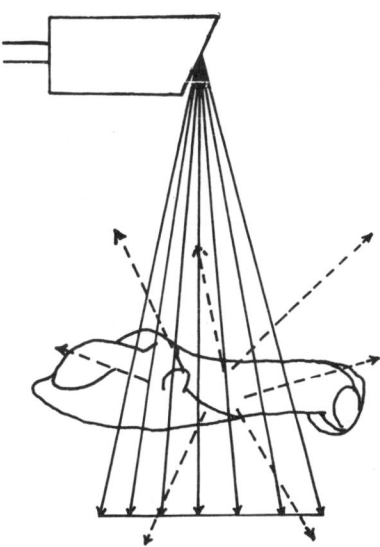 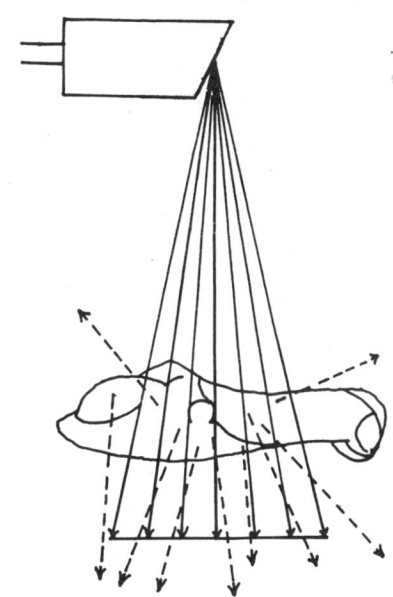

Fig. 4-25 : At low kVp most scatter radiation is backwards.

Fig. 4-26 : At high kVp most scatter radiation is forward. More scatter radiation reaches the film in this case.

SCATTER RADIATION CONTROL DEVICES

Main scatter radiation control devices are beam collimators and grid. **Use of filters also reduces scatter radiation but primary objective is to remove low energy photons from the primary beam so as to reduce patient dose.**

Beam collimators

Collimation refers to the regulation of X-ray beam by beam restricting devices to restrict it to the site of the part of the patient under examination. **Collimation offers following advantages:**

    i) Reduction in scatter radiation and thus improvement in radiographic quality.

    ii) Decrease in patient dose by reducing the area being exposed.

Several types of collimators are available which are placed in the path of X-ray beam as close to the tube as housing permits. Following types of collimators are most commonly used:

 i) Aperture diaphragm.
 ii) Cones and cylinders.
 iii) Variable aperture collimator.

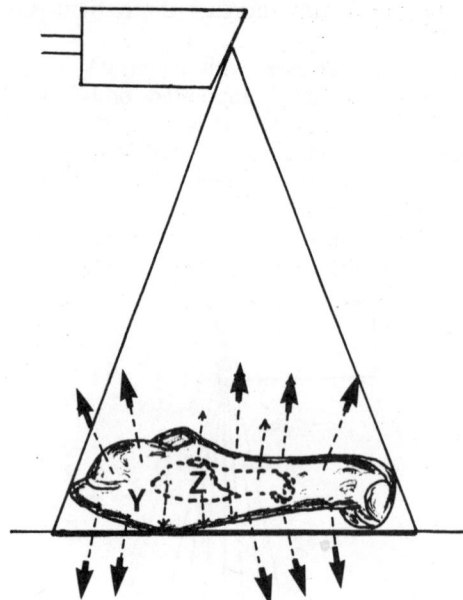

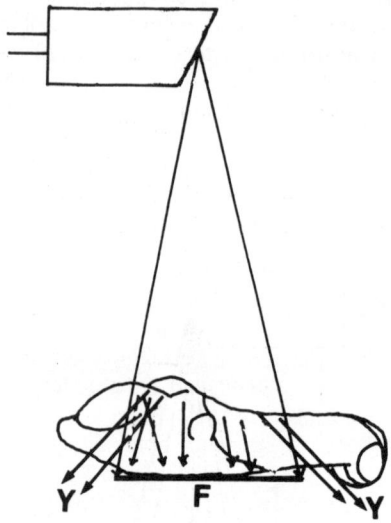

Fig. 4-27 : A thick part (Y) causes more scatter radiation than a thin (Z) part.

Fig. 4-28 : A small field size allows a large angle for some scatter radiation (Y) to pass away from the film (F).

Aperture diaphragm: It is simplest of the collimators and is made of a sheet of lead with a circular, square or rectangular hole in the centre that permits X-ray beam to pass through (Fig. 4-29). Principal disadvantage of an aperture diaphragm is a fairly large penumbra formation at the periphery (Fig. 4-30). another disadvantage is the inconvenience experienced while frequently changing the diaphragm as a separate size of diaphragm is required for each size of the film.

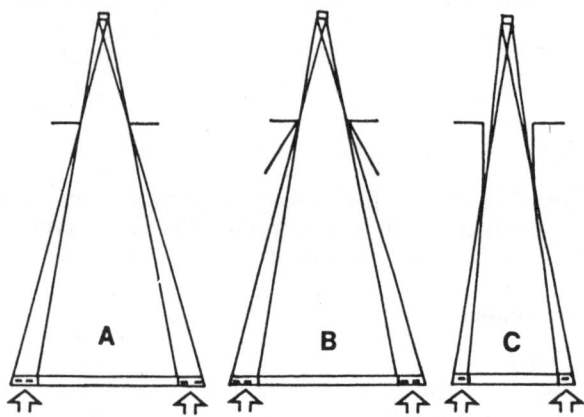

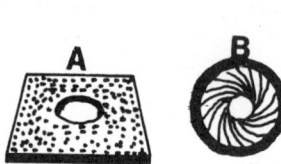

Fig. 4-29 : A = Aperture diaphragm, B = Adjustable aperture diaphragm.

Fig. 4-30 : Penumbra formation (arrows) with aperture diaphragm (A), cone (B) and cylinder (C). Penumbra formation is more with aperture diaphragm and less with a cylinder.

Cones and cylinders: These are conical, or cylinderical metal tubes (Fig. 4-31A) that channel an X-ray beam to the required field size. The base of a cone or a cylinder is made of lead to absorb X-rays. Both of these devices are ineffective in removing penumbra. A cylinder produces comparatively less penumbra because beam collimation takes place at its far and. Since size of cones and cylinders is fixed, these are appropriate only for specific radiographic examinations.

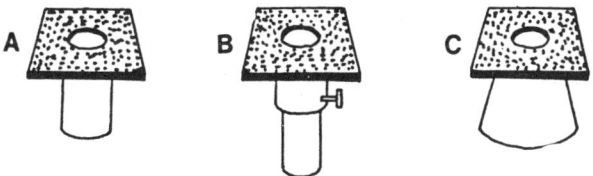

Fig. 4-31A : A = Non-adjustable cylinder, B = Adjustable cylinder, C = Cone.

Variable aperture collimator: This beam restricting device with adjustable lead shutter (Fig. 4-29) is the best and most commonly used in diagnostic radiology. Variable aperture collimator offers following advantages over other beam restricting devices :
   i)   X-ray beam can be adjusted to a variety of rectangular shapes and sizes.
   ii) Exposure field can be illuminated to permit its visualisation.
   iii) Penumbra is greately reduced.
   iv) Rectangular beam obtained with adjustable lead shutter exposes only the area of interest (Fig. 4-31B) thus reducing the patient dose.

There are two sets of adjustable diaphragms placed one above the other (Fig. 4-32) to work in pair. This allows rectangular or square fields of exposure of varying dimensions. Illumination is done with the aid of a light bulb and a relfecting mirror (Fig. 4-33). The light bulb, mirror and collimator shutter are so adjusted as to coincide illuminated field with X-ray beam field. Accuracy

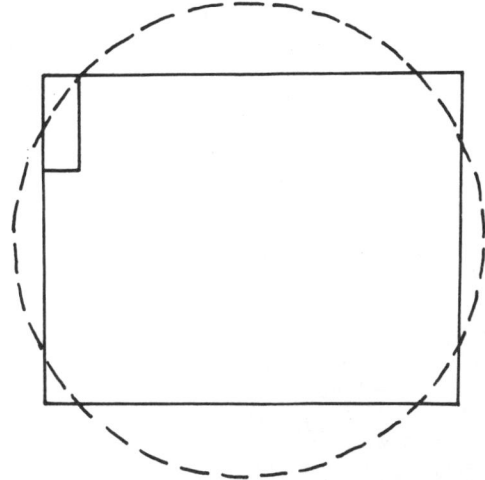

Fig. 4-31B : The field of exposure with adjustable lead shutters is limited to area of interest (rectangular area) than with a cone (circle). This arrangement reduces patient radiation dose and enhances radiographic quality (From Morgan, J.P. and Silverman, S. 1990. Techniques of Veterinary Radiography. 4th edn., Iowa State University Press, Used with permission).

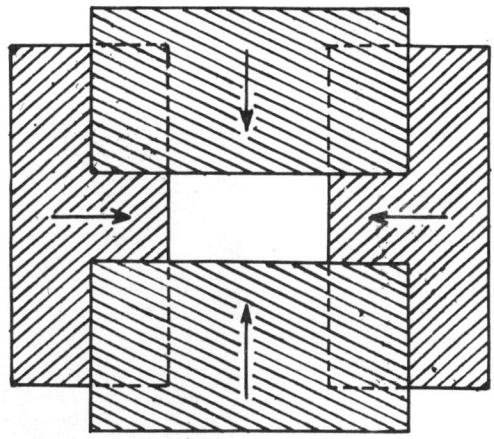

Fig. 4-32 : Top view of adjustable diaphragm (From Curry, T.S., Dowdey, J.E. and Murray. R.C. 1984. Christensen's Introduction to the Physics of Diagnostic Radiology. 3rd edn. Lea and Febiger, Philadelphia. Used with permission).

of the collimator should be checked periodically as adjustment may fall out with constant use. This can be done by a simple method detailed below:

i) Place a cassette on top of table and set the illuminated field smaller than the cassette.

ii) Position 'L' shaped piece of wire at each corner of the illuminated field.

iii) Make an exposure at 3 mAs and 45 kVp.

iv) Process the film.

v) Darkened area created by the exposure should rest inside wire perimeters. If primary beam exceeds this perimeter, malalignment of the mirror is indicated which should be corrected. Proper alignment is shown in fig. 4-34.

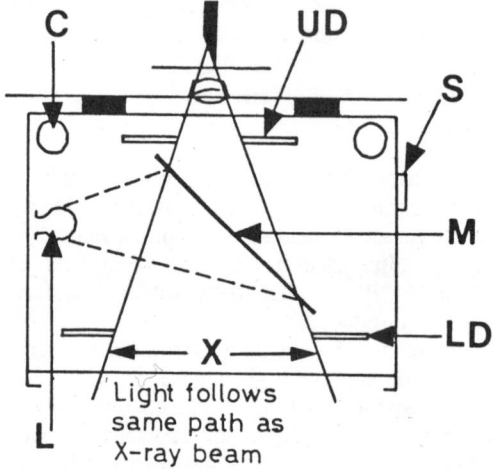

Fig. 4-33 : Diagrammatic representation of light beam diaphragm. C = Controls to open and close diaphragms, UD = Upper set of diaphragms, S = Light switch, M = Radiolucent mirror, LD = Lower set of diaphragms, X = X-ray beam, ·L = High intensity light bulb (From Forster, E. 1985. Equipment for Diagnostic Radiography. M.T.P. Press Ltd, Kluwer Academic Publishers, Netherland. Used with permission).

Fig. 4-34 : Since darkened area of exposure lies within wire perimeters, proper alignment of mirror of light beam diaphragm is indicated.

Grid

Grid is a flat plate containing series of alternating strips of radiodense (lead) and radiolucent (interspacer) material encased in a protective covering of thin aluminium. The interspace material is made of plastic or aluminium and has the main function of supporting lead strips at required angle and position. Aluminium is generally preferred as it is easy to mould, is non-hygroscopic, provides more durability to the grid and absorbs relatively more secondary radiation. Grid is placed between the part to be examined and cassette so as to absorb scatter radiation falling on the film. Since primary radiation is oriented in the same direction as lead strips, it passes through the grid. However, multidirectional scatter radiation is mostly absorbed (Fig. 4-35). **Use of grid, however, results in removal of large quantity of X-rays required to produce desired radiographic density and thus exposure factors have to be increased to compensate for the loss.**

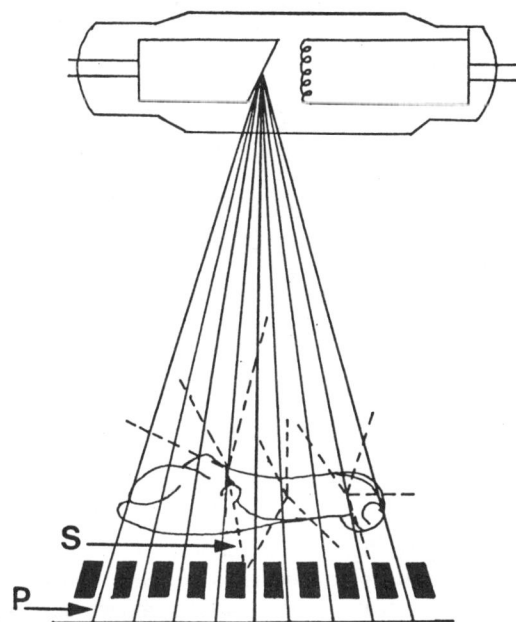

Fig. 4-35 : A grid prevents scatter radiation (S) from reaching the film while primary radiation (P) passes through.

Grid ratio : It is the ratio of the height of lead strips to the distance between the strips (Fig. 4-36). This ratio expresses grid's ability to absorb scatter radiation. **High ratio grids absorb more scatter radiation than low ratio grids but require more perfect centering, higher exposure and much narrower focal range.** In general, grid ratio ranges from 4:1 to 16:1. If exposures are to be made under 90 kVp, usually 8:1 or 10:1 grids are recommended. **Higher ratio grids are recommended for higher kVp ranges.** Use of 16:1 grid increases the patient dose significantly and so are rarely used.. Moreover, the difference between 12:1 and 16:1 grids is not large. A 5:1 grid absorbs about 85% of scatter radiation while 16:1 grid can absorb 97% of it.

Grid frequency: It is the number of lead strips per inch in a grid. Most grids have frequencies in the range of 60-110 lines per inch. **Grids with higher frequencies (more lines per inch) show less distinct grid lines on a radiograph as lead strips are much thinner. Such grids, however, require higher exposure and are less effective in absorbing high energy scatter radiation.** Therefore, as the grid frequency increases, grid ratio has to be increased to maintain same efficiency.

Lead content of grid : Lead content of a grid is expressed in gm per cm^2 and is closely related to grid ratio and frequency (Fig. 4-37). If grid ratio remains constant and grid frequency is increased to improve contrast factor, lead content of the grid must decrease. This is so because making the lead strips thinner does not increase grid frequency as mush as the lead content is decreased. For example, in a 10:1 grid if lines per inch are increased from 71 to 77, the lead content will decrease to 326 mg/cm^2 from 600 mg/cm^2. Lead content is decreased to improve contrast. However, there is a limitation to the extent grid frequency can be increased for the grid to be effective.

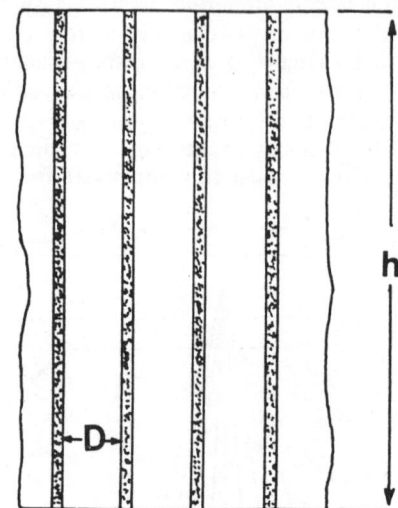

Fig. 4-36 : Cross section of a grid - a diagrammatic representation. D = Distance between two lead strips (0.25 mm), h = Height of lead strips (2.00 mm).

$$\text{Grid ratio} = \frac{h}{D} = \frac{2.0}{0.25} = 8.$$

(Fig. 4-36 to 4-38 are from Curry, T.S., Dowdey, J.E. and Murry, R.C. 1984. Christensen's Introduction to the Physics of Diagnostic Radiology, 3rd edn. Lea and Febiger, Philadelphia. Used with permission).

Grid selectivity : An ideal grid should transmit all primary radiation and absorb all scatter radiation in order to reduce patient exposure and to improve radiographic contrast. However, it is not possible in actual practice. **The ratio of transmitted primary radiation to transmitted scatter radiation is the grid selectivity.** Therefore, more a grid removes scatter radiation, higher is its selectivity. Grid selectivity is related to its lead content and usually **high ratio grids are said to have high selectivity because of higher lead content.** However, as has already been discussed, grid frequency can change lead content without changing grid ratio.

Contrast improvement factor of a grid: It is the ratio of contrast of a radiograph obtained with grid to contrast of a radiograph obtained without grid. **Principal function of grid is to improve radiographic contrast by reducing scatter radiation.** Most grids have contrast improvement factor between 1.5 to 2.5. In other words, radiographic contrast is almost double with the use of grid. Generally, contrast improvement factor of higher ratio grids is greater.

Bucky factor : It is the ratio of incident radiation falling on the grid to radiation transmitted through the grid. **Bucky factor is a measure of total quantity of radiation absorbed by the grid.** Usually high ratio grids have higher Bucky factor as more scatter radiation is absorbed. **Bucky factor of a grid indicates the extent of increase required in the exposure factors with the use of that grid.** It usually ranges between 3-4 for grid ratios 5-12. **Higher Bucky factor results in more radiation exposure to the patient.**

Types of grids

Parallel grid (linear grid) : In a parallel grid, lead strips are placed parallel to each other (Fig. 4-38) . Parallel strips absorb oblique scatter radiation while interspaces allow X-rays, perpendicular to the surface of the grid, to pass through to expose the film. Main advantage of a parallel grid is that X-ray tube can be angled along the length of the grid without grid cut-off. Moreover, lead strips focus at infinity and thus have no specified focal distance. For using parallel grid, focal film distance must be 120 cm or more which is an undesirable feature in veterinary radiography. Parallel grids are, therefore, of limited interest in veterinary radiography.

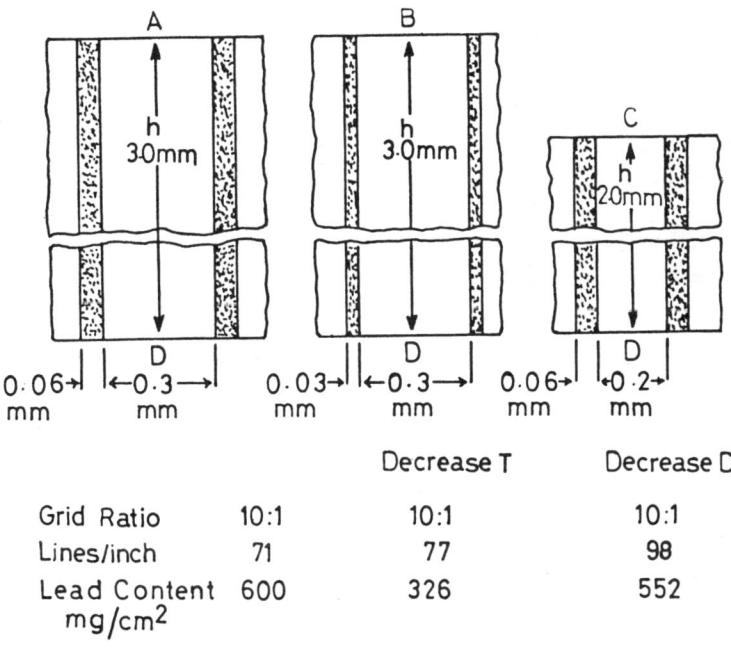

		Decrease T	Decrease D
Grid Ratio	10:1	10:1	10:1
Lines/inch	71	77	98
Lead Content mg/cm^2	600	326	552

Fig. 4-37 : Relationship between grid ratio, lead content and number of lines per inch. T = Thickness of lead strips. If lead strips are made thinner (B) as compared to in A, number of lines per inch increases without affecting grid ratio and there is considerable reduction in lead content. If number of lines per inch is increased by decreasing width of interspaces as in C, the grid ratio remains the same and content decreases marginally. Therefore, there is a limitation on the number of lines per inch a grid may have to be effective.

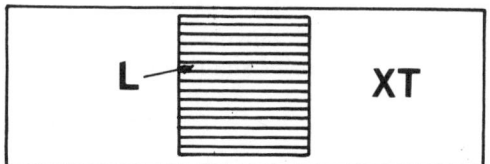

Fig. 4-38 : A parallel or linear grid (L). XT = X-ray table.

Crossed grid : It consists of two superimposed parallel grids placed at right angle to each other (Fig. 4-39) . A crossed grid is more efficient in absorbing scatter radiation. However, exposure factors are increased. Moreover, grid cut off also occurs if it becomes necessary to angle X-ray tube. In other words, X-ray beam must be centered at right angle to a crossed grid.

Focused grid : A focused grid may be parallel or crossed. In this type of grid, lead strips are angled increasingly towards edges so that if planes of the strips are extended in space, these meet along an imaginary line called convergent line in case of a parallel grid or point in case of crossed grid (Fig.4-40). Perpendicular distance between surface of the grid and convergent line or point is called **grid focal distance.** Focused grid has a focal distance range within which it can be used without grid cut-off. A high ratio grid has less focal range than a low ratio grid. Focal distance is marked on each grid.

Fig. 4-39 : A crossed grid.

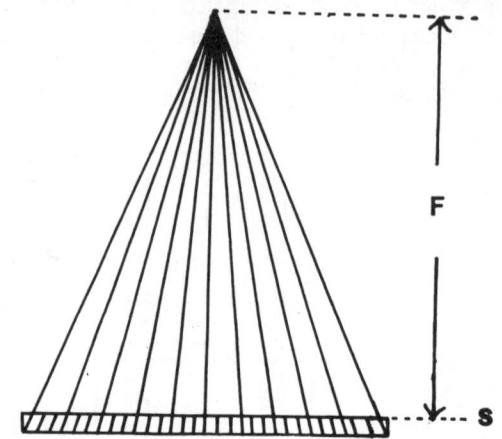

Fig. 4-40 : A focused grid. S = Surface of grid, F = Grid focus distance.

Grid cassette : It is a special type of cassette of which top is built in the form of a grid. Cassette grids are mostly useful when portable machines are used for radiographic work.Most of such grids are focused and have a grid ratio of 5.1 to 8.1. Care must be exercised while constructing cassette to make sure that grid is in close contact with the X-ray film as far as possible. A grid cassette offers following advantages :

i) Centering is determined more easily and accurately when film and cassette remain aligned as a single unit.
ii) Film and grid are maintained in close contact and are easy to handle.
iii) Grid is protected from damage which often occurs when a cassette is placed under a larger size grid and weight of a heavy patient comes over it.

Moving grid : With the use of a stationary grid, grid lines may appear on the radiograph. A moving grid, also called **Potter-Bucky diaphragm, Bucky diaphragm** or **Bucky grid** (Fig. 4-41) eliminates grid lines. It is a focused grid that moves mechanically across the X-ray beam during a radiographic exposure. It moves at a uniform speed adjusted to exposure time. The grid is usually fixed underneath the X-ray table (Fig-4-42) and can vary from a manually operated to a completely automatic reciprocating unit. It is available in all grid ratios. Disadvantages of a moving grid include its cost, possible mechanical failure, noise, limitation of exposure time and relatively more exposure because of increased grid cut-off. **A moving grid increases the patient dose by about 15% than that occurs with a stationary grid of similar physical characteristics.**

Use of grid

A grid is used when thickness of the part to be radiographed measures more than 10 cm. If grid is used, mAs must be increased 3-4 times. An increase in kV is not necessary but can be increased to maintain a relatively low time factor to avoid motion unsharpness of the image. If kV factor is increased by 10, mAs can be reduced by one half.

For proper functioning of the grid, it must be positioned correctly relative to the central axis of the primary X-ray beam. Any error in placement causes additional absorption of primary radiation to cause grid cut-off. **Grid cut-off refers to the loss of primary radiation as a result of undesirable absorption, and images of lead strips are projected wider than these would with ordinary magnification** (Fig. 4-43) . Thus cut-off is dependent upon the geometric relationship between the primary X-ray beam and lead strips of the grid. Grid cut off may be partial or complete and may result in reduced radiographic density or total absence of film exposure due to "cut-off" of the useful X-rays which otherwise would have reached the film. **Grid cut-off is higher with higher ratio grids and short focal distance. In a radiograph with grid cut-off, lighter areas appear in the centre or on one or more margins of the film, or whole radiograph may appear lighter. Following four basic situations produce grid cut-off:**

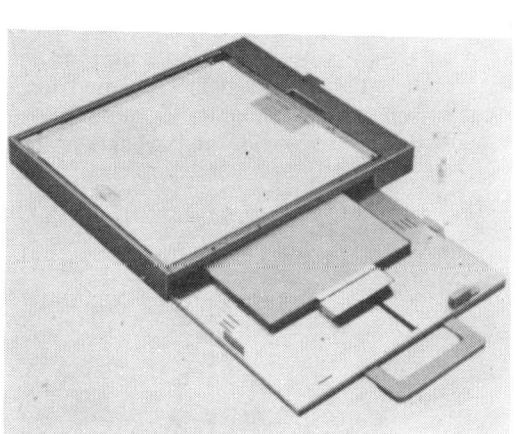

Fig. 4-41 : Moving or Bucky grid for cassette formats of 5" × 7" to 14" × 17" (Courtesy Siemens India Pvt Ltd)

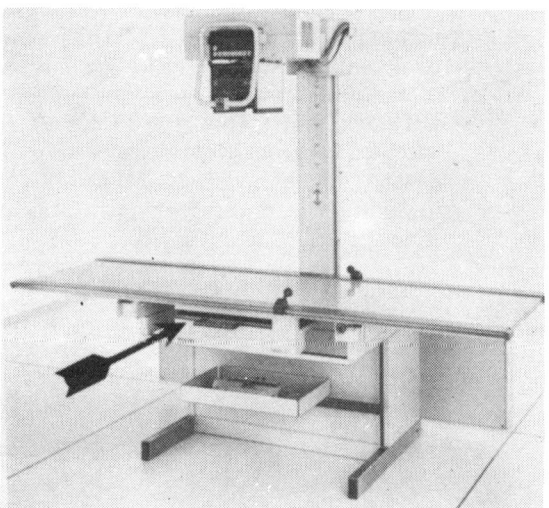

Fig. 4-42 : A moving grid (arrow) fixed underneath the X-ray table (Courtesy Siemens India Pvt Ltd).

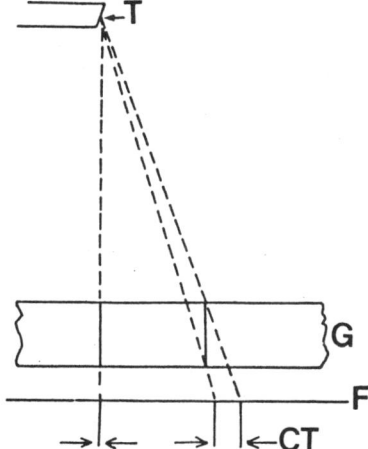

Fig. 4-43 : Grid cut off. T = Target, G = Grid, F = Film, CT = Cut off. Cut off is greatest with high ratio grids (From Curry, T.S., Dowdey, J.E. and Murry, R.C. 1984. Christensen's Introduction to the Physics of Diagnostic Radiology. 3rd edn. Lea and Febiger, Philadelphia. Used with permission).

i) Upside down grid : All focused grids are identified with a label indicating tube side and prescribed focal distance. If the grid is used upside down, peripheral cut-off occurs while centre of the film is exposed normally (Fig. 4-44B) . Higher the grid ratio, narrower will be the exposed area on the film.

ii) Off-level grid (tilted grid) : A grid must be perpendicular to the central X-ray beam. If the grid is tilted, central X-ray beam falls on the grid at an angle and cut-off occurs across the entire radiograph resulting in reduced density (Fig. 4-44C). This type of grid cut-off is common when horizontal X-ray beam is used to produce a radiograph.

iii) Off-centre grid : This is the most common type of grid cut-off and occurs when X-ray tube is positioned lateral to the central line of the grid but from a correct focal distance (Fig. 4-44D) . Any lateral shift results in grid cut-off across the entire grid to produce a uniformly light radiograph. The severity of cut-off is directly proportional to the extent of decentring and grid ratio and inversely proportional to the focal distance. Magnitude of cut-off is less when low ratio grids and long focal distance are used.

iv) Off focus grid : This type of cut-off occurs when X-ray tube is positioned beyond the specified range of focal distance (Fig. 4-44E) . Farther the grid from the specified focal distance, more severe is the cut-off. Grid cut off in such a situation is more severe towards the periphery and cut-off increases with higher ratio grids and when distance from the centre is increased.

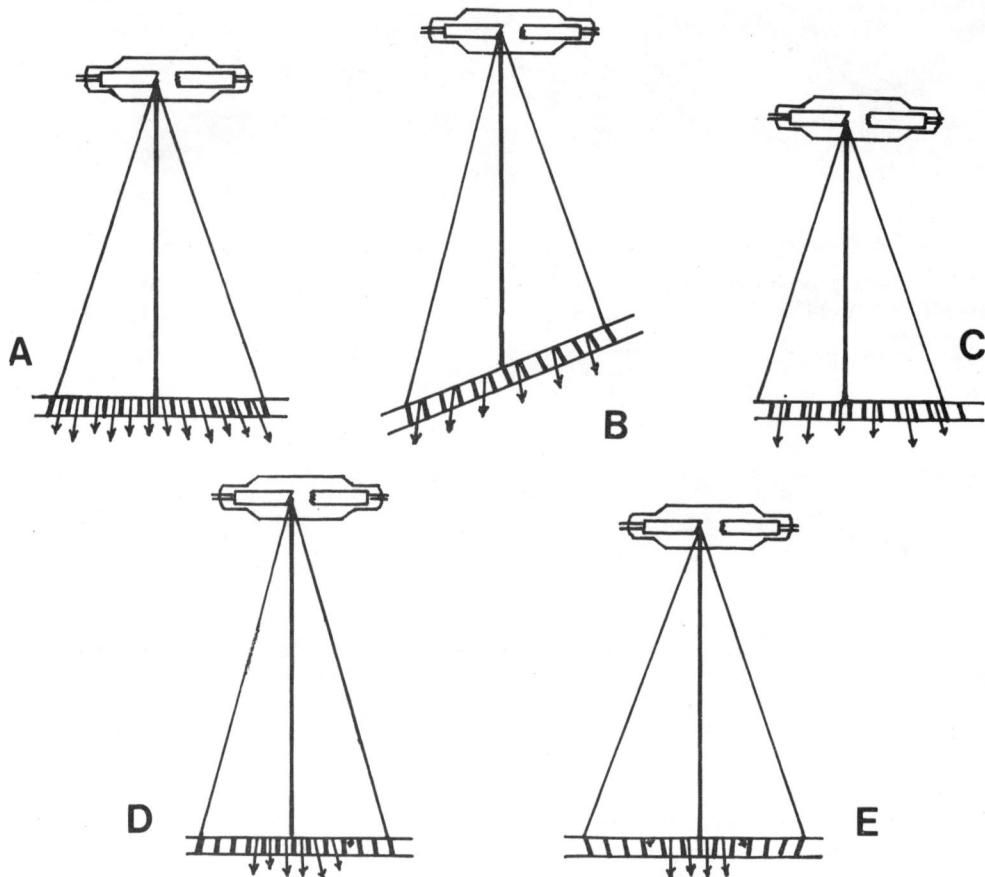

Fig. 4-44 : Four basic situations that produce grid cut off. A = Proper position of grid, B = off level grid, C = Off centre grid, D = Off focus grid, E = Upside down grid (After Bushong, 1975).

Care of grid

Grids are expensive and fragile items and must be handled with great care. If dropped accidentally, grid may be damaged permanently. Spilling of water or contrast material over the grid may present repetitive artifact on a radiograph or may damage interspace material. Such liquids, if spilled, should be immediately wiped off the surface of the grid.

Air gap technique

It is an alternate method to the use of grid to control quantity of scatter radiation interacting with the radiographic film. Scatter radiation arising from the patient being exposed to X-rays disperses in all directions. More nearer is the film to the patient, more it is exposed to scatter radiation. Air gap technique thus uses the simple method of increasing the distance between the patient and film thus reducing the quantity of scatter radiation interacting with the film (Fig. 4-45) with resultant better quality of radiograph. Usually the distance kept between the patient and film for air gap technique is 6-9 cm. Some magnification of the image does occur of which extent, however, is acceptable. The technique has found application in human radiography in the areas of chest and cerebral angiography. It is, however, not that effective when high kVp is used as direction of scatter radiation in such a situation is more forward. Due to this reason, technique has not found favour in large animal radiography. The technique is sometimes referred to as air filtration but is not proper as air does not act as a selective filter of scatter radiation in this method. It is rather the distance between the patient and film which allows a quantity of scatter radiation to escape without interacting with the film.

Filters

Diagnostic X-ray beam consists of a spectrum of different energies. Many of these photons are of lower energy (long wavelength) and do not contribute towards diagnostic quality of a radiograph, and rather increase the radiation dose of the patient. Primary purpose of placing a filter between the patient and X-ray tube is to remove less energetic (soft) X-rays from the primary beam which have no chance to reach the film. **The filtered X-ray beam decreases the exposure dose of the patient and scatter radiation.**These factors inturn increase the radiographic detail. Though mainly low energy photons are filtered, some high energy photons are also attenuated.

Filtration of diagnostic X-ray beam has two components : inherent filtration and added filtration. The sum of inherent and added filtration is total filtration of an X-ray beam.

Glass envelop of the X-ray tube, insulting oil surrounding the tube and the backelite window in the tube housing, all are responsible for inherent filtration, the glass envelop being the main. Added filtration results from placing an absorber (metal filter) in the path of a heterogenous X-ray beam. Thus while inherent filtration can not be controlled, added filtration can be controlled. Almost any material can be used for the purpose but aluminium and copper are most commonly used. Aluminium (atomic number 13) is considered excellent for most diagnostic X-ray tubes and is inexpensive and easy to procure. Copper (atomic number 29) in combination with aluminium is considered better for high energy producing units.

Filtration ability of a material is measured in terms of aluminium that would be required to cause same degree of filtration as the material in question. Inherent filtration of a diagnostic X-ray tube usually ranges 0.5 to 1.0 mm aluminium equivalent. The American National Council of Radiation Protection and Measurements has recommended total filtration at various kVp levels as : < 50 kVp 0.5 mm; 50-70 kVp 1.5 mm and at > 70 kVp 2.5 mm aluminium equivalent. An aluminium filter of 2 mm thickness absorbs most lower energy X-rays from the primary beam. Apart from removing most low energy photons from the primary beam, filters also decrease radiation dose of the patient, by about 80% if a 3 mm aluminium filter is used.

Wedge filters, used extensively in X-ray therapy, have limited use in diagnostic radiology. These filters are made of aluminium or lead acrylic that is 30% lead by weight. A wedge filter may be used in a situation where there is large variation in the thickness of tissues in a part being radiographed. By using a wedge filter, it becomes possible to filter X-ray beam as per the thickness of the part of the area to be radiographed so as to obtain a radiograph of uniform density (Fig. 4-46). Intravenous fluid bags have been used to create a wedge filtration effect while radiographing extremities of small animals. These are placed beneath the tissue of lesser thickness in order to equalise radiographic opacity over the entire limb. These used bags are routinely available in a clinic and can serve as good substitutes to wedge filters of lead acrylic or aluminium.

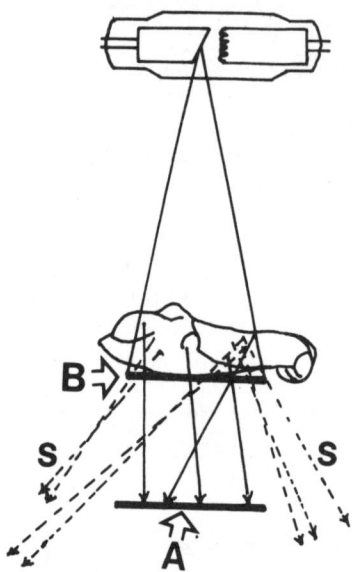

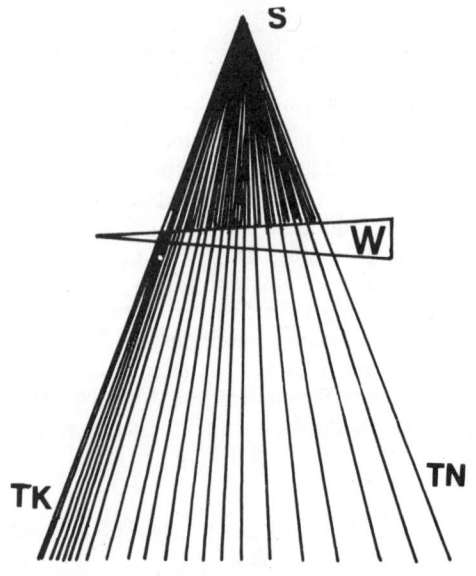

Fig. 4-45 : Air gap technique. If the film is kept at a distance (A) from the object, much scatter radiation (S) escapes away from the film than if the film is near the object (B). Air gap technique thus enhances radiographic quality.

Fig. 4-46 : Wedge filter. S = Source of X-rays, W = Wedge filter, TK = Thick part, TN = Thin part (From Curry, T.S., Dowdey, J.E. and Murry, R.C. 1984. Christensen's Introduction to the Physics of Diagnostic Radiology. 3rd edn. Lea and Febiger, Philadelphia. Used with permission).

REFERENCES

Bookstein, J.J. and Steck, W. 1971. Effective focal spot size. Radiology. **98**, 31.

Bushong, S.C. 1975. Radiologic Science for Technologists. The C.V. Mosby Co., Saint Louis.

Curry, T.S., Dowdey, J.E. and Murry, R.C. 1984. Christensen's Introduction to the Physics of Diagnostic Radiology. 3rd edn., Lea and Febiger, Philadelphia.

Doi, K., Loo, L.N., Anderson, T.M. and Frank, P.H. 1981. Effect of crossover exposures on radiographic image quality of screen-film systems. Radiology. **139,** 701.

Douglas, S.W. and Williamson, H.D. 1980. Principles of Veterinary Radiography. 3rd edn., Williams and Wilkinis Co., Baltimore.

Gillette, E.L., Thrall, D.E. and Lebel, J.L. 1977. Carlson's Veterinary Radiology. 3rd edn., Lea and Febiger, Philadelphia

Gould, R.G. and Hale,J. 1974. Control of scattered radiation by air gap technique: applications to chest radiography. Am. J. Roentgenol.**122**, 109.

Kealy, J.K. 1987. Diagnostic Radiology of Dog and Cat. 2nd edn., W.B. Saunders Co., Philadelphia.

Koblik, P., Hornott, W.J. and 0'Brien, T.F.1980. Rare earth intensification screen for veterinary radiography. An evaluation of two systems. Vet. Rad. **21**,224.

Kramer, R.W. 1993.A simple tissue equivalent "Wedge" filtration technique. Vet. Rad. Ultrasound **34,** 18.

Morgan, J.P., Silverman, S. 1990. Techniques of Veterinary Radiography. 4th edn., Iowa state University Press, Ames.

Meyer, W.1977. Radiography review: Radiographic density. J.Am. Vet. Radiol. Soc. **18**, 13.

Rao, G.U.V., Fatouros, P.P. and James, A.E. 1978. Physical characteristics of modern radiographic screen film system. Invest Radiol. **13**, 460.

Roynold, J.A. 1955. Factors affecting radiographic quality. Vet. Med. **50**, 187.

Selman, J. 1977. The Fundamentals of X-ray and Radium Physics. 6th edn., Charles C. Thomas Publisher, Illinois.

Stromberg, B., Olsson, S.E. and Lundgreen, M. 1978. Rare earth intensifying screens in veterinary radiography. J. Small Anim. Pract. **19**, 689.

Sturm, R.E. and Morgan, R.H. 1949. Screen intensification systems and their limitations Am.J. Roentgenol. **62**, 613.

Thompson, T.T. 1974. Selecting medical X-ray film. Appl Radiol. **3**, 47.

Ticer, J.W. 1984. Radiographic Techniques in Veterinary Practice. 2nd edn., W.B. Saunders Co., Philadelphia.

Trout, E.D., Kelley, J.P. and Cathoy, G.A. 1952. The use of filters to control radiation exposure to the patient in diagnostic roentgenology. Am.J. Roentgenol **67**, 1942.

Wagner, R.F. and Weaver, K.E. 1976. Prospects for X-ray exposure reduction using rare earth intensifying screens. Radiol. **118**, 183.

SELECTED QUESTIONS

1. Name the geometric factors that affect radiographic quality. Explain the effect of FFD on image magnification.
2. What is penumbra? Which factors affect its formation?
3. What qualities should be there in an ideal film base?
4. How a permanent image is formed on an X-ray film?
5. What is a characteristic curve of the film? What are its advantages to a radiographer?
6. Which factors influence film contrast? Name the factors that affect development of the film in relation to film contrast.
7. What do you understand by :
   i) Film speed ii) Film latitude iii) Cross-over exposure.
8. What is the difference between screen and non-screen films?
9. What precautions are required while handling and storing the films?
10. Name the layers of an intensifying screen and list required qualities of the phosphor layer material.
11. What are absorption and conversion efficiencies of the intensifying screens?
12. What is the advantage of intensifying action of the screen?
13. What are the rare earth intensifying screens?
14. List the causes of poor film-screen contact.
15. Why a cassette is used in radiology?
16. Name structural components of a cassette.
17. What are the main visual requirements of a good diagnostic radiograph?
18. Name the factors that influence detail. What is radiographic mottle?
19. What is the difference between radiographic and subject density? How mA affects radiographic density?
20. What is radiographic contrast? Which factors affect subject contrast?
21. How effective atomic number influences contrast?
22. How kVp affects the quality of scatter radiation?
23. What is collimation? Name the devices used for collimation of X-ray beam.
24. Describe the test that can be used to check accuracy of variable aperture collimator.
25. What do you mean by:
   i) Grid ii) Grid-ratio iii) Grid cut-off
26. What is the difference between a linear grid and a crossed grid? Why a linear grid is not preferred in veterinary practice?
27. What effect does a filter have on the X-ray beam?
28. What is the purpose of using a wedge filter?

CHAPTER

5 FILM PROCESSING AND RADIOGRAPHIC FAULTS

SECTION-A

FILM PROCESSING

A.P. BHOKRE

J.R. JINDAL

To obtain a quality radiograph three basic requirements are :
  i)   Correct positioning of the patient and part to be radiographed.
  ii) Correct exposure.
  iii) Correct processing of the film.
    It is often frustrating when all the work done goes in vain due to faulty procedures of processing the film in a dark room. Dark room procedures include:
  i)   Loading of the film in a cassette (Fig. 5.1).

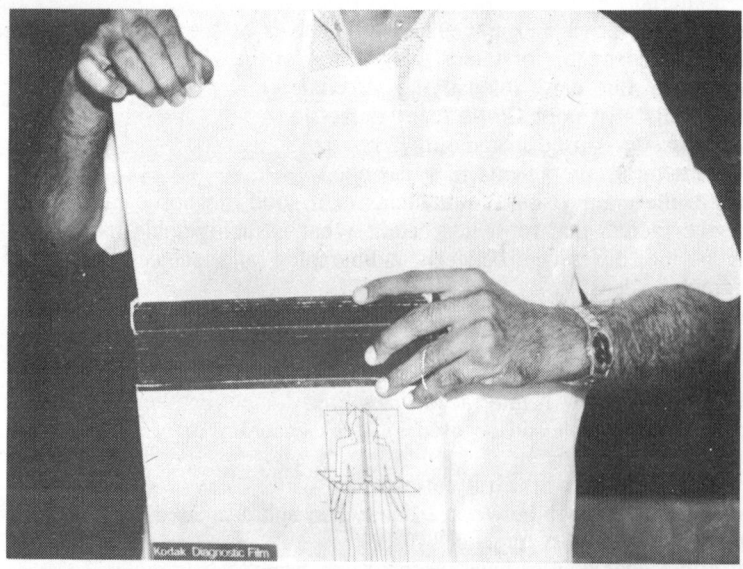

Fig. 5-1A : Taking out the film from the box.

ii)   Unloading of the exposed film in dark room.

iii)   Processing of the film to make latent image visible for viewing.

Errors in dark room lead to unnecessary repeated radiographic examinations, wastage of time and energy and increased cost of clinical examination. Thus it is essential to understand and master dark room procedures which should then be carried out in an orderly manner.

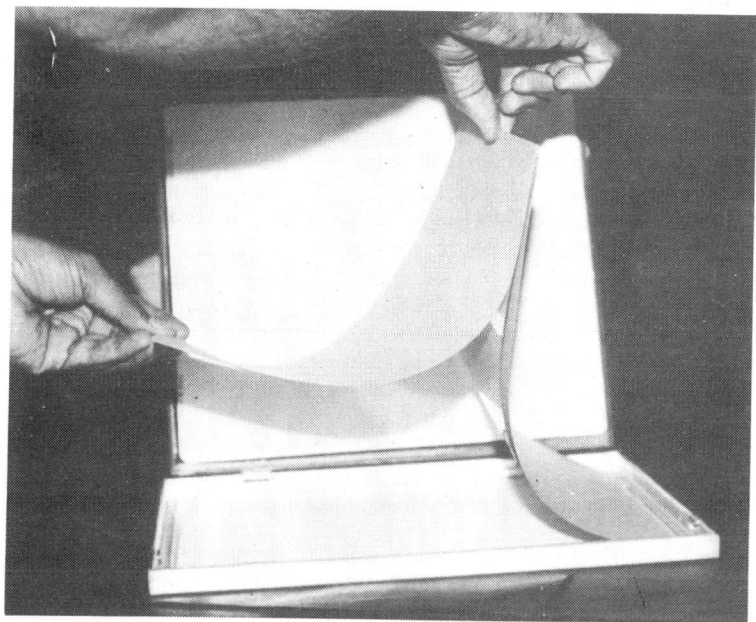

Fig. 5-1B : Placement of the film with wrapper in a cassette.

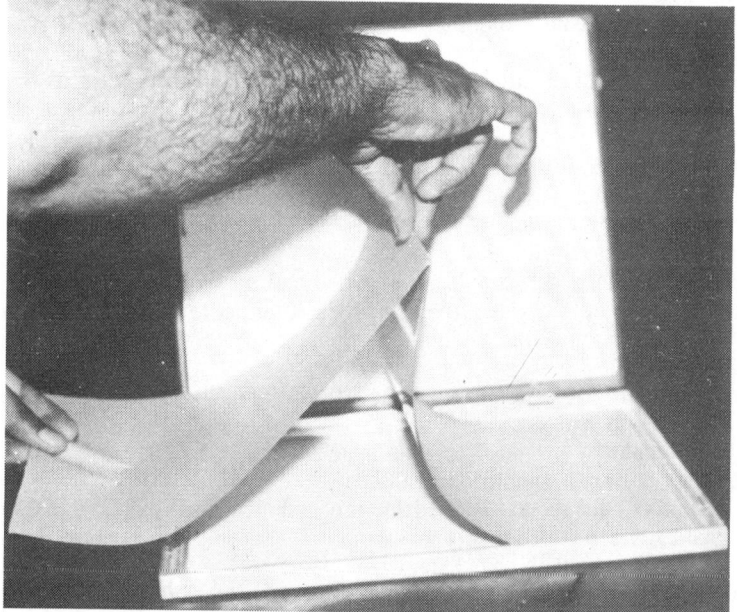

Fig. 5-1C : Releasing the film from its wrapper into the cassette.

DARK ROOM

In veterinary practice, manual processing of the film is preferred as automatic processors are expensive to buy and maintain. Dark room planning should, therefore, suit wet tank manual processing.

Dark room construction

i) Dark room should have sufficient space to accommodate a dry bench (3' × 2'), a wet bench of almost same dimensions and a sink. Too large a room is as undesirable as a small room.

ii) Dark room should be near the X-ray examination area.

iii) The walls should be constructed of solid concrete (15cm thick) or the room should have a lead box inside to store boxes of unexposed films currently in use.

iv) Floor should be impervious to fluids and processing solutions and easy to clean. Linoleum, or any other such material, may be preferred for floor covering as it is easy to clean.

v) The walls and roof may be painted white or cream enamel as such a paint acts as a good reflecting surface for safe light.

vi) There should be provision for sufficient running water.

vii) Room should be well ventilated but light proof. Ventilation holes should be adequately baffled. There should be no space for light to sneak through under the doors. Any light entering the dark room causes film fog.

viii) Entry to the dark room should be preferred through a double door with bolts inside to avoid any accidental opening when processing is going on.

ix) The room should neither be damp nor subjected to extremes of temperatures.

Dark room layout

Layout of the dark room to maintain an orderly flow of work is more important than the size of the room. General plan should be preferably such that after entering the room flow of work is from left to right side. Since unloading of the film is involved first, dry bench should precede 'wet bench'. There should be enough space between dry section and wet section (Fig.5-2) so as to avoid accidental splashing of processing solutions over dry section. If both are adjacent to each other provision should be made for a partition between the two sections (Fig. 5-3) **Everything in the dark room should have a definite place for its storage so that there is no difficulty in obtaining it under minimal illumination conditions.**

   A. **Dry section :** i) Dry bench should have enough space to open the largest available cassette and preferably be of 3' × 2' dimensions. The height of the bench should be around 3 feet so that anyone is able to work comfortably in a standing position.

   ii) There should be provision of cupboards under the bench top to store cassettes, timer, film marking devices etc.

   iii) Top surface of the bench should be of wood and heavy linoleum.

   iv) Brackets should be provided over the bench to store different sizes of film hangers. Hangers of the same size should be grouped together and kept separated from other sizes.

   v) There should be provision for keeping film storage bins or a lead box to store X-ray films currently in use.

   vi) Safe light should be provided to illuminate the area.

   vii) A waste paper basket should be available.

   B. **Wet section** i) This section is used for processing of the film. Usual set up is to install a thermostatically controlled processing unit or a sink (with provision of floating thermometer) in which processing tanks are kept.

   ii) China tiles should be fixed on the walls around processing and washing area.

   iii) There should be a safe light in this section too.

iv) A viewer should be provided near the fixer tank with a drip sink underneath. (wet films are often taken out from the fixer tank, viewed and then again replaced in fixer for complete fixing).

v) There should be provision for hanging a towel in this area so that during processing hands can be washed and dried to load the cassette.

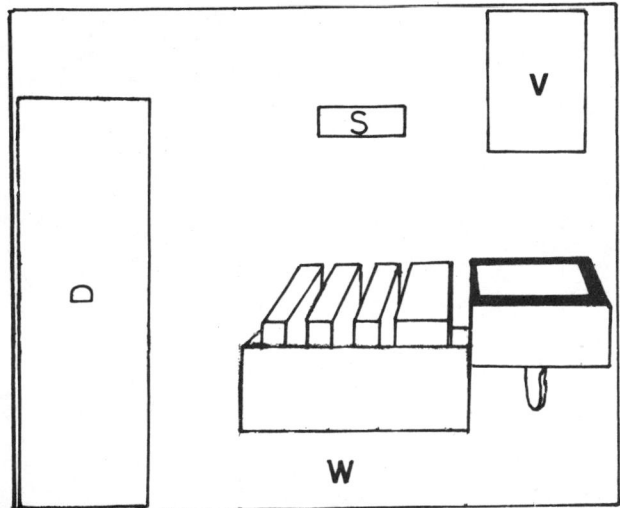

Fig. 5-2 : Hypothetical layout of dark room with sufficient space between the dry and wet sections. D = Dry section, W = Wet section S = Safe light, V = Viewer with a drip sink underneath.

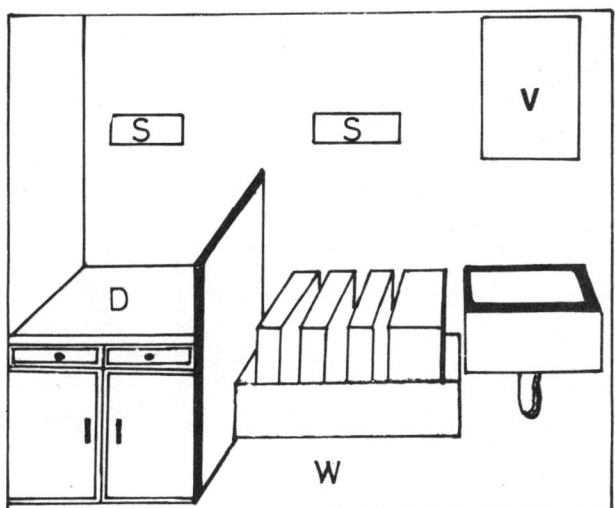

Fig. 5-3 : Hypothetical layout of dark room with a partition between the dry and wet sections. D = Dry section, W = Wet section, S= Safe light, V = Viewer with a drip sink underneath.

C. Film hangers : Three types of hangers are available : channel type, tension type and clip type (Fig. 5-4). In channel type of hangers, processing solutions may be retained in the channels which require careful cleaning for their maintenance. Film should be taken out of these hangers when being dried. In clip type hangers, there is risk of clips scratching another film during processing. A similar risk is involved in tension type hangers. Loading of the film in hangers is shown in figs. 5-5 and 5-6.

Fig. 5-4A : A channel type film hanger.

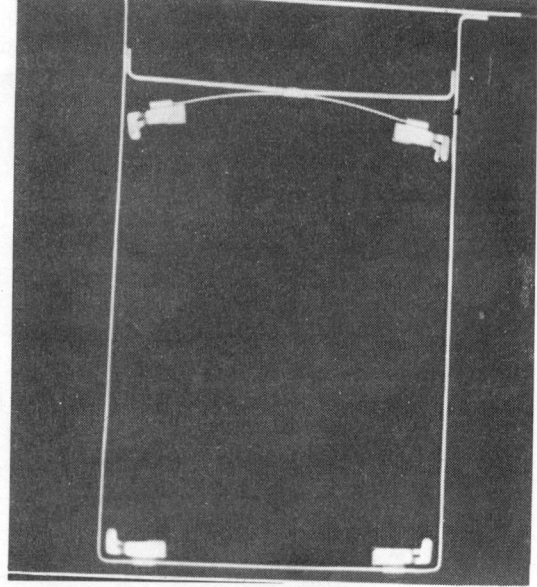

Fig. 5-4B : A tension type film hanger.

Fig. 5-4C : A clip type hanger.

D. Safe Light : It is a box (Fig. 5-7) containing a low watt (10 watt maximum) frosted bulb covered by a specific filter. One type of safe light may not always be suitable for all types of films. For example, changing from calcium tungstate screens to rare earth screens requires change from blue-sensitive film to green-sensitive film which inturn may require change of safe light. It is better to run the following **safe light fog test :**

  i)   Place a film on the dry bench.
 ii)   Place a few coins on top of the film.
iii)   Switch on safe light fixed on top of dry bench.
 iv)   Wait for two minutes.

v) Process the film.

vi) **If image** of coins is present on the film, the film is not safe in the prevailing dark room conditions.

·**Fig. 5-5A :** Holding of the film and channel type hanger.

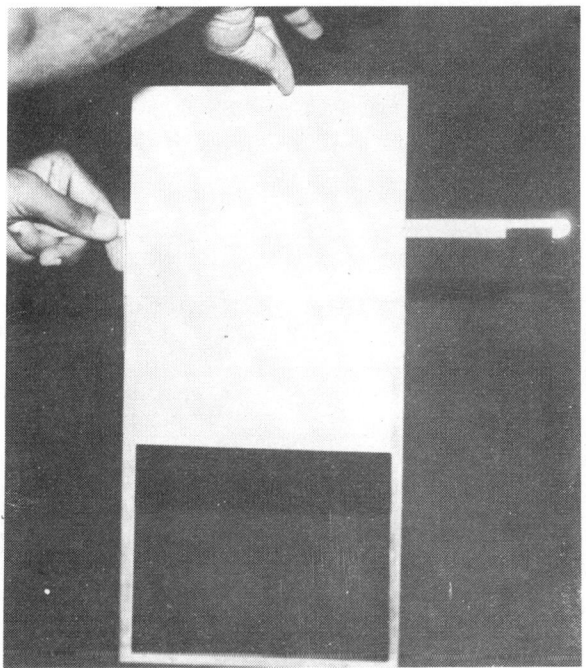

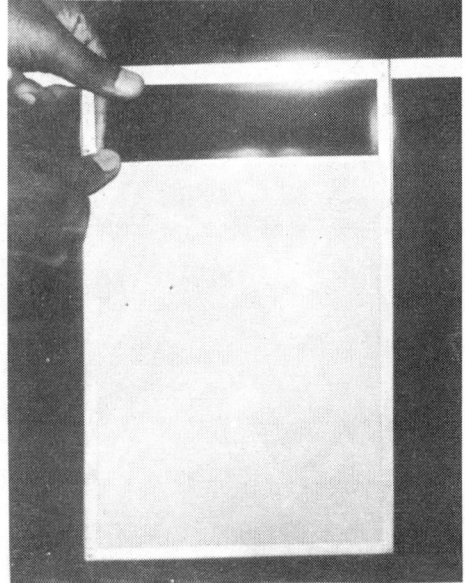

Fig. 5-5B : Insertion of the film in channel type hanger.   **Fig. 5-5C :** Channel type hanger with film in it.

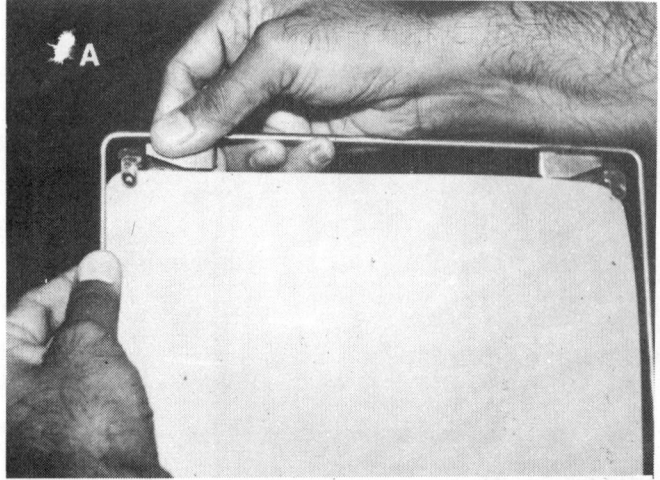

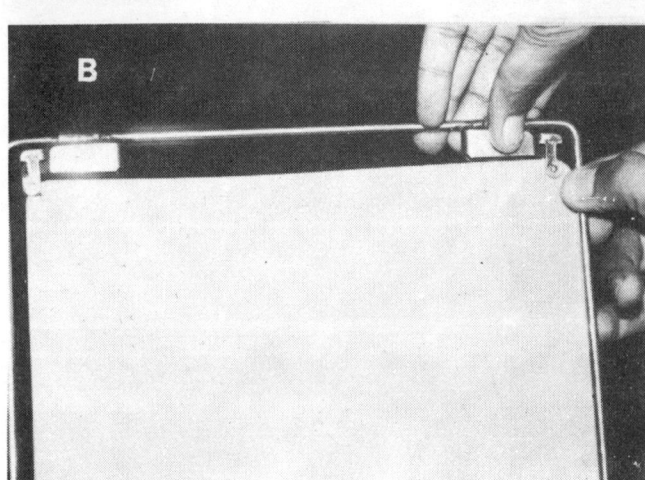

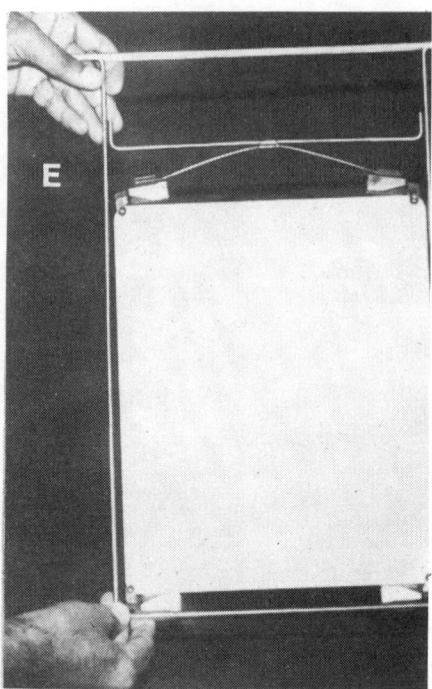

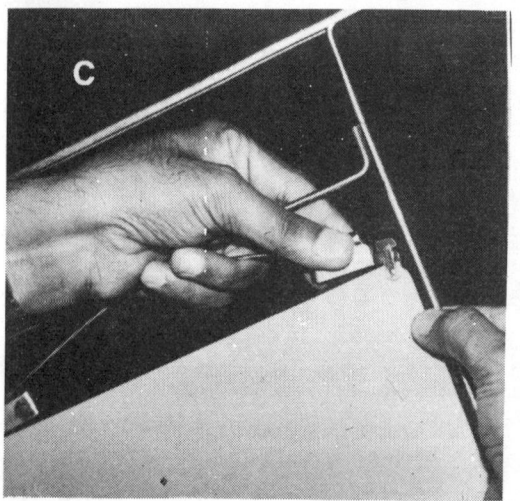

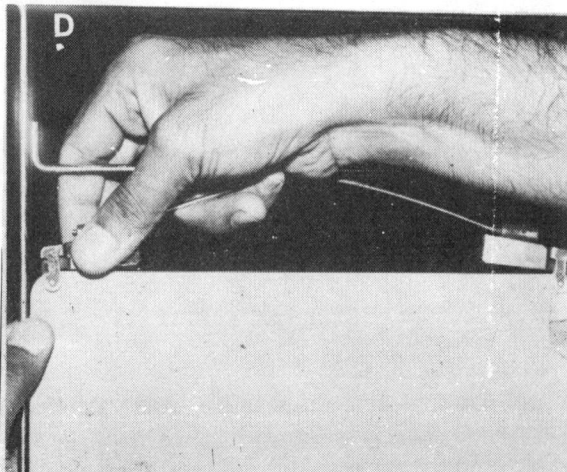

Fig. 5-6 : A to E show steps for loading a film in the tension type hanger.

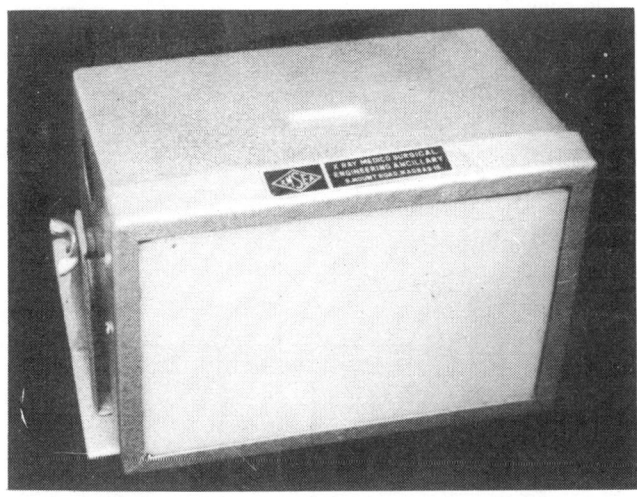

Fig. 5.7 : A safe light suitable for wall mounting.

Following points should be considered regarding safe light :
- i) Safe light must have correct filter.
- ii) It should be atleast 3 feet away from the film.
- iii) Bulb in safe light must be of correct wattage (maximum 10 watts).
- iv) A film should not be left exposed to safe light indefinitely.
- v) Films should be exposed to safe light only during loading or unloading of cassettes and during processing.

FILM PROCESSING

A) Processing tanks (Fig 5-8)

X-ray films are now universally processed in standard vertical tanks which are available commercially of 9, 13 and 22 litres capacity. The material used is either plastic or stainless steel. Four tanks are required i.e. for developing, rinsing, fixing and washing. Two fixer tanks can be incorporated into one processing unit if flow of work is more as time spent on fixing is two times more than that for developing. Washing tank should be of a larger capacity as many films get pooled there. Developing and fixing tanks must always remain covered with lids. Periodic elaborate cleaning of these tanks is not necessary. When solutions are changed tanks should be cleaned with water and long handled brushes. Tanks may occasionally be sterilised using hypochlorite bleaching solution to reduce the risk of bacterial growth.

B) Processing solutions

Processing solutions include developer, rinser, fixer and replenisher for developer :

i) Developer : Developer reduces exposed silver halide crystals of the film to metallic silver, which converts the latent image into a visible one. Developing solutions contain reducing agent, an activator, a restrainer, a preservative and a solvent.

a. Reducing agent : Hydroquinone or metol is used as reducing agent to convert exposed silver halide crystals into metallic silver. It provides contrast in image by developing dark and grey tones on the film.

b. Activator : Sodium carbonate is used as an activator to soften and swell film emulsion so that reducing agent can act on the film. It also provides an alkaline medium for the reducing agent. Phenicide/phenidone can also be used as a reducing agent but is expensive.

c. Restrainer : Potassium bromide as a restrainer controls the activity of the reducing agent so that fogging of the film does not occur.

d. Preservative : Sodium sulphite is used as a preservative to control rapid oxidation of developing agents.

e. Solvent : Water is used as a solvent for chemicals.

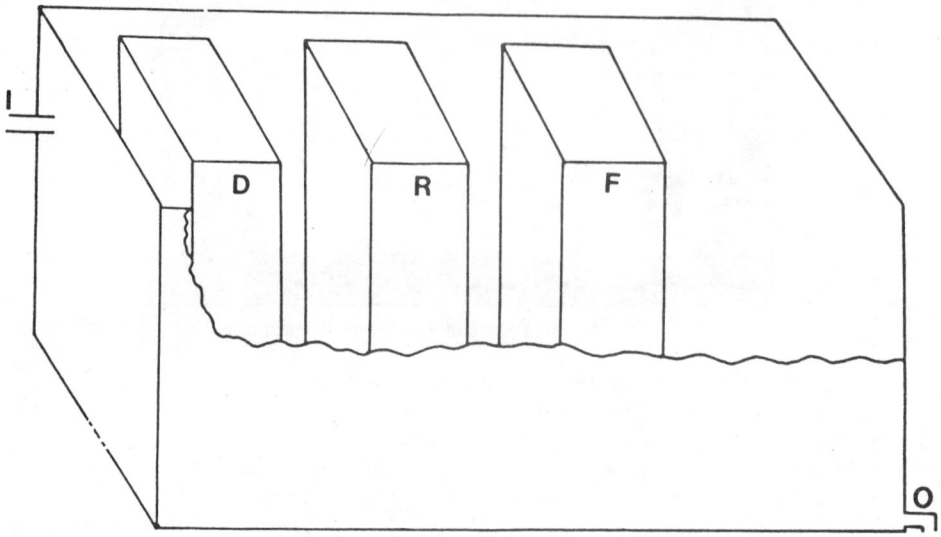

Fig. 5-8 : Diagrammatic representation of processing tanks in a unit. D = Developer, R = Rinser, F = Fixer, I = Water inlet, O = Water outlet.

Each type of developer reduces the exposed grains of silver halide for a specified time and if prolonged time is used fog develops on the film. Each manufacturer, therefore, recommends specific developing time, usually 4-5 minutes at 20°C. It is essential to follow manufacturer's recommendations in this regard. **Higher temperature of developing solution requires lesser developing time and vice-versa.** Time temperature combinations are discussed further in this section. During extreme temperatures, if processing soluition's temperature is not controlled thermostatically, ice or warm water can be added to water around the developing tank to maintain required temperature.

Strong solutions of original developer (replenisher) are available with manufacturers to replace volumes lost during developing process and to maintain even chemical activity as far as possible. It also permits a constant developing time. About 4 litres of replenisher should be added for fifty 14" × 17" films or their equivalent area developed. However, replenshing system does not work indefinitely. Once replenisher has been used in equal quantities to that of original solution, whole developing solution should be made fresh . **Once developing solution has been made, it should never be kept for more than 3 months as its oxidation renders the solution unfit for use.** If some quantity of solution is to be stored, oxidation rate can be reduced by using coloured stoppered bottles. For the same reason, developing tank lid should always remain in position.

ii) Rinser : After an exposed film has been developed it is rinsed so as to stop overdeveloping and to avoid developer being carried over to contaminate fixer. In routine practice, rinsing is done in running water that changes rinse bath about eight times an hour. If running water is not available, **stop bath** containing 128 ml of glacial acetic acid in one litre of water can be used. If running water is used, water should be allowed to drip down the film before being transferred to fixer. In case of stop bath, the film is directly immersed in fixer as soon as it is taken out from stop bath. In both cases, rinsing should be done by agitating the film for 10-30 seconds. At 27°C and using rapid hand processing, 10 seconds rinsing is adequate.

iii) Fixer : Fixer has following functions :

   i)   Removal of unexposed silver crystals so that silver image can be retained as a permanent record.
   ii) Stops development of the film by neutralising developer.
   iii) Shrinks and hardens film emulsion.

Fixer contains following ingradients :

a) Fixing agent : It is a solution containing sodium thiosulphate or ammonium thiosulphate. It acts to remove unexposed silver crystals.

b) Acidifier : Acetic acid or sulphuric acid is used as an acidifier to neutralise developer.

c) Hardner : Ammonium choloride or ammonium sulphide is used as a hardening agent. Hardner shrinks and hardens film emulsion.

d) Preservative : Sodium sulphite is added as a preservative to maintain chemical balance of agents used in fixer solution.

e) Solvent : Water is used as a solvent for chemicals.

Usually fixing requires twice as long time as developing but more time is required for hardening of film emulsion. In practice, clearing time of the film is first recorded. **The clearing time is the time taken by the fixer to convert a cloudy appearance of the film to a clear one.** If clearing time is, for example, 2 minutes, 4-5 minutes will be required for fixation. However, since longer time is required for hardening of film emulsion atleast 10 minutes will be required for fixing. Films are usually kept in fixer for about 20 minutes.

Following points should be considered for film fixing :

   i)   A screen film requires less time for fixing than a non-screen film because of thicker emulsion in the latter.
   ii)   Hardening of film emulsion in fixer requires longer time than clearing.
   iii) Fixer solution replenishers are available. If replenisher is used for fixer, the amount added should be the same as that of developer replenisher added. Whole of the fixer solution should also be changed when developer solution is changed. If replenisher is not used, whole of the solution should be changed when clearing time of the film is twice to that of original time.
   iv)   As more and more silver gets dissolved in fixer, the fixing and hardening properties are weakened.
   v)   Maintain temperature of fixer as that of developer. A warmer solution fixes faster than a cooler solution but at temperatures higher than 24°C, hardening properties are weakened.
   vi)   Agitation of the film in fixer reduces fixing time.
   vii) Fixer concentration should be as recommended by the manufacturer.
   viii) A longer fixing time (even overnight) does not affect the film but washing time is increased.

FILM WASHING

Final washing of the film is done after it has been fixed. Washing removes excess fixer and residual silver. It should be done in a large tank with provision of running cold water.

Following points should be considered for washing of the film :

   i)   If water in the tank is changing 8-10 times an hour, 20 minutes washing is sufficient.
   ii) If fixation time has been prolonged, washing time should be increased.
   iii) Film can be left overnight in washing but drying time of the film is increased because of swelling of film emulsion.
   iv) Non-screen films require longer washing time than screen films (recall that same is the case for fixing because of thicker emulsion in non-screen films.)
   v)   If washing is not proper, films get discoloured with passage of time.

TEMPERATURE OF PROCESSING SOLUTIONS

The chemical activity of processing solutions is closely related to the temperature of the solutions, as has already been discussed. It is important that temperature of developer, fixer and rinsing and washing tanks is maintained at the same dagree. **Large variations in temperature of various solutions can cause reticulation i.e cracking of the film emulsion which produces a network of lines over the image.** A floating thermometer (Fig. 5-9) should be used to monitor temperature if processing unit is not thermostatically controlled.

FILM DRYING

After washing, the film is drained and hung up to dry. Films should not come in contact with each other and air should be able to freely circulate around them. It is essential that no dust is present in the room as otherwise films will get spoiled. Drying cabinets with heating provision can be used to quicken the drying process.

FILM MARKING

Film identification is necessary to maintain proper records. Marking should indicate place, case number, date of radiographic examination, age of the animal and radiographic view. Information can easily be incorporated on the film by attaching lead markers (Fig. 5-10) with the aid of adhesive tape on the cassette corner before exposing the film. Film can also be marked with white ink after it has been dried. However, by writing afterwards a film can not be used as a legal document, should the necessity arise in a veterolegal case or for insurance purposes.

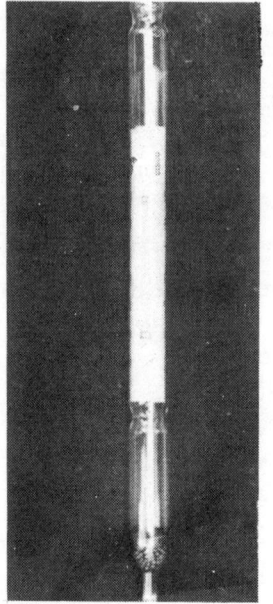

Fig. 5-9 : A floating thermometer.   **Fig. 5-10 :** Lead markers for film identification.

FILM PROCESSING TECHNIQUES

Film can be processed either at a recommended temperature or by using time-temperature combination. The manufacturers recommend temperature at which solutions are to be used. It is usually 20°C and recommended developing time is 4-5 minutes. Processing results are best if this method is used. In case temperature is within permissible limits, a time-temperature combination can be used. The film will take shorter time to develop in warmer solution and vice versa (Fig. 5-11). Recommended time-temperature combinations are : 8½ minutes at 15.5°C, 6 minutes at 18.5°C, 5 minutes at 20°C, 4½ minutes at 21°C and 3½ minutes at 24°C.

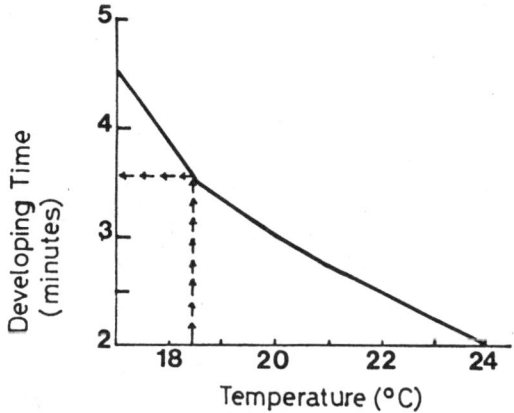

Fig. 5-11 : Time-temperature chart for developing the film. Arrows indicate that if temperature of developer is 19°C, developing time will be 3½ minutes.

Processing steps
   i) Check the level of processing solutions.
   ii) Stir the solutions.
  iii) Check temperature of solutions.
  iv) Select correct size film hanger.
   v) Switch on safe light, switch off white light.
  vi) Open cassette on dry bench and take out the film by grasping a corner with thumb and finger.
  vii) Fix film in hanger.
 viii) Close cassette.
  ix) Place film in developer. Set the timer for required developing time. Agitate film for few seconds to remove air bubbles. Check that films do not touch each other or sides of the tank.
   x) Clean hands, reload cassette with a fresh film.
  xi) At the end of developing time, lift the film out and allow developer to drain.
  xii) Rinse film (10-20 seconds).
 xiii) Transfer film to fixer, agitate it for a few seconds. Replace developing tank lid.
 xiv) Once film has been cleared (about 2-3 minutes in fresh fixer solution), white light can be switched on. Film can be viewed in an emergency case for provisional diagnosis. Replace film in fixer and replace fixer tank lid. Keep the film in fixer for 10-20 minutes.
  xv) Wash film in running water for about 20 minutes.
 xvi) After washing, allow the film to dry.

Although not a recommended procedure yet sight developing may be necessary under some situations in veterinary practice. In such a case, film is taken out of developer after one minute and viewed in safe light. If film density is high, remove the film, rinse and transfer to fixer in an attempt to salvage some information from the film. Usually high densities in a short developing time are produced when exposure factors are not controlled adequately e.g., while using portable X-rays machines.

AUTOMATIC PROCESSING

Automatic processors (Fig. 5-12) are used if flow of radiographic work is high. Such processors eliminate dark room errors and thus avoid otherwise possible repeated radiographic examinations. There is better quality control and cleaner dark room operations. Film processing is also very fast. However, automatic processors are very expensive to buy and maintain and require large quantities of original solutions and replenishers. Such processors are not being currently used in India in veterinary practice.

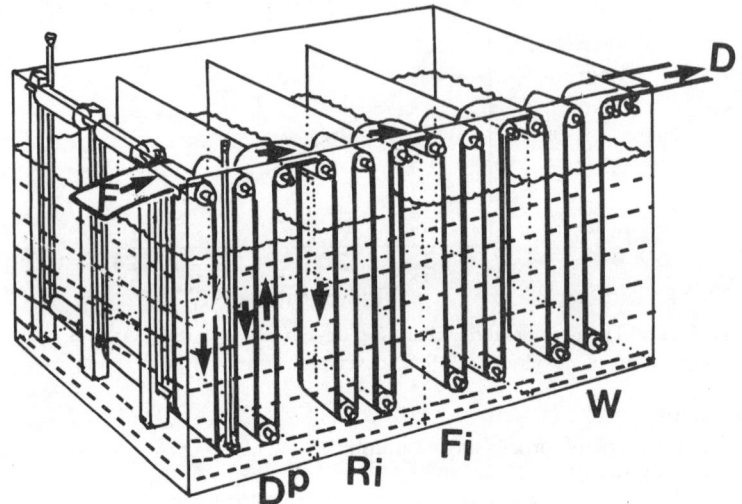

Fig. 5-12 : Diagrammatic representation of automatic processor unit. F = film, D = To drying cabinet, DP = Developer, R = Rinser, Fi = Fixer, W = Washing.

SILVER RECOVERY FROM FIXER

Since unexposed silver halide crystals are removed by fixer, used fixer solution contains silver. This silver can be removed by using metallic replacement or electrolytic or chemical precipitation technique. Due to this reason, used fixer solution should not be thrown away and be sold in the market to recover some cost. Silver recovery potential of 20-25 X-ray films of 14" x 17" size is about 30gms while 60-65 films of 8" x 10" are required for the same potential.

SECTION - B

RADIOGRAPHIC FAULTS

MOHINDER SINGH
S. THILGAR

It is often erroneously thought that good equipment, proper positioning and correct exposure are all that required to make a quality diagnostic radiograph. This belief and neglecting of other minor details often spoil an otherwise good radiograph. Neck restraint chains, tapes or clothes tied over the site to be exposed, dirty skin, unclean accessories, improper beam collimation and accidents in the processing room (e.g. dropping of material on open cassettes) are also worth consideration. To avoid radiographic faults and artifacts, proper care of the accessory equipments and a thorough understanding of the radiographic and processing techniques are necessary. Every error noticed and immediately rectified in routine enhance the quality of a radiograph. Various possible radiographic defects and their remedies are discussed in this section. These errors are least if the animal is prepared properly, correct positioning is done, correct exposure factors are used, care is exercised while loading and unloading the film in a cassette and finally the film is processed properly.

DARK RADIOGRAPH (HIGH DENSITY)

Causes a) Over-exposure (Fig. 4-21).
  i) High kV.
  ii) High mA.
  iii) Long exposure time.
 b) Short focal-film distance.
 c) Use of a wrong screen-film combination. Use of ultra fast film or ultra fast screen with a technique chart developed for par speed film or screen.
 d) Over-development:
  i) Too high temperature of developer.
  ii) Too long developing time.
  iii) Inadequate dilution of developer.
  iv) Exposure of film to visible light prior to/or during development.
Prevention : a) Reduce kV, mA or exposure time.
 b) Use recommended focal-film distance.
 c) Use proper screen-film combination.
 d) Avoid over-development.
  i) Adjust the temperature of developer.
  ii) Use correct developing time.
 iii) Use properly mixed and diluted developer.
  iv) Avoid exposure of film to visible light.

WHITE OR LIGHT RADIOGRAPH (LOW DENSITY)

Causes : a) Under-exposure (Fig. 4-21).
  i) Low kVp.
  ii) Low mA.
  iii) Shorter exposure time.
 b) Increased focal-film distance.
 c) Wrong screen-film combination.
 d) Under-development.
  i) Too low temperature of developer.
  ii) Short development time.
  iii) Developer is exhausted.
  iv) Developer is diluted.
  v) Inadequate mixing of developer.
 e) Two films placed in the same cassette.
Prevention : A) Use proper kVp, mA or exposure time.
 b) Reduce focal-film combination.
 c) Use proper screen-film combination.
 d) Avoid under-development.
  i) Adjust temperature of developer.
  ii) Correct the developing time.

iii) Change the developer.

iv) Ensure adequate mixing of developer.

e) Ensure proper loading of the cassette.

FOG

i) Radiation fog (Fig. 5-13)

Causes : a) Exposure of film to radiation during storage.

b) Exposure of film to radiation during transport and delivery.

c) Exposure of film during radiographic examination.

d) Exposure of film to scatter radiation.

Prevention : a) Do not store film near a source of radiation. Store film in a lead-lined box.

b) Ensure proper protection during transport and delivery.

c) Keep loaded cassette away from X-ray room during radiographic examination.

d) Use grid on parts thicker than 10cm.

ii) Chemical Fog

Causes : a) Overdevelopment of film.

b) Exhausted or contaminated developer.

Prevention : a) Correct temperature of the developer and time of development.

b) Replace the developing solution.

iii) Light Fog

Causes : a) Light leakage into the dark room.

b) Light leakage through cassette and transfer box.

c) Light leakage through broken lead film storage box.

d) Faulty safe light.

    i) Bulb too bright.

    ii) Cracked or fadded filter.

e) Prolonged inspection of film during development.

f) Turning light 'on' before proper fixation of the film.

Prevention : a) Ensure perfect light leak proof processing room.

b) Use light leak proof cassette and transfer box.

c) Replace the broken lead film storage box.

d) i) Use proper wattage bulb.

    ii) Replace cracked or faded filter.

e) Avoid frequent and prolonged inspection of film during development.

f) Allow adequate time for fixation of film before turning on the light.

iv) Film Fog

Causes : a) Expired or old film.

b) Films stored under too high temperature and/or humidity.

c) Presence of ammonia or other fumes in processing room.

d) Excessive pressure on film during storage or handling in dark room.

Prevention : a) Do not use expired film.

b) Store film in proper storage conditions.

c) Remove source of fumes from the processing room.

d) Avoid pressure on film during storage and handle the film gently in dark room.

BLACK SPOTS

Causes : a) Two films stuck together during fixation.
   b) Splash of developer on film before processing.
   c) Dust or liquid on the processed film (Fig. 5-14).
   d) Linear scratches on film.
   e) Light leakage into film storage box or into poorly fitting cassette.
   f) Dropping of water droplets from hangers on the processed film during drying process.
   g) Sharp pressure on film or bending of film before development.

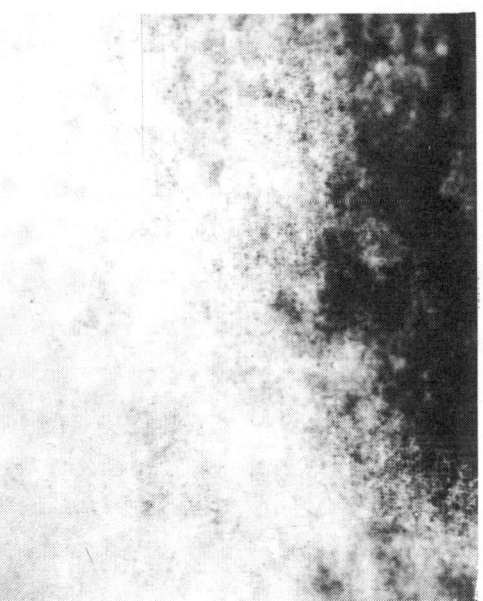

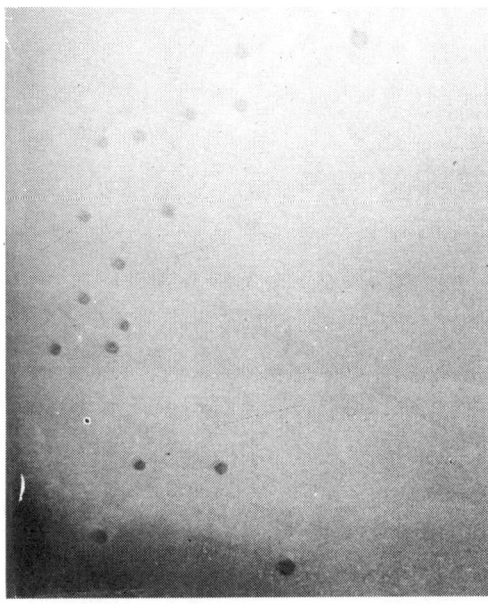

Fig. 5-13 : Radiation fog.

Fig. 5-14 : Black spots due to splashing of liquid on unprocessed film.

Prevention : a) Keep proper distance between two films during fixation.
   b) Handle the film with clean and dry hands and clean the working bench surface in dark room.
   c) Protect film from dust or liquid.
   d) Handle film gently and carefully.
   e) Repair or replace storage box or cassette.
   f) Remove all the water droplets from the hanger.
   g) Handle film carefully and avoid its folding.

WHITE SPOTS

Causes : a) Pitted screen (Fig. 5-15).
   b) Dust, grit or any other material with radiopaque base present on film, screens, grid or the animal (Fig. 5-16).
   c) Crescent marks due to folding of film before exposure.
   d) Splash of water or fixer on film (Fig. 5-17) before processing.
   e) Trapping of air bubbles on film's surface (air bells) during development.
   f) Scratches on emulsion (Fig. 5-18).

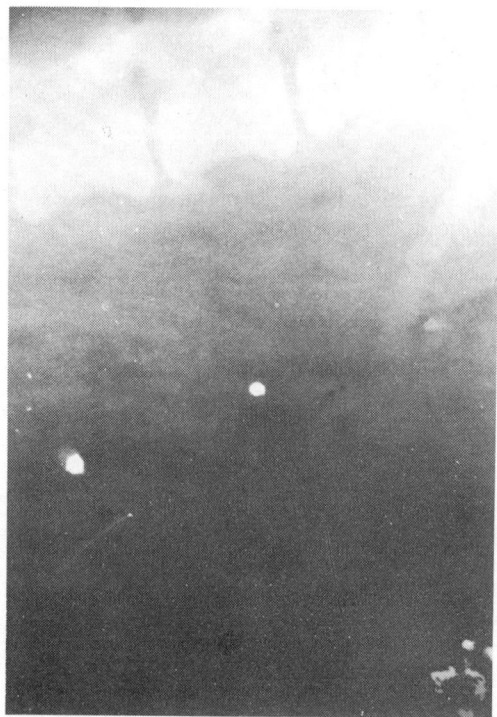

Fig. 5-15 : Radiographic faults due to pitted screen.

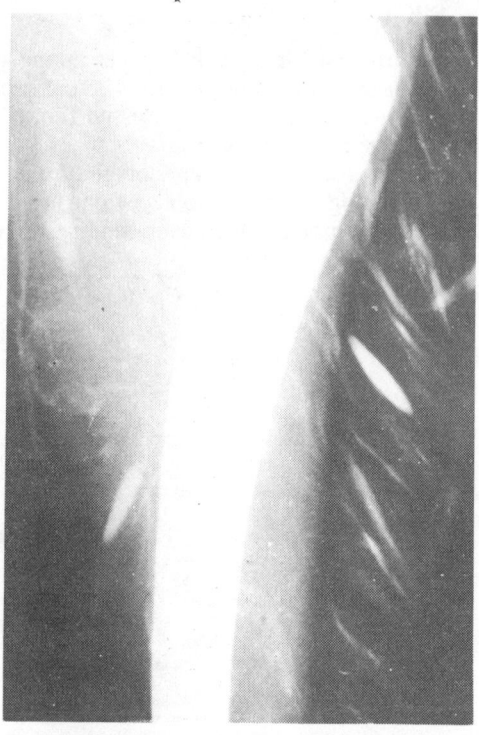

Fig. 5-16 : Radiodense material attached to hair/wool of the animal.

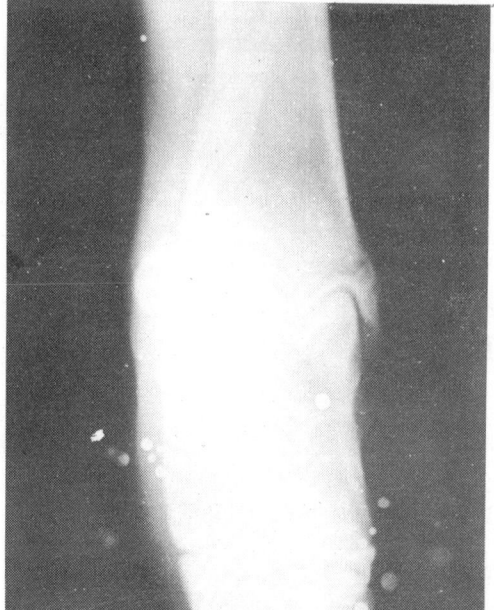

Fig. 5-17 : White spots on the radiograph due to splash of fixer on unprocessed film.

Fig. 5-18 : Scratches on film emulsion.

Prevention : a) Replace screen.
  b) Remove foreign material from the body of the animal and clean or replace intensifying screens or grid.
  c) Handle film gently and carefully.
  d) i) Clean hands before unloading the film from cassette.
  ii) Clean working bench surface.
  e) Agitate the film occasionally in the developer.
  f) Handle film carefully during processing.

CRESCENT SHAPED BLACK MAKS

Cause : Sharp bending of the film before development (Fig. 5-19).
Prevention : Handle the film carefully.

BLACK STATIC MARKS (ARBORESCENT STREAKS)

Causes : a) The marks (Fig. 5-20) develop due to static electricity discharge as a result of improper handling of the film during its removal from the box or interleaving paper and during loading and unloading of the cassette.
  b) Low humidity in the room.
Prevention : a) Handle the film carefully and avoid its friction with other surfaces i.e. interleaving paper, intensifying screen etc.
  b) Maintain correct humidity.

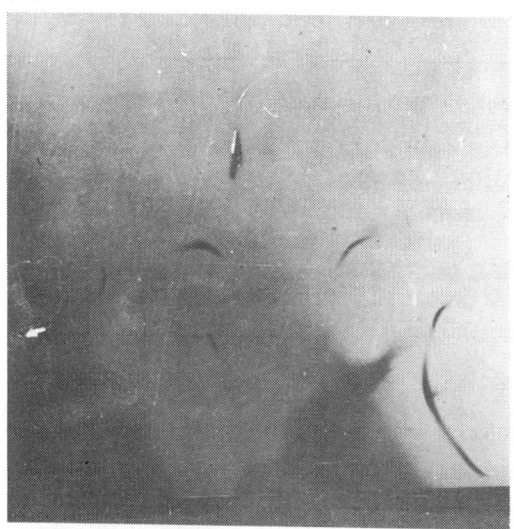

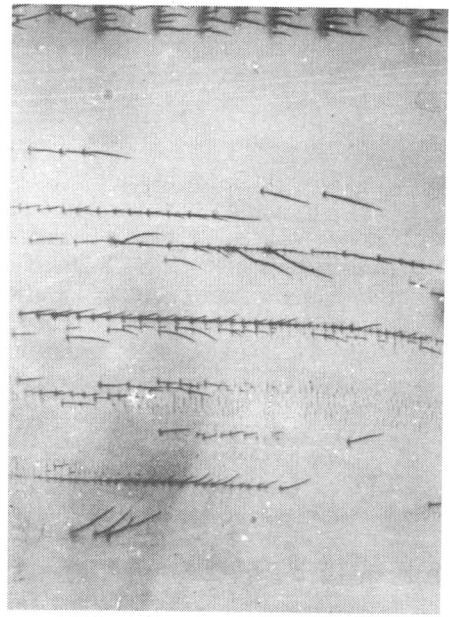

Fig. 5-19 : Crescent shaped black marks due to sharp bending of the film before development.

Fig. 5-20 : Black static marks.

STREAKS ON FILM

i) Chemical streaks

Causes : a) Insufficient film agitation during processing.
  b) Dirty film hangers.

    c) Insufficient rinsing.

    d) Running down of water drops on semi-dried film.

    e) Water splashes.

    f) Dirty water used for washing of film.

Prevention : a) Ensure proper agitation of developer and fixer solutions and of film.

    b) Use clean film hangers.

    c) Do proper rinsing.

    d) Remove water droplets from film or use film drier.

    e) Replace washing water and increase flow rate in washing tank.

ii) Mechanical streaks

Causes : a) Dark scratches (Fig. 5-21) occur on film while loading or unloading the cassette.

    b) White scratches occur during processing when hanger of a film scratches across the surface of another film.

Prevention : a) Handle the film carefully during loading or unloading.

DISTORTED OR BLURRED RADIOGRAPHIC IMAGE

Causes : a) Blurred image due to motion of patient, X-ray tube or cassette during exposure.

    b) Distorted image due to i) poor film-screen contact, ii) improper centring of the primary beam, and iii) central beam being not perpendicular to the cassette.

Prevention : a) Avoid motion.

    b) i)   Clean intensifying screen.

        ii)   Correct inadequately moulded screen.

        iii) Check proper centring of the central beam.

GRID FAULTS

Causes : a) Grid lines due to wrong focal-film distance outside range of grid radius.

    b) Grid lines on the film due to off-centring from mid-line.

    c) Central beam is not perpendicular to the grid.

    d) Grid lines on both edges of film due to use of reverse grid.

Prevention : a) Use correct focal-film distance as recommended for a particular grid.

    b) Centre the primary beam correctly.

    c) Direct the central beam in such a way that it is perpendicular to the grid.

    d) Check the side of grid before exposure for its proper placement.

FROSTY AREAS ON FILM

Cause : Improper final washing of film.

Prevention : Wash the film properly with fresh water.

YELLOW RADIOGRAPH

Causes : a) Use of exhausted or diluted fixer.

    b) Insufficient fixation (Fig. 5-22).

Prevention : a) Replace the fixer solution.

    b) Ensure proper fixing of the film.

BRITTLENESS OF PROCESSED FILM

Causes : a) Excessive drying temperature or prolonged drying time.

    b) Excessive fixation.

Prevention : a) Use proper temperature and time, if film is dried in a film dryer.
b) Fix the film properly.

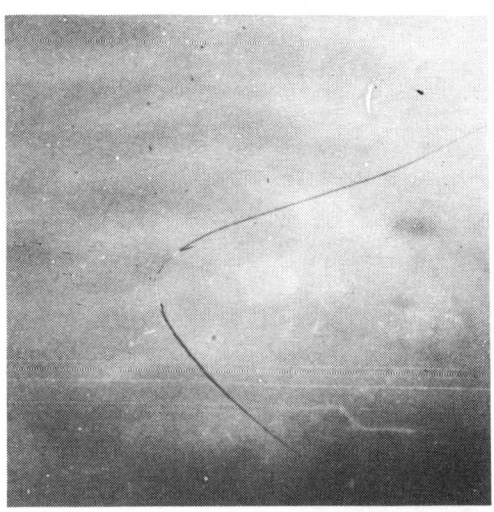

Fig. 5-21 : Dark scratches on the radiograph due to faulty loading/unloading of the cassette.

Fig. 5-22 : Insufficient fixation of the film.

FINGER MARKS

Causes : Handling of the film with dirty, greasy or contaminated hands. Hand developed film contaminated with fixer shows white marks; if contaminated with developer, the marks appear black.

Prevention : Clean the hands and dry before handling the film.

RETICULATION

Causes : a) Wide variation in temperature of developer and fixer solutions
b) Exhausted fixer solution.

Prevention : a) Control temperature of solutions.
b) Replace the fixer.

WHITE HORIZONTAL AREA ON TOP OF THE FILM

Cause : Low level of developer in tank.
Prevention : Maintain proper level of the developer.

DARK HORIZONTAL AREA ON TOP OF THE FILM

Cause : Low level of fixer in the tank (Fig. 5-23).
Prevention : Maintain proper level of fixer.

BLANK FILM

Causes : a) Film not exposed.
b) Central beam not properly directed and centred on the cassette.

Prevention : a) Check exposure switch and X-ray machine.
b) Ensure proper centring of the X-ray beam.

FRILLING OF GELATIN

Cause : Use of excessive hot solutions for processing (Fig. 5-24).

Prevention : Control the temperature of processing solutions.

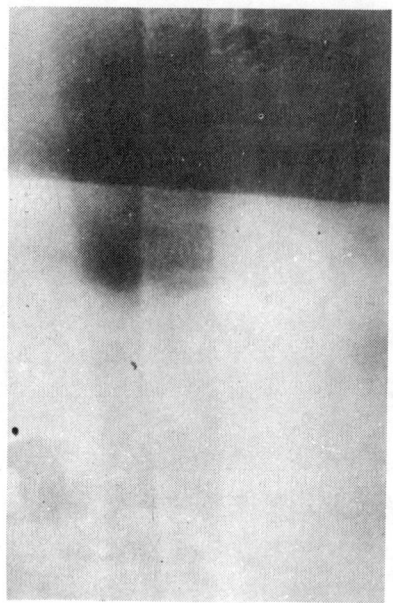

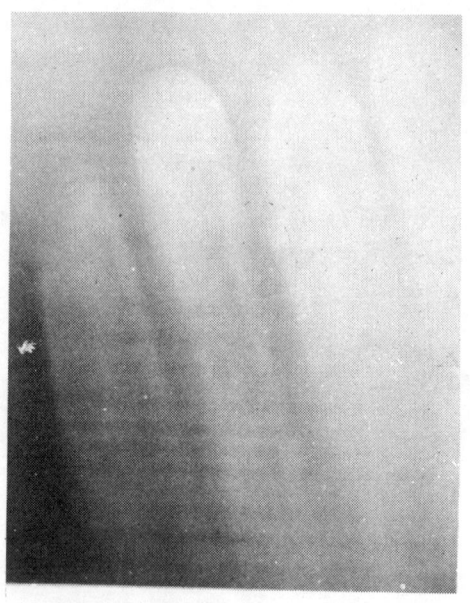

Fig. 5-23 : Dark horizontal area on top of the film due to low level of fixer.

Fig. 5-24 : Frilling of gelatin due to excessive hot processing solutions.

REFERENCES

Butler, H.D. 1986. Seven steps to better radiographs. Vet. Med. **81,** 347.

Douglas, S.W. and Williamson, H.D. 1980. Principles of Veterinary Radiography. 3rd edn., Lea and Febiger, Philadelphia.

Eastman Kodak Company. 1980. The Fundamentals of Radiography. 11th edn., Rochester, New York.

Gillette, E.L., Thrall, D.E. and Lebel, J.L. Carlson's Veterinary Radiology. 3rd edn., Lea and Febiger, Philadelphia

May and Baker Radiographic Manual. 1966. The Processing of Radiographic Film-chemistry and mechanics. England.

Reynolds, J.A. 1955. Factors affecting radiographic quality. Vet Med. **50,** 187.

Selman, J.1977. The Fundamentals of X-ray and Radium Physics. 6th edn., Charles C Thomas Publishers, Springfield, Illinois.

Sweeny, R.J. 1983. Radiographic Artifacts : Their cause and control. J.B. Lippincott Co., New York.

Ticer, J.W. 1984. Radiographic Technique in Veterinary Practice. 2nd edn., W.B. Saunders Co. Philadelphia.

Van Der Plaats, G.J. 1980. Medical X-ray Techniques in Diagnostic Radiology. 4th edn., Martinw Nijhoff Publishers, London.

SELECTED QUESTIONS

1. What are 'dry section' and 'wet section' in a dark room?
2. What is a "safe light"? What is safe light fog test?
3. What is the purpose of developer? What is its composition?
4. What do you understand by 'stop bath'?
5. What are the constituents of a fixer? What is clearing time?
6. Why temperature of developer, fixer, rinsing solution and washing tank should be maintained at the same degree?
7. What are the advantages of film marking?
8. List the causes of the following radiographic faults :
   i) Over-development, ii) Light fog, iii) Black static marks iv) Reticulation
9. List the processing steps from developing to drying of the film.

```
-DARK ROOM-
ENTRY OF
LIGHT AND CLUMSINESS
PROHIBITED
```

```
-X-RAY FILM-
EVEN MINOR CAN VIEW IT
```

CHAPTER

6 ESTIMATION OF RADIOGRAPHIC EXPOSURE FACTORS

GAJRAJ SINGH
BHARAT SINGH

Formulation of an exposure technique chart enables selection of correct predetermined exposure factors for the anatomical part to be exposed in order to ensure same radiographic quality on every radiographic examination. A technique chart also obviates wastage of X-ray films which may occur due to inappropriate setting of machine. A number of variable factors affect the exposures, therefore, a comprehensive understanding of different inter-relationships between various factors is necessary so that available X-ray machine can be used to its full potential.

The exposure technique chart developed for one X-ray machine may not produce satisfactory results with another X-ray machine. This is due to the differences in the various X-ray machines, types of accessory equipments and processing procedures used. It, therefore, becomes essential to develop technique chart indepenently for a particular type of X-ray machine being used.

This chapter deals with variable factors which may affect exposure factors. These factors include:

   i) Make and type of X-ray machine
  ii) incoming line voltage
 iii) Kilovoltage
  iv) Milliamperage and exposure time
   v) Focal film distance
  vi) Grid type
 vii) Speed and type of film and intensifying screens
viii) Thickness and nature of part being exposed
  ix) Temperature and time of developing

Since all the mentioned variables affect the quality of a radiograph, if any factor is altered it must be compensated by making alterations in the other variables so as to produce a quality radiograph.

Make and type of X-ray machine

A combination of kVp, mA and time setting that produces a satisfactory radiograph with one machine may not necessarily produce the same quality radiograph with another machine. Exposure factors used for half-wave rectified machine are generally not applicable to full-wave rectified machines.

Incoming line voltage

A fluctuation in the line voltage will not allow consistent output and thus negating all other variables.

Kilovoltage

Kilovoltage determines the penetrability of the X-ray beam and thus affects the radiographic density. Exposure latitude, degree of variation from the correct exposure factors, that still produces a diagnostic radiograph varies with the range of kVp used. **High kVp technique increases the exposure latitude and reverse is true with lower kVp** (Table 6-1). For most radiographic examinations, a long scale of contrast is required to visualise subtle differences in tissues densities. **It is, therefore, desirable to use high kVp range which produces a long scale contrast in comparison to short scale contrast produced by low kVp range.** A lower kVp range may be used to advantage in certain bone examinations requiring recording of relatively fewer tissue density differences. However, it is possible only when the available X-ray equipment is powerful enough to compensate for low kVp with a high mA and a reasonable exposure time.

TABLE 6-1: Exposure Latitude

kVp range	Exposure latitude
46-55	±2 kVp
56-65	±4 kVp
66-75	±6 kVp
76-85	±8 kVp
86-95	±10 kVp

The variations in kVp become necessary according to the thickness of the part to be radiographed so as to maintain proper radiographic density. **For each centimeter increase in the part thickness, it is necessary to add 2 kVp up to 80 kVp; 3 kVp between 80 to 100 kVp and 4 kVp above 100 kVp. Similarly decreased kVp may be used to compensate for decreasing part thickness.**

Use of higher kVp is desirable in veterinary practice as it offers following advantages:

i) Higher exposure latitude in comparision to lower kVp range (Figs 6-1 and 6-2).
ii) Longer scale of contrast.
iii) Shortest possible exposure time and thus elimination of risk of motion of the animal during exposure.
iv) Reduced radiation exposure of personnel.

There is a general relationship between kVp and mAs setting i.e. **a change in kVp requires a change in mAs and vice versa to maintain similar radiographic density.** The change in kVp settings required in various ranges when mAs is doubled or reduced to half is shown in Table 6-2. For example, if at 61 to 70 kVp normally used mAs is 10 and if it is reduced to 5, kVp will have to be increased by 8. If mAs is increased to 20, kVp will have to be decreased by 8.

TABLE 6-2: kVp change required with change in mAs (refer to text for details)

kVp range	kVp change required when mAs doubled or halved
41-50	±4
51-60	±6
61-70	±8
71-80	±10
81-90	±12
91-100	±14
101-110	±6

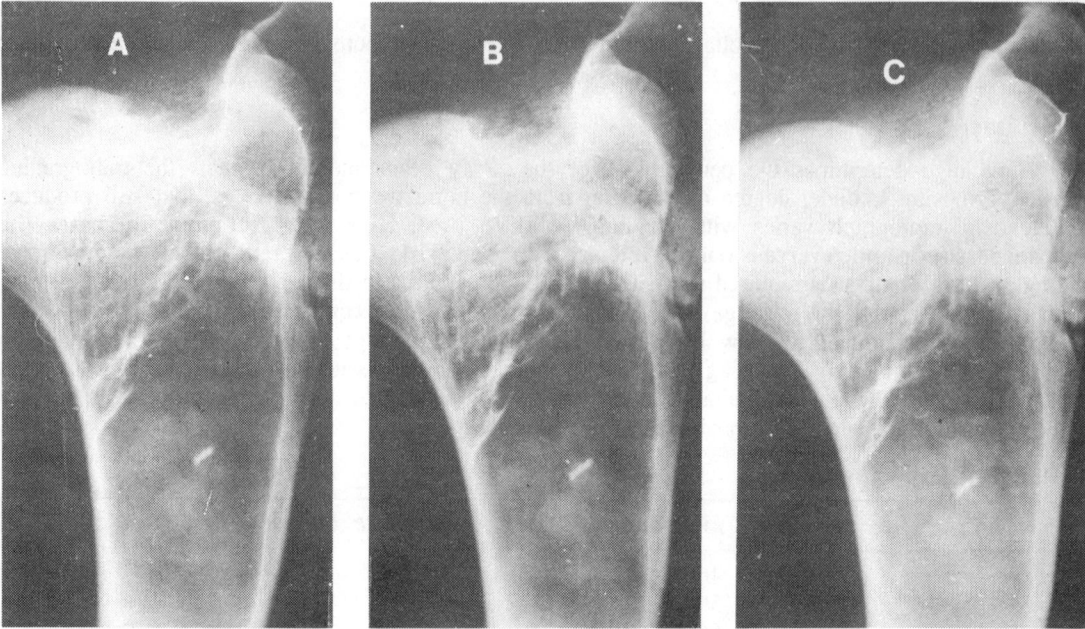

Fig. 6-1 : Higher kVp range offers a wide exposure latitude. Radiographs A, B and and C were made with kVp 78, 70 and 62 respectively, with all other factors being constant. All appear to be of same quality.

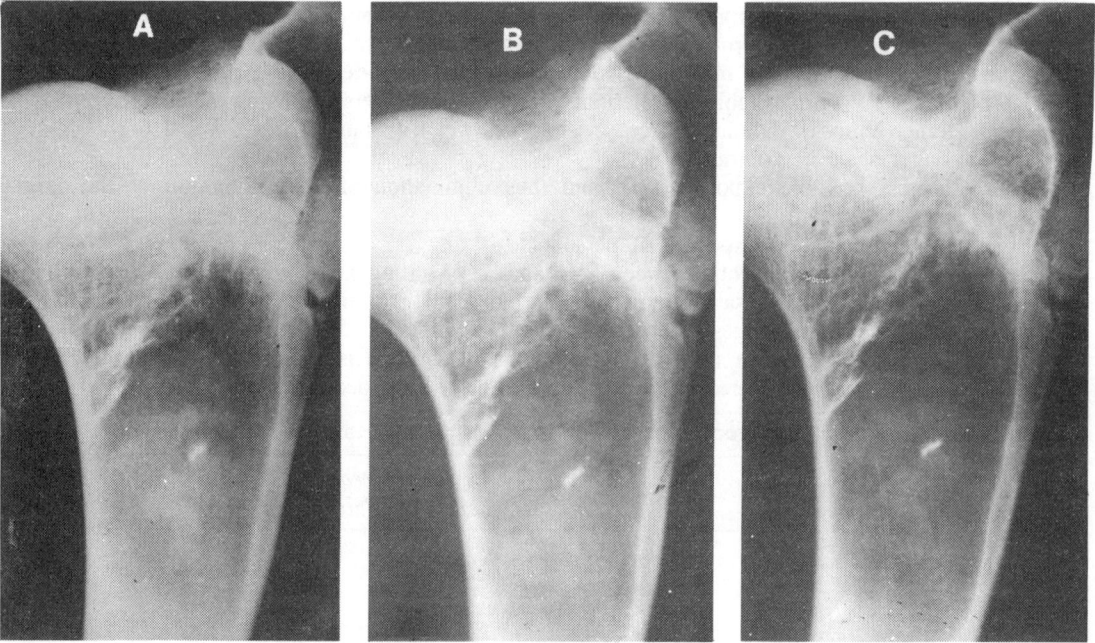

Fig. 6-2 : At lower kVp range, exposure latitude is less. Radiographs A, B and C were made with kVp 48, 50 and 52 respectively, with all other factors constant. Small variations in kVp have pronounced effect in this range.

Milliamperage and exposure time

Since milliamperage (mA) and exposure time are closely related, it is better to consider these together. An increase in mA allows shortening of exposure time to maintain satisfactory radiographic density. Therefore, concept of milliamperage-second (mAs) has been developed. **An exposure made at 200 mA for 1/10 second or at 400 mA for 1/20 second will produce the same radiographic density as in both cases mAs will be 20.** However, shorter exposure time is preferred in veterinary practice, as has already been discussed.

Focal-film distance

Focal film distance (FFD) is the distance between the focal spot in the target of the X-ray tube and X-ray film. It is desirable to use constant FFD for every radiographic examination. In veterinary practice, 90-100 cm FFD is considered a good compromise between the distance and exposure factors. Increase in FFD decreases the total number of X-rays available to expose the film. The relationship between FFD and radiographic density has been discussed in chapter 4. In brief, the radiographic density is inversely proportional to the square of the FFD. In other words, if FFD is doubled only one-fourth of the total number of X-rays will be available to expose the film because the same field of X-ray beam is spread over an area of four times. When FFD is to be changed, a simple calculation given below may be made to determine the mAs required to maintain adequate radiographic density:

$$\frac{(\text{new FFD})^2}{(\text{old FFD})^2} = \frac{\text{new mAs}}{\text{old mAs}}$$

Example: if mAs at 75 cm FFD is 20 what will be mAs at 100 cm FFD?

$$\text{Ans:} \quad \frac{(100)^2}{(75)^2} = \frac{X}{20}$$

$$X \text{ (new mAs)} = 35.6$$

Grid type

A grid (refer to chapter 4 for details) prevents scatter radiation from falling on the film and thereby improves radiographic contrast. However, use of a grid requires increase in exposure factors. This increase depends on the ratio of grid. The extent by which the factors should be increased may be determined by trial exposure.

Table 6-3 provides a rough guide for magnitude of increase in the exposure with the use of grids.

TABLE 6-3: The extent of change required in mAs with use of grid

Grid ratio	Increase exposure (mAs) by a factor of
5 : 1	2
8 : 1	3
12 : 1	4
16 : 1	4.5

Note: Grid ratio of 8:1 is most commonly used in veterinary practice, however, in areas of greater thickness higher ratio grid may be used.

The increase in exposure factors may be accomplished by increasing the kVp or mAs, however, mAs is generally increased because most techniques utilise the maximum milliamperage permissible with the available machine, the exposure time being increased to increase mAs.

Speed and type of the film and intensifying screens

The importance of type and speed of the film and intensifying screens has been discussed in chapter 4. In brief, **screen film in combination with intensifying screens requires about five times less exposure as compared to non-screen film to produce same radiographic density.** The choice of film and screen type is governed by the speed. A par speed film requires approximately double the exposure than a high speed film to maintain same radiographic density. However, par speed film usually produces radiographic detail that is satisfactory for average diagnostic use. The detail or slow speed film requires twice as much exposure as par speed film but offers advantages for radiographic examination of fine structures, e.g. periarticular tissues. Similarly high speed screens require approximately one half the exposure to produce same radiographic density as par speed screens. Since difference in radiographic density is not significant between high speed and par speed screens, the use of high speed screens is usually recommended for most radiographic procedures in veterinary practice because of the decreased exposure requirement.

The use of rare earth intensifying screens decreases the exposure factors by a factor of 12 when compared with par speed film and screens. However, radiographic mottle becomes objectionable and, therefore, a combination of rare earth screen and film that increases the speed maximum by a factor of 8 is generally recommended.

The thickness and nature of part being exposed

In addition to the part thickness, the nature of part and presence or absence of any pathological lesion within the part being exposed also influence the exposure. A part of same thickness may contain air (thorax), soft tissue (abdomen) or bone (limb, vertebrae, skull). Because of variations in tissue density, following guidelines are recommended to modify machine settings:

A. Without pathological lesions :

i)	Skull vertebrae, pelvis	- increase kVp 5-10
ii)	Soft tissue in the cervical region	- decrease kVp 5-10
iii)	Thorax	- half mAs
iv)	Immature animal	- half mAs
v)	Heavy muscled animals	- double mAs
vi)	Plaster cast	- double mAs
vii)	Contrast studies	- either increase kVp 5-10 or double mAs

B. With pathological lesions: The pathological changes may be grouped into two categories: a) Those in which the radiographic density is less than suspected and b) those in which the radiographic density is greater than suspected.

The conditions that cause a decrease in the radiographic density because of an increase in subject density require increase in exposure either by increasing kVp by 5 to 10 or by doubling the mAs. The pathological conditions that require increased exposure include: soft tissue calcification, bone lesions resulting in increased density, ingesta filled gastrointestinal tract and abnormal fluid quantity.

The conditions that cause an increase in radiographic density due to a decrease in the subject density require decreased exposure. One may either decrease kVp by 5 to 10 or half the mAs. These conditions include megaoesophagus, localised destructive bone diseases, generalised bone diseases, pulmonary cyst and presence of large volumes of gas.

Temperature and time of developing

Temperature and time for developing the film must be kept constant. Low temperature and/or too short developing time generally produce a radiograph that appears to be underexposed. Similarly too high temperature and/or longer developing time produce a dark film that appears to be overexposed.

FORMULATION OF EXPOSURE TECHNIQUE CHART

To formulate a technique chart, it is necessary to maintain some variables constant. These variables include : incoming line voltage, focal film distance, type of films and intensifying screens being used, use of grid and dark room procedures. Variations in thickness of the body part necessitate a change either in kVp or mAs. Different technique charts need to be developed depending on the species of the animal and use of grid and intensifying screens. Since most exposures are made using screen films, and a grid is used only when the thickness of the part being examined is more than 10 cm, a technique chart should first be deviced by using screen film without a grid. Technique chart can be developed by using either variable kVp or variable mAs by a series of trial exposures on a normal average sized animal. It is difficult to describe the procedure to be followed for the first exposure. Previous experience of a person is of great help. Setting of kVp and mAs also depends on the individual X-ray machine. The following procedure, however, may be helpful :

(i) Measure the thickness of the part to be examined in centimeters, multiply it by two and add 40 (standard factor). The number so obtained (say X) represents the approximate kVp at FFD 100 cm, mAs 8, with par speed film and high speed intensifying screen.

(ii) Make three trial exposures, using lead blockers, on a single film with kVp X and mAs 4, 8, 16 and 12.

(iii) Develop the film and check radiographic quality of all three images.

(iv) If none is of good quality and film is overexposed, reduce either kVp by 10 or mAs by one-half of the original. If film is underexposed, increase the factors by same order.

(v) Make again three trials on a film by varying the kVp or mAs.

(vi) If machine does not permit increase in kVp, increase mA or exposure time.

(vii) Note the exposure factors providing best results.

(viii) Develop the chart on structures of different thickness by varying the kVp i.e. either by decreasing or increasing the kVp. For each centimeter of thickness, add to or substract from kVp 2 upto 80 kVp, 3 between 80-100 kVp and 4 above 100 kVp. Using the above method, formulate exposure factors for all parts of the body.

Variable kVp technique chart may not be possible for all X-ray machines because of lack of selection range for 1 or 2 kVp. In such a case, modify the procedure to a variable mAs by using kVp-mAs conversion rules (Table 6-2). However, in such a case, relatively longer exposure times are required which certainly are not advantageous. Once the variable kVp and variable mAs exposure technique charts have been developed, a combination of both can be considered. In general, following points must be considered while formulating exposure technique charts :

(i) Use shortest possible exposure time.

(ii) Use highest possible kVp.

(iii) Use highest possible constant mA.

(iv) Use constant FFD.

REFERENCES

Cahoon, J.B. 1970. Formulating X-ray Technique, 7th edn., Duke University Press, Durham, North Carolina

Douglas, S.W. and Williamson, H.D. 1980. Principles of Veterinary Radiography, 3rd edn., Williams and Wilkins Co., Baltimore

Gillette, E.L., Thrall, D.E. and Lebel, J.L. 1977. Carlson's Veterinary Radiology. 3rd edn., Lea and Febiger, Philadelphia.

Morgan, J.P. and Silverman, S. 1990. Techniques of Veterinary Radiography. 4th edn., Iowa State University Press, Ames.

Ticer, J.W. 1984. Radiographic Techniques in Veterinary Practice, 2nd edn., W.B. Saunders Co., Philadelphia

Wakson. J.C. 1954. Considerations in formulating X-ray exposure factors. Vet. Med. **49,** 435.

SELECTED QUESTIONS

1. List the advantages of using higher kVp in veterinary practice.
2. What is mAs?
3. What is FFD? What is its relationship with mAs?

CHAPTER

7 ESTABLISHING A RADIOLOGY SECTION

HARPAL SINGH
I.S. CHANDNA

It is not possible to lay down rigid guidelines for establishing a radiology section since requirements may vary from establishing a unit in a small space in a town hospital to an independent unit in a teaching institute. The ultimate shape of the unit will also depend upon the finances available and previous experience of the persons consulted for establishing the unit. Considerable improvement can be expected if a visit is arranged to an established unit with almost similar requirements and observations of persons working in such a unit are sought before planning is done. **Following points should be considered while planning the unit:**

   i)   Anticipated present work load and future requirements in the next 10-15 years.
   ii) Financial resources available.
   iii) Type of equipment to be installed for radiographic work.
   iv) Handicaps of similar established units and scope of improvements.
   v)   Requirements of radiation safety enforcing agencies e.g., of Bhaba Atomic Research Centre, Bombay.

Proper design of a radiology section requires combined skill and experience of the following:

   i)   Radiologist and radiographer to provide necessary information about the present and anticipated future requirements.
   ii) Equipment manufacturer to provide information regarding space requirement of equipment for its best utilisation within the frame work.
   iii) Safety enforcing agency representative to provide information regarding safety requirements.
   iv) Architect to design the construction of the unit as per requirements pooled by first three sources.

Various states of India are now increasingly accepting polyclinics concept at district headquarters. Therefore, recommendations made in this chapter take into consideration teaching institutes and polyclinics. It is understood that general outlines need to be provided and final set up will vary according to local requirements. **The overall set up of radiology section should consider the following requirements:**

   i)   Space
   ii) X-ray machine
   iii) Accessory equipment
   iv) Staff

SPACE

Radiology section should be located near the clinic and surgery area. Table 7-1 shows minimum room requirements for a teaching institute and a polyclinic (Fig. 7-1 and 7-2).

Large animal X-ray room

Sufficient space and free mobility are the key requirements in designing large animal X-ray room. Apart from being comfortable, larger space also reduces radiation exposure of personnel due to decreased scatter radiation. Exact space requirement will depend upon the type of X-ray machine to be installed. It is important that provision is kept for restraining devices e.g. a travis (Fig. 7-3) and a casting trolley. Entrance to radiology section should face the clinic for an easy approach. The height of the entrance gate should be such that it suits the height of the species of animals to be handled and a truck or tractor trolley can enter for unloading recumbent animals, if need arises. A separate entrance gate should be provided for personnel working in the radiology unit.

Table 7-1: Minimum room requirements for X-ray facilities.

	Facility	Teaching Institute	Polyclinic
1.	Large animal X-ray room	Yes	Yes. Same area can be
2.	Small animal X-ray room	Yes	Used for small animals
3.	Dark room	Yes	Yes
4.	Film file room	Yes	Yes
5.	Radiologist's office	Yes	Not necessary
6.	Radiographer's office	Yes	Yes
7.	Interpretation room	Yes	Not necessary
8.	Control panel area	Yes	Yes
9.	Waiting room	Yes	Not necessary
10.	Radiological museum cum teaching hall	Yes	Not necessary

The floor of the X-ray room should not be slippery. The walls should be made of bricks or solid concrete, 15 cm thick concrete wall is better as it has a lead equivalent of 1.5 mm thickness and thus minimises radiation hazards. Wood is not recommended as it does not provide any radiation protection. The room should be well ventilated but direct sunlight should be avoided as it may interfere during centring with a collimator.

If the same X-ray machine is to be used for radiography of both small and large animals, a separate table should be provided for small animals.

Small animal X-ray room

Except for the space and any special requirement, criteria for small animal X-ray room construction would be similar to that of large animal X-ray room. If fluoroscopic facility is also to be installed, room construction should be such that it can be easily darkened.

Irrespective of the type of radiographic work involved, prominent warning signs (Fig. 7-4) should be displayed at the entrance gates of the radiology section to the effet that there is radiation hazard inside and that expectant mothers and children under 18 years are not allowed to enter.

Dark room

Dark room should be adjacent to the X-ray room to save time. When separate large animal and small animal radiographic facilities are planned, dark room should be centrally located to economise establishment facilities. Details of dark room requirement has been discussed in chapter 5.

Film file room

Radiographic record is a valuable property of the department or clinic. Legal rules that apply to medical records also apply to radiographs and radiographic reports. Needless to say that each film must be properly identified. Such records are also useful teaching aids. Therefore, filing of these records should be systematic for easy retrieval. Film file room should be near the radiologist's office with minimum size of 8' x 10'. The room should have sufficient provision of storing racks or cabinets for film boxes so that films can be systematically stored yearwise and species-wise.

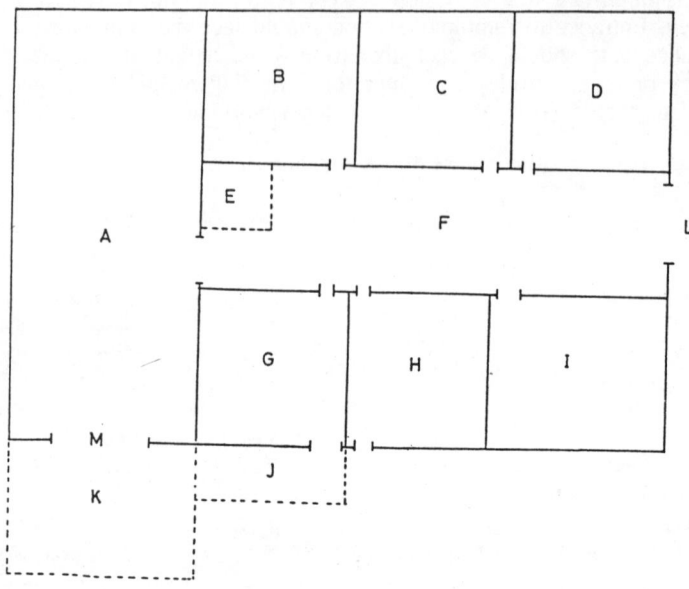

Fig. 7-1 : Possible layout for a radiology unit in an institute. A = Large animal X-ray room, B = Processing room (dark room), C = Radiologist's office, D = Film file room,   E = Area for control panel for large animal X-ray machine, F = Corridor, G = Small animal X-ray room, H = Radiographer's office, I = Conference cum museum room, J = Waiting area, K = Waiting ·area for large animals, L = Main enterance of the building, M = Entry gate for large animals.

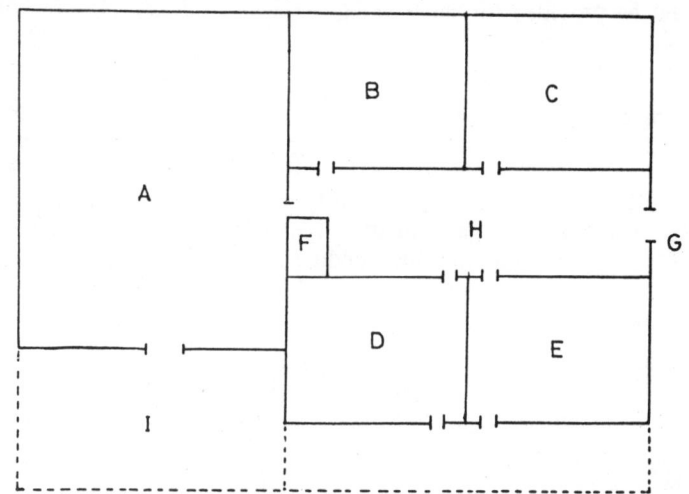

Fig. 7-2 : Possible layout for a radiology unit in a polyclinic. A = X-ray room, B = Processing room, C = Film file room, D = Radiographer's office, E = Doctor's Office, F = Control panel area. G = Main gate of the building. H = Corridor, I = waiting shed for animals.

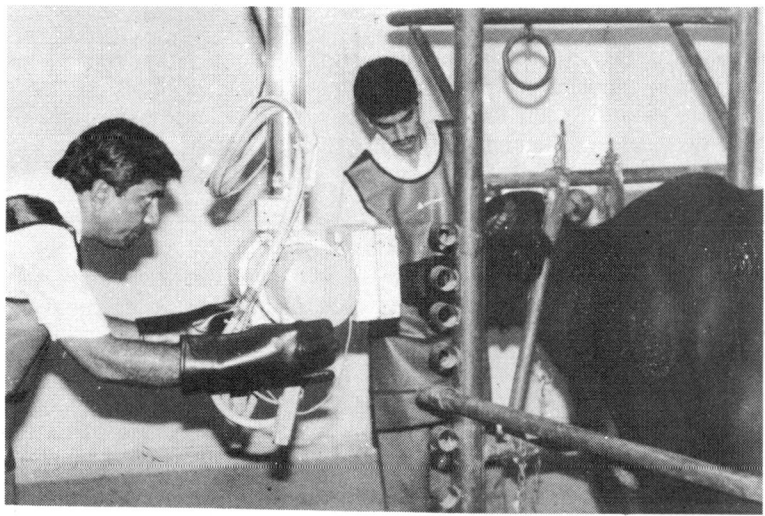

Fig. 7-3 : Use of travis for controlling the animal in X-ray room.

Fig. 7-4 : Radiation warning sign.

Rooms for radiologist and radiographer

If there is no separate provision for radiographic interpretation room, the office of the radiologist should be spacious so that illuminators can be fixed at a convenient place. Such a room is not necessary in a polyclinic as clinician incharge himself discharges the duties of a radiologist. The office space for radiographer should be located near the X-ray room and dark room.

Control panel area

Separate areas should be provided for control panels of small and large animal X-ray machines. This area should be located in such a way that it is seldom in line with the primary X-ray beam. There should be a protective partition with a lead glass window (30 cm x 30 cm) to view the animal and machine during exposure. This is necessary to observe movement of the animal or X-ray tube during exposure which produces motion artifacts on processed films.

Interpretation room

If space and finances permit, a separate radiographic interpretation room should be planned. Otherwise, radiologist's or clinician's office, or museum hall can be used for this purpose. In any case, film interpretation should be left for a place where there is least disturbance. Provision should the made for atleast two viewing illuminators and a spot light.

Radiological museum cum teaching hall

In teaching institutes, there should be provision for radiological museum cum teaching hall. It should have provision for fixing a number of viewing illuminators just above the writing desks. There should also be provision to systematically store radiographs to be used as teaching aids. Space should be enough to accommodate 20-30 students. Cushioned boards can be fixed on the walls to display photographs of interesting radiographs with sufficient details of the case underneath.

Waiting area

There should be provision for waiting area outside the X-ray room and at such a location that there is minimum radiation hazard and minimum obstruction in the flow of work. In a polyclinic, waiting area of the clinic can itself serve the purpose.

X-RAY MACHINES

Different types of X-ray machines and tube stands have already been discussed in chapter 3. An X-ray machine suitable for small animal radiography may not be suitable for large animal radiography. Therefore, it is most essential that needs of installing radiographic machine are fully identified before planning of the unit is done. Future requirements should also be anticipated. Detailed information should be sought concerning the required equipment from various manufacturers. It is always better to seek advice from established radiological units to learn about the working performance of the particular machine.

For large animals, higher milliamperage X-ray machine should be preferred. Three pahse stationary or ceiling suspension X-ray equipment of atleast 300 mA and 125 kV with atleast 1/120 second timing is suitable for most examinations. Machine of 800 to 1000 mA are also available. A single phase mobile unit of 100 mA and 100 kV with a switch time of 0.05 to 5.0 second range may also be considered additionally if finances permit. Such a machine is especially useful for radiographic examination of extremities. For small animals, a three phase control rated 200 to 300 mA, 125 kV and atleast 1/60 second timing machine provides most required range of exposures.

ACCESSORY EQUIPMENT

By selecting right type of accessory equipment deficiencies can be offset, safety can be enhanced and working life of the machine can be extended. Accessories such as films, cassettes, grids, screens, collimators and processing equipments have been discussed in chapter 4. Remaining requirements are discussed in this chapter.

Positioning aids

Positioning blocks and cassette holders are not expensive and are easy to construct. These positioning aids help to obtain constant views of a part and personnel to keep their hand away from the primary beam. Following positioning aids should be available in the radiology unit :

Positioning wooden blocks: The X-ray tube usually does not drop to level for the primary beam to be centered on the foot of standing animal. It thus becomes necessary to elevate the foot by placing underneath a wooden block (Fig. 7-5) of suitable size and height. Slots can be cut in this block to hold the cassette (Fig. 7-6).

Sand bags: These (Fig. 7-7) are used in case of small animals or small ruminants. Sand bags are flexible enough to mould to suit the position of the animal and heavy enough to help maintain the desired position. To make such bags, fine sand is placed in a plastic or cotton bag upto 2/3rd of its capacity and this bag is then placed inside another bag of suitable material except plastic.

Foam blocks: Foam blocks (Fig. 7-8) of various sizes and thickness are used to provide support to the animal or area of interest. Foam is radiolucent and relatively inexpensive. Their placement should be such as to avoid being in the direct path of primary X-ray beam as otherwise air density will be produced on the radiograph. Foam can also absorb liquids which after drying produce radiodense shadows on the radiograph.

Fig. 7-5 : A positioning wooden block.

Fig. 7-6 : A positioning wooden block with slots(s) for cassette placement.

Fig. 7-7 : Sand bags.

Fig. 7-8 : Foam blocks of different sizes.

Cassette holders: Use of cassette holders (Figs. 7-9 and 7-10) reduces radiation hazards to personnel as these help them to be away from the primary X-ray beam. It becomes easier to place the cassette in a correct position for successive radiographs and chances of possible injury to the personnel from the animal are also reduced. Additionally, chances of motion of the cassette during radiographic exposure are greatly reduced. A long adjustable stand or 'leg' of the cassette holder (Fig. 7-10) may be made from metal or conduit pipe that slides through a sleeve. The stand can be fixed in a desired position with a thumb screw. Cassette holders are easy to construct from wood or aluminium.

Fig. 7-9 : One type of cassette holder.

A weight bearing cassette holder can also be constructed from wood (Fig. 7-11). It can be used for radiography of the foot in a standing animal or for ventrodorsal projections. Such a holder not only provides protection to the cassette but it also becomes easier to change the cassette, if required.

Lead blocker

The use of a lead blocker allows more than one radiographic exposure of the same film. Such blockers are thus important from economy point of view especially when extremities are to be radiographed. The procedure also enhances radiographic quality due to reduction in the field size. A lead blocker is a sheet of lead (Fig. 7-12) placed over the part of the cassette not desired to be exposed. After an exposure has been made, blocker is shifted to the exposed part of the film and another part of the film is exposed. A lead blocker should be atleast 2 mm thick.

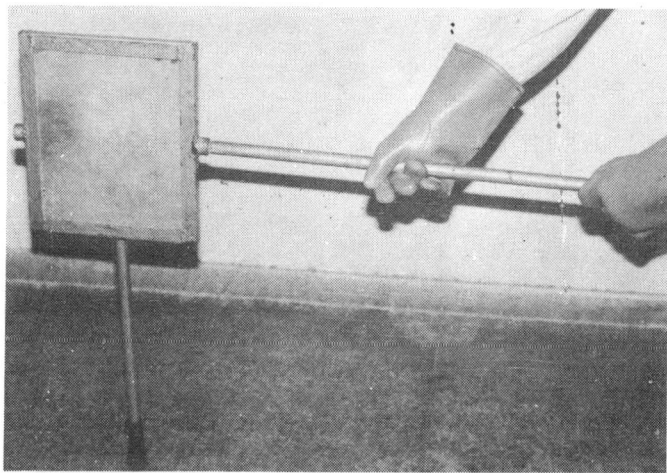

Fig. 7-10 : A cassette holder with a stand.

Fig. 7-11 : A weight bearing cassette holder. S = slot for cassette placement.

X-ray Table

A variety of X-ray tables is available for use in diagnostic radiology. Most important requirement for such a table in veterinary practice is its rigidity. Most tables are used in conjunction with a floor mounted tube stand (Fig. 7-13). However, the X-ray table can also be used in conjunction with a ceiling tube column or mobile X-ray unit. Most tables used for small animal radiography measure 196 x 61 cm with a height of 82 cm from the floor, X-ray tables are either positioned on the smooth surface of the floor or in channels laid down in the floor. In the latter type, electrical connections and support plates can be attached to the table from beneath. Surface of some tables can tilt (Fig. 7-

14) so that it can move at right angle to the floor. In others, no motion is permitted. Former tables permit shifting of the X-ray tube and flouroscopic screen with the table. An X-ray table usually has provision for a bucky tray beneath its surfcace which also holds the grid. Such a system makes it easier to change the film without disturbing the animal. Table top should be stain free, smooth and washable.

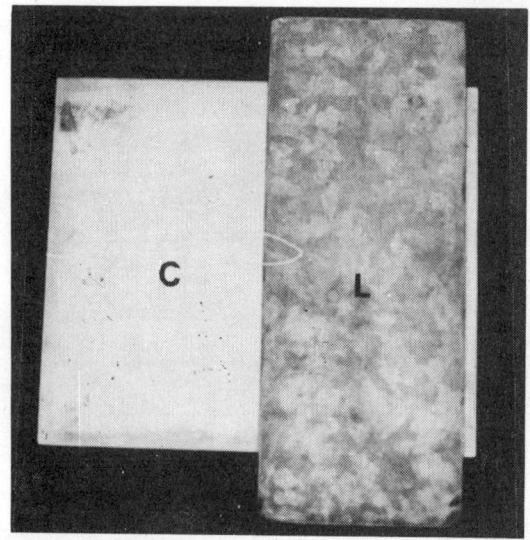

Fig. 7-12 : Lead blocker (L) for the cassette (C).

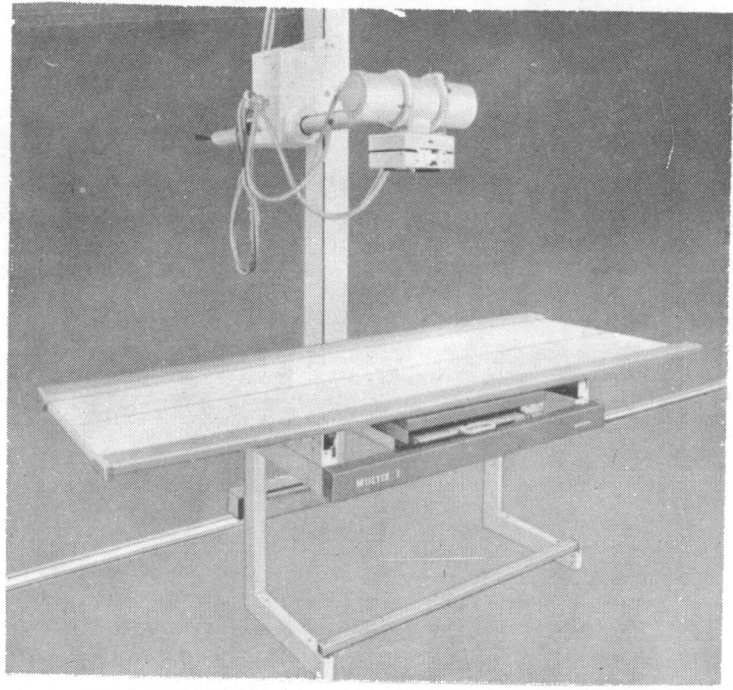

Fig. 7-13 : An X-ray table with a floor mounted tube stand (Courtesy Siemens India Pvt Ltd).

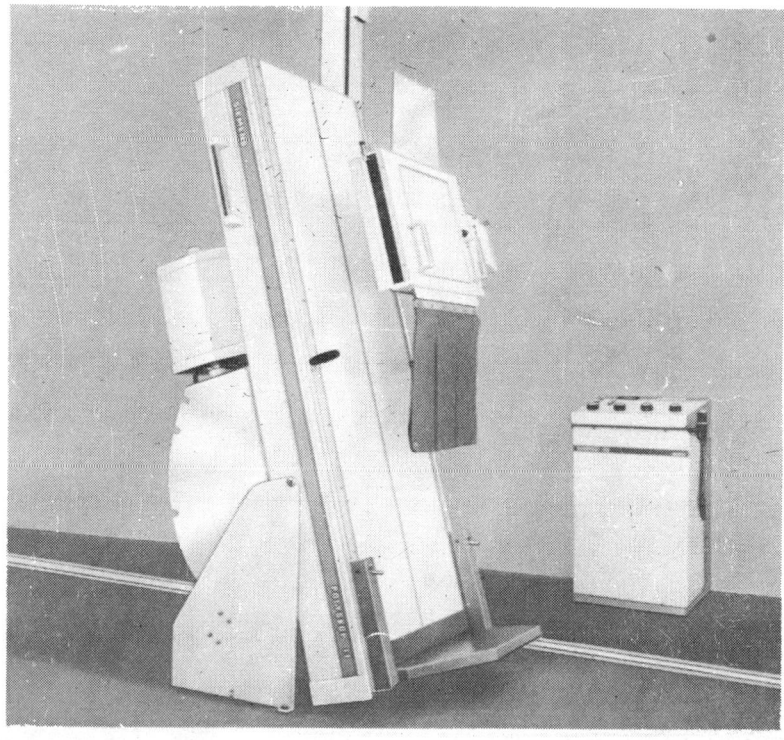

Fig. 7-14 : A multiposition X-ray table with bucky and grid assembly (Courtesy Siemens India Pvt Ltd).

Measurement calipers and centring aids

A caliper (Fig. 7-15) is required to measure part thickness for determining exposure factors. Accurate centring of the primary X-ray beam is important to produce good quality radiograph. A piece of wooden stick or expanding rule (Fig. 7-16) can be used for the purpose. Modern X-ray machines incorporate expanding rule for adjusting FFD and also for centring the film.

Viewing illuminators

For proper radiographic interpretation, radiograph should be viewed on an illuminator. Such viewers usually accommodate films upto the size of 14" x 17". A variety of illuminators is available from the manufacturers. Some accommodate single film while others can accommodate two or more films (Fig. 7-17).

STAFF REQUIREMENT

In teaching institutes, radiology needs to be developed as a discipline. So services of trained radiologist should be available. In polyclinics, doctor incharge himself has to be a radiologist too. If not already trained, short term training from an established centre is mandatory. In either case, services of a radiographer and a dark room assistant should be available. In those clinics in the field where work load is not much, a dark room assistant is enough to assist the doctor who then does the duties of a clinician, a radiologist and a radiographer.

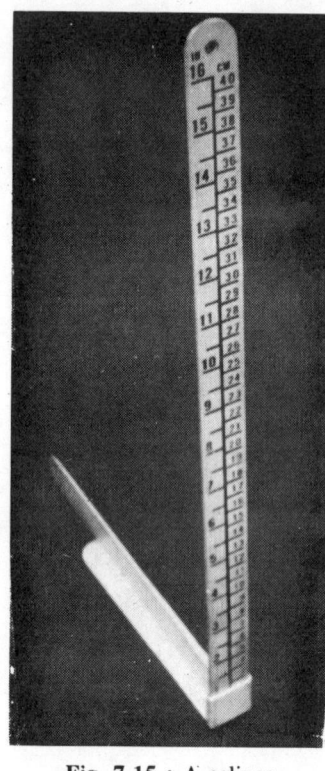

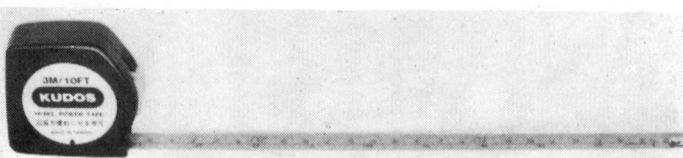

Fig. 7-16 : An expanding rule.

Fig. 7-15 : A caliper.

Fig. 7-17 : An illuminator for two films.

REFERENCES

Bargai, U. 1987. A complete mobile radiology unit for large animal field work. Vet. Rad. **28**, 66.

Douglas, S.W. and Williamson, H.D. 1980. Principles of Veterinary Radiography., 3rd edn. Williams and Wilkins Co., Baltimore.

Forster, E. 1985. Equipment for Diagnostic Radiography. M.T.P. Press Ltd, Kluwer Academic Publishers, Netherland.

Phillips, D.F. 1987. Radiology in your practice: Choosing the right equipment. Vet. Med **82,** 587.

Ticer, J.W. 1984. Radiographic Techniques in Veterinary Practice. 2nd edn., W.B. Saunders Co., Philadelphia.

Webbon, P.M. and Ramsey, L.J. 1983. Survey of X-ray machine in veterinary practice, Vet. Rec. **112,** 224.

SELECTED QUESTIONS

1. What main points would you consider while planning a radiology section?
2. What is the advantage of using positioning aids?
3. Why cassette holders should be used while making radiographic exposures?

CHAPTER

8 RADIATION HAZARDS AND SAFETY

V.K. SOBTI
N.R. PUROHIT
N.S. SAINI

The radiation exposure to man usually results from diagnostic as well as therapeutic radiation devices. Use of such devices in veterinary practice exposes veterinarians and their coworkers to definite risks. It is, therefore, essential that biological effects of radiation and related concepts are well understood. **It is the responsibility of the veterinarian incharge to enforce radiation safety measures in the radiological section and to educate other personnel about the potential hazards of radiation.**

BIOLOGICAL EFFECTS OF RADIATION

The concepts of ionising radiation are discussed in chapter 2. Once absorbed by the tissues, all types of radiation produce changes within the living tissues. The X-ray beam while traversing the tissue forces the electrons to be ejected from the atomic lattice. The atom is thus left with surplus positive electrical charge. The cells within the tissue come to a state of high chemical reactivity which can initiate biological effects. The cellular injury causes pathological and physiological changes leading to "radiation sickness" and other manifestations of radiation damage to the body. The radiation effects may be somatic or genetic. While somatic effects are harmful to a person in his life time, genetic effects affect generations. Sometimes years may pass between the radiation exposure and the evident damage. This factor complicates the situations as it may be difficult to establish a cause and effect relationship. Leukaemia and malignant tumours are common somatic effects. Genetic effects may not be manifested for several generations. Biological effects of radiation are briefed here.

Direct and indirect effects of radiation

After the transfer of radiant energy to the atoms and molecules in the form of excitation and ionisation, the resultant chemical changes in the molecules can be produced by direct and indirect effects of radiation. **The direct effect appears due to absorption of energy by the molecules. Indirect effects are caused by the products of radiation decomposition (radiolysis) of water and other solutes of the body.** After radiolysis of water in the cells, there is formation of free radicals with unpaired electrons (Fig. 8-1). In this figure, H^0 and $(OH)^0$ are the free radicals. These free radicals and the H_2O_2 formed by them are highly reactive and mutagenic. This may be one of the chief mechanism by which damage is caused by the ionising radiation as 80% of the biological system is water. Since these free radicals possess large amount of energy, they can readily break chemical bonds in vitally important macromolecules of the body such as proteins, nucleic acids and lipids. Moreover, the biological effects of radiation are amplified by oxygen which is always present in the cells. Proteins are present in the cell as part of the membrane, enzymes, hormones and antibodies. The radiation affects the secondary and tertiary chains or may cause a break in H^+ or disulfide bonds which maintain the secondary or tertiary structures of proteins. Alteration of the protein structure, however, does not seem to be a major damaging effect produced by radiation in a living system.

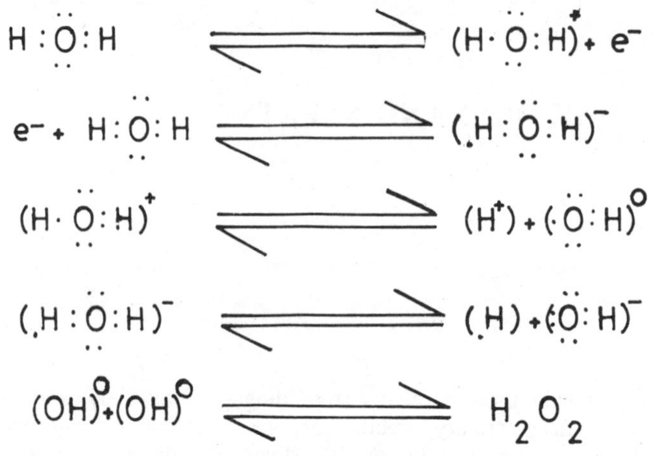

Fig. 8-1 : Formation of free radicals with unpaired electrons after radiolysis of water in the cells.

The various types of damages produced by the radiation in a DNA molecule are : change of base (deamination), loss of base, H^+ bond breakage between chains, single strand break, double strand break, cross-linking with helix and cross linking with other DNA molecules and proteins. Pyrimidines are more radiosensitive than purines. Thymine seems to be the most sensitive. Since the order of bases in DNA determines the genetic information, any alteration is most likely to change the genetic information. These changes may lead to mutations, disturbed normal cell proliferation or other cellular activities. The free radicals also attack the double bond between carbon atoms of the fatty acids. In carbohydrates, depolymerisation is the most common effect.

Radiation sensitivity of different body cells

Radiation sensitivity refers to the loss of reproductive capability of the proliferating cells. Since, the radiation damage is usually mitotic linked, **cells regularly proliferating seem to be most radiosensitive.** Thus the **cells which do not proliferate, such as nerves and muscles are relatively radioresistant,** though some functional changes may occur following high dose of radiation. Stem cells of the haemopoietic systems and cells of the gut, skin and testes are highly radiosensitive. These cells, in the order of decreasing sensitivity can be listed as : lymphoid cells, epithelial cells of the small intestine, haemopoietic cells, germinal cells, epithelial cells of the skin, connective tissue cells, cartilage and growing bone cells, cells of the brain and spinal cord and cells of the skeletal muscles and mature bone.

Vegetative intermitotic cells (germinal layer of the skin, megaloblasts) have high mitotic rate and are considered highly radiosensitive. The differentiating intermitotic cells (promyelocytes, basophilic erythroblasts), which are daughter cells of vegetative cells, are somewhat less radiosensitive. The reverting postmitotic cells (hepatic and renal epithelial cells) do not divide or divide at low rate and are relatively radioresistant.

Susceptibility of different species to the radiation

The susceptibility of various animal species to ionising radiation differs with limits. The LD 50/30 is influenced by the source of radiation and the dose rate (**the LD 50/30 is the dose that will kill 50% of the population within a period of 30 days**). The LD 50/30 of X-rays or gamma radiation has been observed to be highest in the rabbit (7.5 Gy) followed by in the rat (7.1 Gy), mouse (6.4 Gy), hamster (6.1 Gy), Guinea pig (4.5 Gy), man (3.0 Gy), dog (2.5 Gy), ass (2.5 Gy), pig (2.5 Gy), and goat (2.4 Gy). (Gy = Gray, 1 Gy = 1 Joule of energy per kg of absorbing material and 1 Gy = 100 rads).

Early vs late radiation effects

The effects of radiation on various organs and tissues may: a) appear within days or weeks after exposure, or b) appear slowly i.e. manifested only after a relatively long induction period (months or years). In general, the effects of the first category are associated with extensive killing of the cells and the acute radiation sickness. Those of the second category arise through undefined mechanism and resemble the degenerative and neoplastic phenomenon of growing old.

A. Early effects of radiation: The intensive irradiation of the entire body severely depletes radiosensitive cells in many organs simultaneously. The combined effect produces **"radiation sickness"** (acute radiation syndrome). The resulting symptoms depend upon the radiation dose rate given and are summarised in Table 8-1. The clinico-pathological changes include immediate and severe lymphopenia with a slow recovery in survivors. Major acute radiation damage occurs within stem cell pool. As the stem cell population is depleted, it no longer provides cells to replace those continually being lost from the functional pool. **The death caused by radiation damage to haemopoietic system is the result of severe neutropenia which permits the development of fulminating infections in the body.**

TABLE 8-1: Symptoms of acute radiation syndrome

Time after exposure	Supralethal dose 1000 rads or 10 Gy	Medium lethal dose 500 rads or 5 Gy	Sublethal dose 200 rad or 2 Gy
First day First and second week	Nausea and vomition Nausea, vomition diarrhoea, fever, inflammation of throat, rostration, dehydration, emaciation leading to death.		
Third week		General malaise, haemorrhage, loss of appetite, pallor, diarrhoea, fever, inflammation of throat, emaciation leading to death in 50% of the victims.	Loss of appetite and hair, inflammation haemorrhage, diarrhoea. No death in absence of complications

Adapted from Upton and Kimball, 1968.

The degree of injury to different tissues by whole body radiation can be described as under:

a) Lymphoid tissues : The lymphocytes of lymph nodes, thymus and spleen are highly radiosensitive. Within 15 minutes after a moderate dose of total body irradiation, there is marked reduction in cell division among lymphocytes and many show necrotic changes. Regeneration may occur within 2 to 5 hours of the exposure, but large doses permanently inhibit cell regeneration and lymphoid tissues show marked shrinkage. Secondary changes due to formation of abscesses, ulcers, haemorrhages or other lesions in the organs drained by a particular lymph node may occur.

b) Bone marrow: The precursors of red blood cells, granulocytes and platelets are radiosensitive. Erythroblasts are most sensitive while myelocytes are somewhat less sensitive. Within a week or two after radiation exposure, most of the radiosensitive elements disappear from the marrow leaving an aplastic marrow containing resistant cells (fat cells, reticular cells, connective tissue cells, macrophages) and haemorrhages.

c) Cardiovascular system : The heart, and large arteries and veins are radioresistant. The endothelium of the capillaries, however, is radiosensitive. The occlusion of small capillaries leading to reduced blood supply to the tissues is one of the main effects of radiation.

d) Digestive system: **Most of the clinical symptoms arise as a result of changes in the digestive system.** Histological and functional changes are seen soon after the exposure. Ulcers and erosions in the buccal cavity result from large doses and may be convenient portals of entry for bacterial and viral pathogens. The pharynx and oesophagus are somewhat radioresistant but heavy radiation doses may lead to necrosis and sloughening of the epithelium.

The stomach is quite radiosensitive and degenerative changes are apparent within 30 minutes. Pepsinogen secreting cells are adversely affected. Parietal cells are functionally disturbed resulting in decreased hydrochloric acid production. The net effect is gastric ulcers and haemorrhages. The duodenum is the most sensitive part of the small intestine. The crypt cells show nuclear fragmentation and their mitotic activity stops. Large doses of radiation may shorten the villi. The large intestines are, however, less sensitive but are prone to ulceration. The net effect of small and large intestine injuries is loss of fluids and their decreased absorption. Electrolyte imbalance and secondary infection may contribute to fatalities.

e) Skin: Germinal layer of the epidermis is adversely affected. Higher doses of radiation cause cell death and increased cellular differentiation at the expense of cell division. These cells migrate to the superficial layer of the skin and desquamate. Normal tight ridged structures of the dermis become loose. Radiation erythema is a most prominent sign in individuals with light colour skin. Maximum dose tolerated by the skin in a single exposure is 10 Gy. At higher doses, dermatitis and subsequently ulcers develop.

f) Respiratory system: Radiation pneumonitis, hyalinisation and fibrosis of the lungs may be observed.

g) Urinary system: Parenchymal cells of the kidneys are radioresistant. The damage to the kidneys occurs due to injury to the blood vessels. The resultant ischaemia produces hypertension. The ureters and the urinary bladder are radoresistant.

h) Organs of vision: A dose similar to the one causing damage to the skin will also induce inflammatory reactions in the conjunctiva and sclera. In experimental animals, radiation has been observed to cause cataract at doses of 3-10 Gy, depending upon the animal species.

i) Male reproductive organs: The male accessory organs, epididymis and vas deferens are radioresistant. In testes, leydig cells, sertoli cells and interstitial tissues are radioresistant.

Spermatogonia are highly sensitive while mature sperms are morphologically resistant but at higher doses may not be able to fertilise ova. If fertilisation occurs, the implantation fails. A dose of 0.5-1 Gy causes massive cell depletion of testes in man and animals while the sterility sets in at doses above 2-4 Gy.

j) Female reproductive organs: Ova and the granulosa cells are highly radiosensitive.

k) Other organs/systems: The brain, spinal cord, peripheral nerves, liver, gall bladder, bones, tendons and muscles are highly radioresistant.

B. Delayed effects of radiation: A typical feature of radiation sickness is that at a very late period (several months in rat and mice, 10-20 years or more in man), various changes may 'again' appear in an organism which has apparently recovered from the radiation damage. These include shortening of the life span, leukaemia, malignant tumours and cataract. In man, there may also be induration and atrophy of the skin, connective tissues and lungs.

RADIATION PROTECTION

Keeping in view the magnitude of the radition hazards, radiation safety has assumed greater significance in routine clinical practice. However, safe radiation procedures are often not followed because of insufficient knowledge of biological effects of radiation and principles of radiation safety and also because of inadequate radiation safety equipment. **Sometimes, radiation safety practices are neglected on the ground that too much time and effort is required. None of these excuses, however, justifies the unsafe use of radiation.**

General principles of radiation safety

A. Increasing the distance between the radiation source and personnel: Distance is an important factor in reducing exposure. **Doubling the distance from the source will reduce the radiation exposure by a factor of four.** The personnel should, therefore, stay as far away from the tube as possible. For this purpose, following points should be kept in mind.

a) No individual other than the operator and those essentially involved in the procedure should be in the X-ray room when the exposures are made.

b) Wherever possible, the animal should not be manually restrained for radiography. Chemical restraint combined with supporting and physical restraint devices should be employed to the maximum extent.

c) To avoid exposure to the primary beam, a cassette holding device should always be used in large animal radiography. Apart from permitting the individuals to stay out of the primary beam, these devices also increase the distance between the source and the individuals.

d) The personnel restraining the animal manually or holding the cassette with cassette holders should be rotated so that the same person is not repeatedly exposed.

e) The operator should be in a shielding booth or behind a shielding screen or atleast 6 feet away from the X-ray source when the exposure is being made.

f) No part of the body should be exposed to the primary X-ray beam. **During fluoroscopic procedures operator should not expose any part of the body or hands directly to the primary beam even after wearing protective gloves and aprons. The amount of scatter radiation received by the persons restraining the animal could be several hundred times more than that received by the patient during exposure of an X-ray film.** Fluoroscopy should never be used as a substitute for a non-motion radiographic procedure.

B. Use of protective barriers: These barriers are designed to seek protection against scatter radiation and not against the hard primary beam. The scatter radiation has lower energy than the primary X-ray beam. The lead shielding material in the gloves and aprons reduces the dose of scatter radiation well below one twentieth of the scater radiation dose. The higher energy of the primary beam, however, may be only attenuated to one fourth by the gloves.

a) Aprons: Aprons (Fig. 8-2) should have a minimum of 0.25 mm lead equivalent for voltages upto 100 kV. The material usually used is lead rubber covered with cloth or plastic impregnated with metallic lead. **Protective aprons should never be folded but kept flat since lead shielding material tends to separate after repeated bendings.**

b) Gloves and goggles: The lead equivalent of gloves (Fig. 8-3) should not be less than 0.33 mm for voltages upto 100 kV. The lead gloves should preferably have 0.5 mm lead equivalent. The gloves should be checked periodically by radiography for cracks which can easily be missed on visual inspection. Lead goggles (Fig. 8-4) may be used during fluoroscopy examinations.

c) X-ray room and equipment: The X-ray facilities should be located away from the flow of traffic and those places where public is likely to be inadvertently exposed. It is also essential that X-ray equipment is periodically checked for possible leakage. Warning signs must be displayed near the location of X-ray units regarding potential hazards. The wall of the X-ray room should be atleast 22 cm thick and should be of concrete into which iron may be introduced. Where there is a possibility of the X-ray beam being consistently directed horizontally, the wall should have a lead lining sandwiched between plywood.

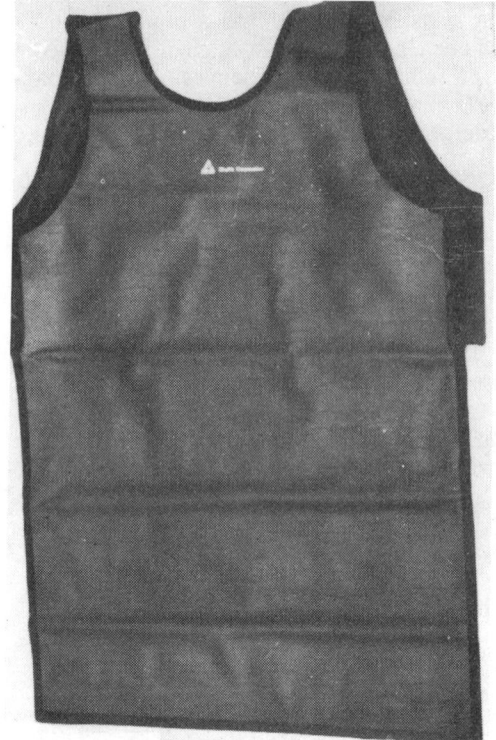

Fig. 8-2 : Lead apron.

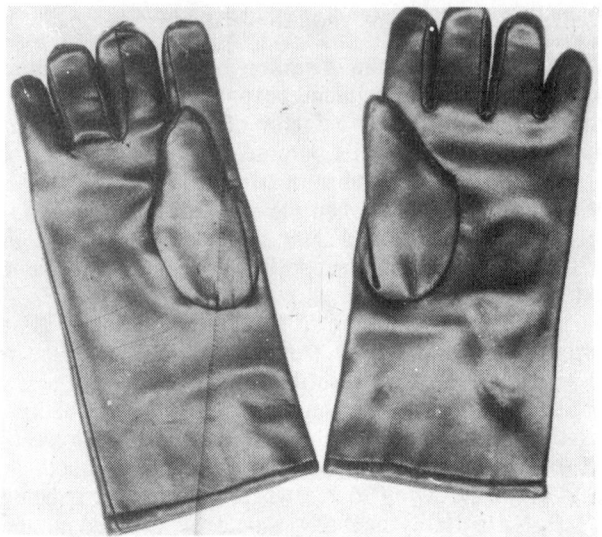

Fig. 8-3 : Lead gloves.

Fig. 8-4 : Lead goggles.

C. Reduction of exposure factors and unnecessary radiography: Correct exposure factors must be used in the first attempt. Repeated exposures due to technical errors definitely enhance the exposure level of the personnel. The X-ray film should be processed correctly and properly to avoid repetition of the process. The use of fast film-screen combinations and a decreased focal spot-film distance reduces X-ray exposure factors. Many times owners demand radiography when it is not justified, unnecessary exposures should be avoided.

D. Use of radiation monitoring devices: Radiation monitoring devices should be worn all the time by the individuals involved in radiographic work. Ideally, one film badge should be at the belt level to monitor whole body exposure and the other above the protective apparel, at the neckline, to estimte exposure to the skin of the head and neck and eyes. In case of a heavy workload, use of protective glasses or face shields may be considered. Even ordinary eye glasses provide protection from the scatter radiation. Film badges (Fig. 8-5) or thermoluminescent dosimeters can be obtained from Division of Radiation Safety, Bhabha Atomic Research Centre, Trombay, Bombay, India. At periodic intervals, these monitoring devices should be sent back to this research institute for claculating radiation dose. For the average radiological work, the personnel monitors should be assessed every month. For practices with low radiology work load, the replacement of monitor badge once in three months is adequate. Longer periods of use increase the possibility of erroneous results either due to fogging of the film inside the badges or due to technical errors.

Fig. 8-5A : Radiation safety monitoring badge.

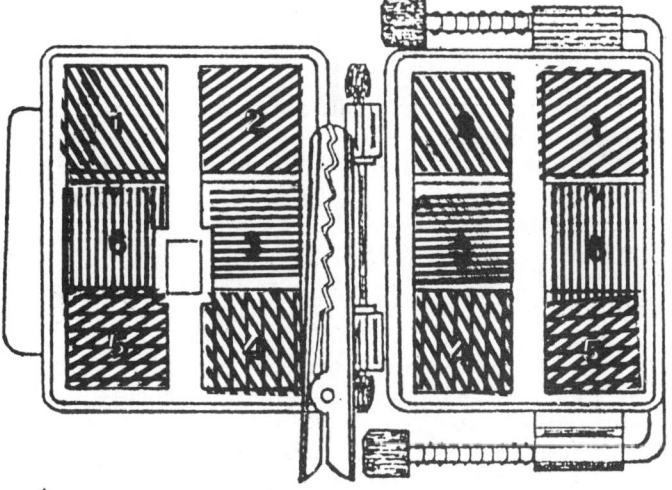

Fig. 8-5B : Structural details of fig. 8-5A. 1 to 6 are different filters. 1 = Open window for alpha rays, 2 = Plastic for beta rays, 3 = Cadmium for slow neutrons, 4 = Thin copper for diagnostic X-rays, 5 = Thick copper for gamma rays, 6 = Lead for fast neutrons and gamma rays.

E. The X-ray beam filtration, collimation and proper shielding of the tube head: a) Filtration: Aluminium filters of atleast 2.5 mm thickness should be used to absorb the soft X-rays to reduce the amount of scatter radiation to the personnel. Without filtration, the exposure dose can increase three to four times.

b) Collimators: Fixed size cone and cylinders and adjustable light beam collimators are used to limit the primary X-ray beam. Adjustable light beam collimators have the advantage of limiting the primary X-ray beam to the exact cassette size regardless of the focal spot-film distance. To improve the diagnostic quality as well as to reduce the scatter radiation exposure to the personnel, the collimator should restrict the beam to the area of clinical interest.

c) X-ray tube head: It should be well shielded from all sides except the exit window.

F) Age and sex of the involved personnel: Persons under 18 years of age should not be involved in radiographic exposures because of the sensitivity of the growing tissues and inclusion of age factor in the formula for determining maximal permissible occupational dose. Because of the extreme sensitivity of human embryos at certain stages of development, pregnant woman or potentially pregnant female should also not be involved.

The recommendations of the International Commission of Radiological Protection regarding maximum permissible dose (MPD) for persons involved in radiation work are summarised in table 8-2. For general public, MPD is equal to 1/10th of the stated dose. **The maximum permissible dose is not a level of exposure that will cause observable symptoms of radiation damage. It is a level at which any appreciable injury is unlikely to be manifested in the life time of an individual. On the other hand, if the recommended radiation safety procedures are followed, a large number of animals can be radiographed without receiving a significant amount of radiation.**

TABLE 8-2: MPD for veterinarian and the radiographic staff.

Body part	One week	one year
Whole body, gonads, bone marrow, lens of the eye	100 m rem	5,000 m rem
Hands forearms and feet	1,500 m rem	75,000 m rem

[MPD (annual) = 5 rems (N-18), where N is the age in years]

Summary of the radiation precautions

1. Never hold the X-ray cassette with hands during an exposure. Cassette holders or general anaesthesia should be used.
2. Personnel not required for assistance should leave the immediate area.
3. Lead aprons and gloves should be worn by all individuals assisting X-ray examination.
4. Primary beam restricting devices should be used.
5. Primary beam filtration should be done.
6. Personnel used in radiographic work should be rotated at intervals, if possible.
7. Pregnant woman and persons under 18 years of age should not be involved in radiographic work.
8. Fast speed screens (e.g. rare earth screens) may be used to reduce exposure factors.
9. Correct X-ray exposures and proper dark room techniques should be used to avoid repeated unnecessary exposures.
10. Radiation monitoring devices should be used.
11. A notice should be posted prominently near the X-ray unit indicating potential hazard and need to wear protective apparel.

12. Radiation survey by qualified experts should be carried out at periodic intervals to check safety precautions and possible leakage from the equipment.

Units of measurement

Unit which measures the energy of an X-ray is called **electron volt (eV)** and is often expressed in thousands (keV). **Roentgen (R)** is the unit for exposure to X-rays or gamma rays i.e. it measures the quantity of ionisation. It is defined as a unit of radiation exposure that will liberate a charge of 2.58×10^{-4} coulombs per kilogram of air. For clarity purposes, **R is that quantity of X-rays or gamma radiation which produces one electrostatic unit (2.08×10^9 ion pairs/cm^3) in 1 cc of dry air after its ionisation at 0°C and 760 mm Hg.** Since R is exclusive for X-rays or gamma rays and does not describe dose to tissue, the unit **Rad** (radiation absorbed dose) is used as the unit of absorbed dose following exposure to any type of ionising radiation. **One rad is equal to the radiation necessary to deposit energy of 100 ergs in 1 g of irradiated material (100 ergs/g).** When soft tissues are exposed to X-rays or gamma rays, rad and R. are nearly equivalent. The biological effects of various types of radiations differ a lot. In order to equate all types of radiation in terms of biological effects, the unit **Rem (roentgen equivalent man)** was evolved. **One rem = rad × quality factor.** Quality factor relates to the biological effectiveness of the given radiation. The quality factor for X-rays and gamma rays is one and for alpha particles 20.

REFERENCES

Bushong, S.C. 1975. Radiologic Sciences for Technologists. The C.V. Mosby Co., Saint Louis.

Douglas, S.W. and Williamson, H.D. 1980. Principles of Veterinary Radiography. 3rd edn., Williams and Wilkins Co., Baltimore.

Gillette, E.L., Thrall, D.E. and Lebel, J.L. 1977. Carlson's Veterinary Radiology. Lea and Febiger, Philadelphia.

Jones, T.C. and Hunt, D. 1983. Veterinary Pathology. 5th edn., Lea and Febiger, Philadelphia.

Kangstrom, L.E. and Kilibus, A. 1972. Radiation safety in small animal radiography. Acta. Radiol. (Suppl.) **319**, 147.

Morgan, J.P. and Silverman, S. 1990. Techniques of Veterinary Radiography. 4th edn., Iowa State University Press, Ames.

O'Riordan, M.C. 1968. Occupational exposure to X-rays in veterinary practice. Vet. Rec. **82**, 22

Park, R.D. and Lebel, J.L. 1987. Adam's Lameness in Horses. 4th edn., Lea and Febiger, Philadelphia

Pizzarello, D.J. and Witcofski, R.J. 1975. Basic Radiation Biology. 2nd edn., Lea and Febiger, Philadelphia.

Rendeno, U.T. and Watrons, B.J. 1980. Radiation Safety. Mod. Vet. Pract. **61**, 730.

Rust, J.H., Trum, B.F. and Kuhn, V.S. G. III. 1954. Physiological observations following total body irradiation of domestic animals with large doses of gamma rays. Vet. Med. **49**, 318.

Ryan, G.D. 1981. Radiographic Postitioning of small Animals. Lea and Febiger, Philadelphia.

Ryan, G.D. and Deigl, H.J. 1969. Safety in large animal radiography. J. Am. Vet. Med. Assoc. **155**, 898.

Ticer, J.W. 1984. Radiographic Techniques in Veterinary Practice. 2nd edn., W.B. Saunders Co., Philadelphia

Upton, A.C. 1960. Ionising radiation and ageing. Gerontologia. **4**, 162.

Upton, A.C. and Kimball, R.F. 1968. Radiations biology. In: Principles of Radiation Protection. Ed., Morgan, K.S. and Turner, J.F. John Wiley and Sons, New York.

Van Der Plaats, G.J. 1980. Medical X-ray Techniques in Diagnostic Radiology. 4th edn., Martinus Nijhoff Publisheers, London.

Yarmonenko, S.P. 1988. Radiobiology of Human and Animals. English translation by G.Leib. Mir Publishers, Moscow.

Zontine, W.J. 1980. Role of radiation safety in equine practice. Proc. Am. Assoc. Eq. Pract. **26**, 449.

SELECTED QUESTIONS

1. List various types of damages cuased by radiation in a DNA molecule.
2. Why nerves and muscles are relatively radioresistant?
3. What do you understand by LD 50/30?
4. What are the early effects of radiation on the digestive system and reproductive organs?
5. Why persons under 18 years of age and expectant mothers should not be involved in radiographic procedures?
6. What are delayed effects of radiation?
7. What do you understand by maximum permissible dose of radiation?
8. List important radiation safety precautions to be followed in a radiology section.
9. Define the following:
   i) Roentgen unit ii) rad iii) rem.

```
-X-RAY ROOM-
ENTRY AGE
ABOVE 18 YEARS
```

```
- X-RAYS -
EXPECTANT MOTHER ?
DONOT ENTER
THIS AREA
```

CHAPTER

9 PRINCIPLES OF ALTERNATE IMAGING AND SPECIAL TECHNIQUES

Conventional radiography continues to be the mainstay of diagnostic imaging in veterinary practice in most third world countries. Because of financial constraints, alternate imaging modalities such as ultrasound, computed tomography, scintigraphy and magnetic resonance imaging are yet to be established in leading institutes. Fluoroscopic facilities, though available, are mostly used in experimental studies. The present chapter on alternate imaging deals in brief only the basics of various alternate imaging techniques.

SECTION-A

FLUOROSCOPY

SUKHBIR SINGH

Fluoroscopy, sometimes incorrectly termed screening, is the dynamic radiological study of the body parts. Fluoroscope is a device which is used to view an X-ray image on a fluorescent screen instead of on a film. The X-rays after passing through the body parts are transformed into visible light which is observed on a screen coated with fluorescent material, either directly or through an intensifying device. The fluoroscope has been a very valuable diagnostic tool ever since it was invented in 1896 by Thomas A. Edison.

Immediately following the discovery of X-rays, fluoroscopy proved to be more useful and reliable than radiography. It was mainly because of poor quality of the film emulsion. In the later years, the radiography took over the fluoroscopy leaving the latter far behind and just for the study of moving parts and spot radiography.

Since the intensity of the image formed on the screen depends on the number of X-rays available for interaction with the crystals of fluorescent material, its applicability is limited to man, smaller domestic ruminants and pet animals. The increased thickness of the patient requires increased exposure to X-rays. In fluoroscopy, the tube currents of 3 to 5 mA are used whereas for the radiographic examination, the tube currents of hundreds of mA are generally used. Though the current has been substantially reduced to 1-3 mA in the image-intensified fluoroscopy yet the patient dose still remains higher than in radiography. The visibility of the fluoroscopic image remains so poor that it can not be viewed with cone vision. Thus maximum use of rod vision is required to study the less illuminated image which necessitates adapting the eyes to darkness before viewing. These days, an automatic brightness control is available with the modern fluoroscopic equipment that maintains the image brightness by varying the kVp or the mA.

FLUOROSCOPIC EQUIPMENT

The simplest type of fluoroscopic unit consists of an X-ray tube, a fluoroscopic table and a fluoroscopic screen. The other parts are mainly to protect both the patient and fluoroscopist. The principle of fluoroscopy is shown in fig. 9-1.

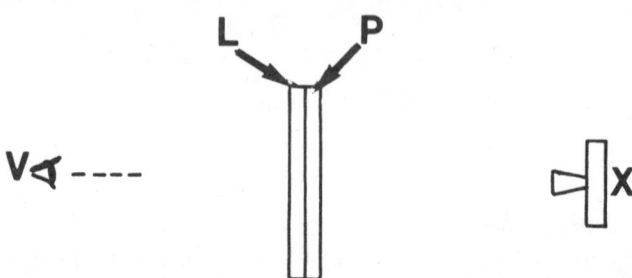

Fig. 9-1 : Principle of fluoroscopy. V = viewer, L = Lead glass, P = Phosphor layer, X = X-ray tube (From Forster, E. 1985. Equipment for Diagnostic Radiography, MTP Press Ltd, Kluwer Academic Publications, Netherland. Used with permission),

X-ray tube: A large number of X-ray machines are used for both the conventional radiography and fluoroscopy. In such machines, a switch contols the switching off or on of the fluoroscopy and automatically marks at the required currents for both the procedures. The X-ray tube is placed below the table, and the tube target must be atleast 45 cm below the table top so as to reduce the patients's input dosage. The fluoroscent screen is attached on top of table so that the tube always follows the movements of the screen and is also centered to it. The tube is mobile along the length of the table and across it.

The X-ray collimator with adjustable shutters helps in regulating the size of the X-ray beam and the image on the screen. The machine is operated with the help of either exposure button or a foot switch. The fluoroscopic unit is also provided with a timer that automatically switches off the unit after five minutes of viewing. It thus helps in reminding the fluoroscopist about the radiation exposure.

Fluoroscopic table: The standard fluoroscopic table is equipped with a motor-driven tilting mechanism. The table top is made of a radiolucent material. Many fluoroscopic tables have an over-table tube and an under-table Potter-Bucky tray so that these can also be used for routine radiography (Figs. 9-2 and 9-3).

Fluoroscopic screen: A fluoroscopic screen resembles an intensifying screen except that it contains different phosphor: silver activated zinc cadmium sulphide. It emits yellow-green light which is in the maximum sensitivity range of the eye. The crystal size is also quite large, i.e. 30-40 microns in diameter, so as to obtain the brightest possible image. The fluorescent screen is slightly thicker (0.45-0.65 mm) than the intensifying screen. The phosphor is coated on a thin card board in a suitable binding material. This coated card is then sandwiched between a plastic sheet on the patients's side and a protective sheet of transparent lead glass on the fluoroscopist's side. The lead glass and the hinged lead strips between the viewer and unit protect the operator from radiation. The different layers of the screen are shown in Fig. 9-4.

The light photons that reach the eye and ultimately the retina are so small in number that these can only be seen by the night (rod) vision. Therefore, the image intensifier (Fig. 9-6) is now a days used with the fluoroscopic unit to increase the image brightness without increasing the radiation dose of the patient. The provision of a closed circuit television system reduces the dose even further. The combined use of image intensifier and closed circuit television has improved the output image to an extent that it can be seen by cone vision and the patient dose is also greatly reduced.

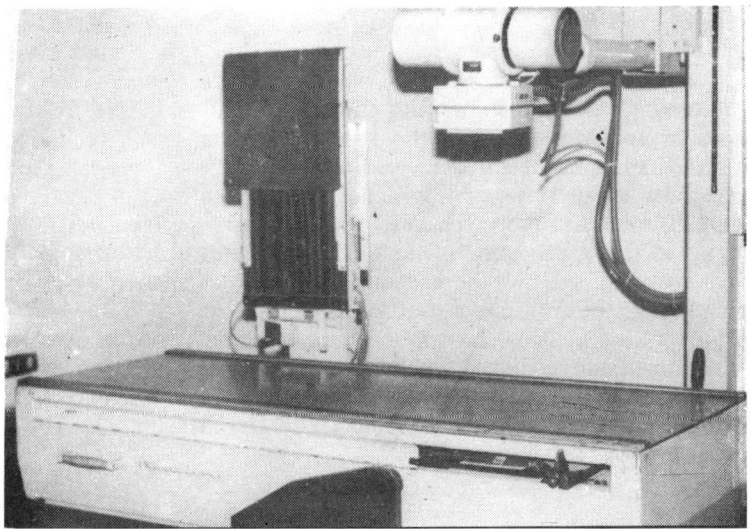

Fig. 9-2 : An X-ray unit with fluoroscopic facilities.

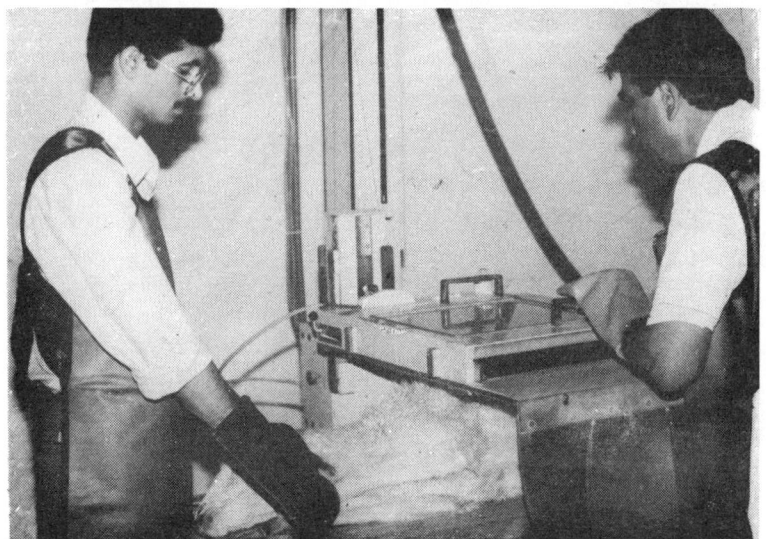

Fig. 9-3 : Fluoroscopic equipment in use.

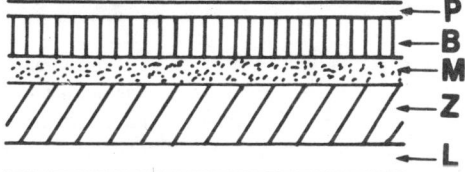

Fig. 9-4 : Cross section of fluorescent screen. P = Plastic sheet, B = Card board, M = White magnesium oxide, Z = Zinc cadmium sulphide, L = Transparent lead glass.

USES OF FLUOROSCOPY

The uses of fluoroscopy in ruminants are limited. It can be of some value only in sheep, goat and very young calves apart from small animals. The fluoroscopy is for immediate visualisation of the radiographic image on the fluorescent screen. It is a technically simple procedure for consultation and teaching. It also saves time and expenses of exposing and developing X-ray film. **The greatest contribution of fluoroscopy is that it permits clinical evaluation of the dynamics of the body such as the peristalsis and movements of the joints.** The fluoroscopy, or its modern counterpart the image intensified fluoroscopy, helps the veterinarians to thoroughly evaluate the upper gastrointestinal tract, especially the oesophageal obstruction or diverticulum using a barium meal.

The fluoroscopy also helps in proper placing of a catheter in the bronchus for bronchography. It is also a valuable aid in the investigation of the thoracic contents and in positioning of catheters in the heart and great blood vessels for the examination of the circulatory system. Fluoroscopy has also been used for the study of birth postures of sheep foetus.

One of the main purposes of fluoroscopy is for proper positioning of a body part for a radiograph. Therefore, most fluoroscopes are equipped with spot film device (Fig. 9-5) to record the fluoroscopic image. The adjustment of field size, the movement of the stored film between the patient and the screen and switching on of the X-ray tube to radiography are all automatic. This way two or four radiographs can be taken on a single film.

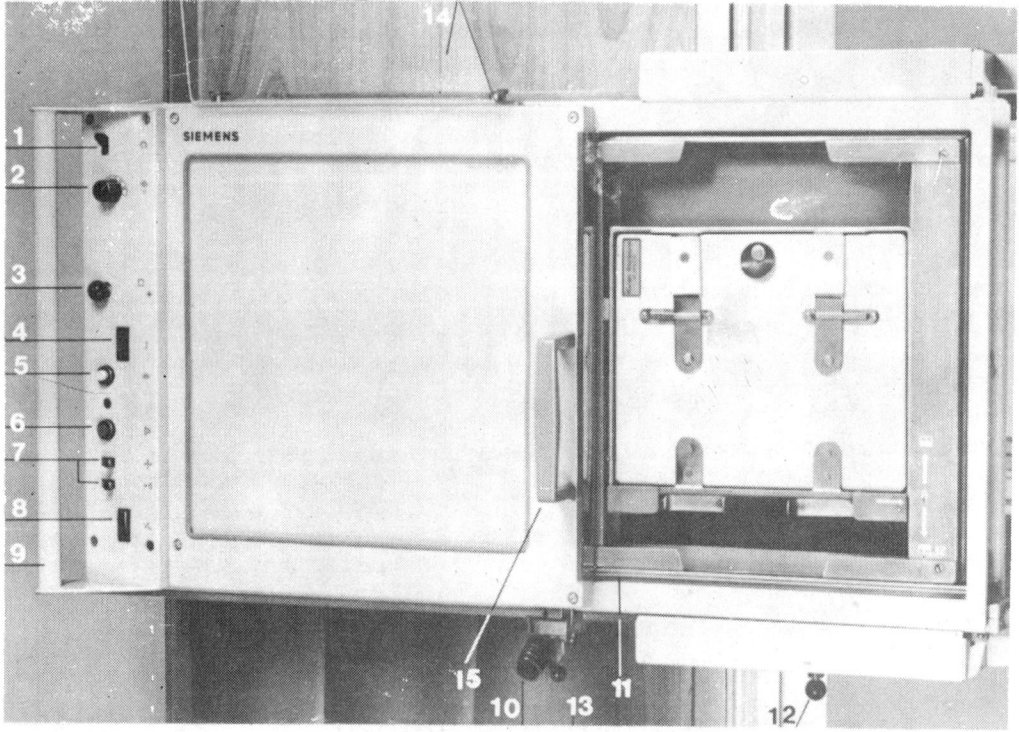

Fig. 9-5 : Spot film device. 1 = Room light switch, 2 = Fluoroscopic kV remote adjustment; 3 = Motorised collimator control, 4 = Motorised longitudinal excursion of the table top, 5 = Fluoroscopy release, 6 = Exposure release, 7 = Switches for electromagnetic locking of spotfilm device, 8 = Unit tilting, 9= Guide handle, 10 = Handle for moving cassette frame with stops for choosing film division, 11= Cassette frame for cassette sizes 8" × 10" and 10" × 12", 12 = Release for sliding the spotfilm device into the parked position, 13 = Scatter radiation guard, 14 = Cough guard, 15 = Hand grip (Courtesy Siemens India Pvt Ltd).

SHORTCOMINGS AND HAZARDS

The radiation hazards and other disadvantages are greater in fluoroscopy than in conventional radiography. Since the intensity of the image is very low, smaller details on the screen are most likely to be escaped. Though the exposure rate at the table top must not exceed 10 roentgen per minute yet an extended investigation generally results in higher exposure dose to the patient. **The radiation hazards to the fluoroscopist and other staff are more even after adopting all preventive measures.** The fluoroscopy was also handicapped in keeping permanent records of dynamic images before the facilities of close circuit television and video recording were available.

SAFETY PRECAUTIONS

The fluoroscopist and his staff must take all the precautions such as wearing of lead gloves and apron as in case of conventional radiography. The smallest possible field should be examined during fluoroscopy using the unit intermittently with limitation of time i.e. 5 minutes. The fluoroscopy of the head should be avoided. Aluminium filter of 3 mm thickness should be used at the aperature of the tube so as to reduce the exposure dose of the patient. The kilovoltage should be increased so as to reduce the mA while producing a brighter image. If there is a slot for a bucky tray below the table top, it must be shielded by a lead drape. **Fluoroscopy should never be used as a substitute for non-motion radiographic examination.** The fluoroscopist and other staff must wear the film badges to monitor their exposure dose.

SECTION-B

IMAGE INTENSIFICATION

D.M. MAKHDOOMI

Two important limitations of fluoroscopy are lack of brightness of the image and increased radiation hazard. Image intensification through an image intensifier overcomes these problems to a large extent. An image intensifier is a complex electronic unit which opens several possibilities:

i) The image produced for viewing is 1,000-5,000 times brighter than that obtained by a conventional fluoroscopic unit.
ii) Brightness and contrast of the image can be electronically controlled.
iii) It is possible to record the motion of organs through a recording system.
iv) Spot film camera can be used.
v) Lower mA settings required with an image intensifier reduce radiation exposure of the patient and personnel.
vi) Since image can be viewed on a television, examination can be done in a lighted room.

An image intensifier is shown in fig. 9-6 and fig. 9-7 is its diagrammatic illustration. It basically consists of four components:

i) Input phosphor and photocathode
ii) Electrostatic focusing lenses
iii) Accelerating anode
iv) Output phosphor

An X-ray beam that carries useful information and exits from the patient enters the intensifying tube to interact with input phosphor layer (zinc cadmium sulphide crystals or cesium iodide). The X-ray energy is converted to visible light photons. These light photons strike the photocathode (photoemissive metal usually composed of cesium and antimony compounds) to emit photoelectrons directly proportional to the intensity of light falling on the photocathode. A potential difference between the photocathode and anode accelerates these photoelectrons towards the anode. Near the anode is situated the output phosphor (silver activated zinc cadmium sulphide). The photoelectrons interact with output phosphor. Since these are accelerated electrons, 50-75 times light photons are emitted than originally present on the input phosphor. A thin layer of aluminium placed on the fluorescent screen allows high energy photoelectrons to pass through towards the output phosphor layer but prevents retrograde movement. This layer also serves as a ground to remove spent electrons which otherwise would produce a negative charge on accumulation. The whole assembly is encased in a lead lined metal container to protect personnel from stray radiation. The container has a lead glass window over the output phosphor to absorb the radiation passing through the intensifier.

There are several means of viewing the image formed by an image intensifier. It can be viewed through a mirror optical system that magnifies the image through mirrors and lenses. However, field of view is small and only one or two persons can view it at a time. A television monitoring system allows more persons to view the image and to control its brightness and contrast. The system, however, is more expensive. A video recording attachment allows recording of the motion of organs. A spot film camera can also be attached to the image intensifier.

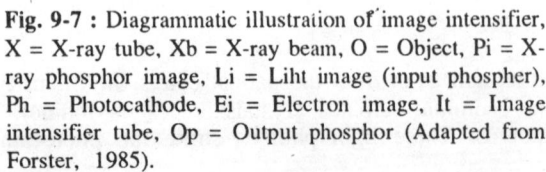

Fig. 9-6 : An X-ray image intensifier (Courtesy Siemens India Pvt Ltd).

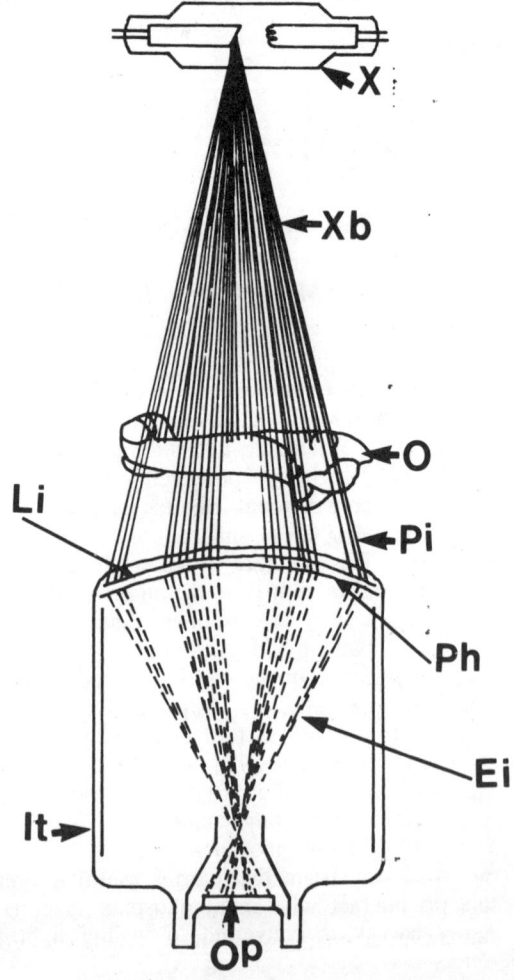

Fig. 9-7 : Diagrammatic illustration of image intensifier, X = X-ray tube, Xb = X-ray beam, O = Object, Pi = X-ray phosphor image, Li = Liht image (input phospher), Ph = Photocathode, Ei = Electron image, It = Image intensifier tube, Op = Output phosphor (Adapted from Forster, 1985).

SECTION-C

XERORADIOGRAPHY

S.K. CHAWLA

It is a method of X-ray imaging in which a visible electrostatic pattern is produced on the surface of a photoconductor. Amorphous selenium is used as the photoconductor on a xeroradiographic plate. When exposed to radiation, selenium becomes locally conductive and partly dissipates its uniform charge. The remaining charge pattern on the selenium forms a latent electrostatic image which by certain procedures is converted into a visible image. Selenium does not conduct current when shielded from X-rays or light.

Xeroradiographic plate consists of an exceedingly smooth aluminium plate coated with a thin layer of selenium. Since selenium is an insulator in the dark, during charging the plate can be considered as a parallel capacitor with selenium layer and aluminium acting as the dielectric (Fig. 9-8). Sensitivity of the plate to X-rays depends upon the selenium thickness and energy (kVp) of the X-ray beam. There is a interface of aluminium oxide between the selenium and aluminium, and a thin layer of cellulose acetate to protect the selenium surface. Though aluminium is a coductor, aluminium oxide is an insulator. The purpose of the interface is to prevent charges induced in the aluminium from migrating into the selenium and dissipating the positive charge induced on the selenium surface. Overcoating prevents lateral conduction of charges which would degrade the electrostatic image. It extends the life of a xeroradiographic plate by ten times.

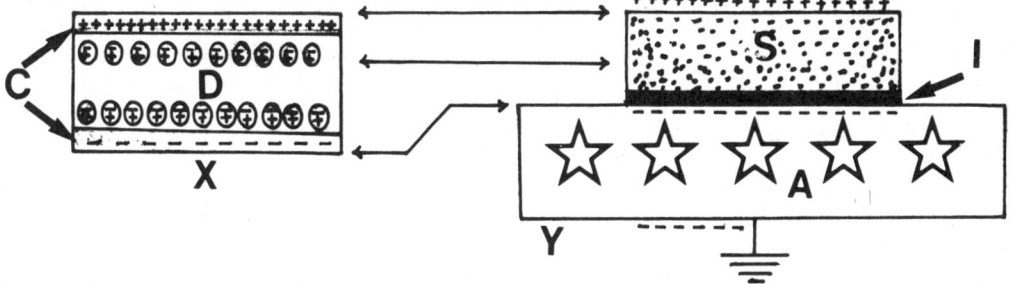

Fig. 9-8 : Charging of a xeroradiographic plate (Y) is similar to charging of a parallel plate capacitor (X). C = Capacitor plates, D = Dielectric, S = Selenium; I = Interface, A = Aluminium substrate (Figs. 9-8 to 9-10 are adapted from Curry et al., 1984).

The first step in the xeroradiographic process is to sensitise the selenium layer by applying a uniform electrostatic charge to its surface in the dark. It is done by making the plate to move at a uniform rate under the stationary charging device which may be a scorotron or corotron and produces a charge on the surface of photoconductor. Thus the layer of selenium on one side becomes charged with a uniformly distributed positive charge (600-1200V) which is retained as long as it remains in the dark due to the great resistance of the layer. The charged plate is then enclosed in a cassette which is light-tight and is rigid enough to provide mechanical protection to the delicate plate. The plate cassette combination is used just as one would use an X-ray film in its cassette.

When the charged selenium plate is exposed to X-rays, electron hole pairs are formed. These charged particles come under the influence of the positive charge on the selenium layer and negative charge in the aluminium substrate. The electrons migrate to the plate surface and discharge the positive charge originally laid down. The positive holes move through the selenium towards the substrate where they are neutralised by the induced negative charge. The amount of discharge of the positive charges on the surface of the plate is proportional to the intensity of X-rays which penetrate the patient. Thus an image arises with localised electrostatic contrasts which correspond to the localised contrasts of the incident radiation image. The remaining charge pattern on the plate, called as the electrostatic latent image (Fig. 9-9), accurately reflects the pattern of attenuation of X-rays caused by the part xeroradiographed.

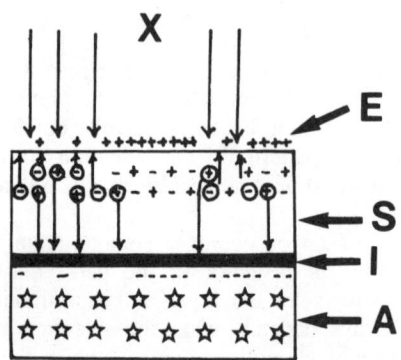

Fig. 9-9 : Formation of electrostatic latent image (E) on a xeroradiographic plate. X = X-rays S = Selenium, I = Interface, A = Aluminium substrate.

The latent electrostatic image is made visible (developed) by exposing the surface of the selenium plate to the fine charged powder particles called toner which usually is a pigmented thermoplastic material of dark blue colour (called type 5 or 5B). Particle size of the material is few microns. The exposed xeroradiographic plate is placed on top of a dark box into which an aerosol of charged toner particles is sprayed through a nozzle. The electrical charge on the toner particles is produced by friction between the toner and the wall of the nozzle, the process being called triboelectrification or contact electrification. Toner particles with both a positive and a negative charge are produced in roughly equal numbers. The tiny negative charge toner particles are attracted to the plate surface in proportion to the intensity of the residual charge and form a contrast of the uncharged areas which are black.

The powder picture on the surface of the xeroradiographic plate is then transferred to a special paper and fixed there to form a permanent image. An electrostatic transfer process is used to attract the toner away from the plate and onto the paper. The paper is coated with a slightly deformable layer of plastic such as a low-molecular weight polyethylene material. When it is pushed against the powder image under relatively high mechanical pressure, the toner particles become slightly embedded in the plastic. The paper is then peeled off the plate, and the loosely held powder image is made into a permanent bonded image by heating the paper to a temperature of about 475°F. The heat softens the plastic coating on the paper and allows toner particles to sink into and become bonded to the plastic. The toner particles do not melt or flow. After fixing, also called fusing, the imaging portion of the xeroradiographic process is completed. The completed image is delivered from the processor ready for viewing.

On a xeroradiographic image, the areas which receive little X-ray exposure appear light blue. If a charged plate is inadvertently exposed to room light and then developed, the paper will be almost devoid of toner. Conversely, a charged but unexposed plate will produce a uniformly deep blue print.

After transfer of the toner to the paper, some of it remains on the plate surface. All of the toner must be removed before the plate can be reused. This is done by exposing the plate to a light source that reduces the bond holding residual tone to the plate. A preclean corotron then exposes the plate to an alternating current which serves to neutralise the electrostatic forces holding the toner to the plate. The residual toner is then brushed off from the plate using a clean brush. The plate can then be reused. It is not charged during storage.

The xeroradiography has found application in soft tissue imaging e.g. in radiographic examination of the mammary glands, muscles, tendons and ligaments. **Main advantages of xerordiography include enhanced visualisation of the borders between images of different densities (edge effect), low contrast which enables differentiation between fat, muscles and bones and wide exposure latitute.** It is also a dry process. Major disadvantage is that the technique can not be used for very thick parts as very high exposure is required.

SECTION-D

SUBSTRACTION TECHNIQUE

B.M. JANI

Substraction technique is a photographic method that eliminates unwanted images and makes it easier to visualise important radiographic information on a radiograph. The principle is illustrated in fig. 9-10. It is mostly employed to delete bone and surrounding densities from an angiogram in order to clearly visualise the pattern and changes in the vascular supply. A typical example is a cerebral angiogram. Overlying bony structures in such an angiogram makes it difficult to critically evaluate minute blood vessels. If somehow all the bony structures are made to disappeear, leaving only the vascular details, even minute vessels will be clear, This is done by destroying the density differences of the bony structures in a careful controlled manner. This will require a survey radiograph of the skull, an angiogram exactly in the same plane and without motion and a negative of the survey radiograph (substraction mask). In the survey radiograph, the bone will appear white and so will it be in an angiogram. However, the density of bone on the mask will be black. When mask is exactly superimposed over the angiogram, a very high density of the bony structures will appear and it will be easier to visualise even minute specified vessels. For this purpose it is essential that no motion existed between the time survery radiograph and angiogram were made. Since injection of contrast materials do cause slight motion, exact superimposition of the mask and angiogram is

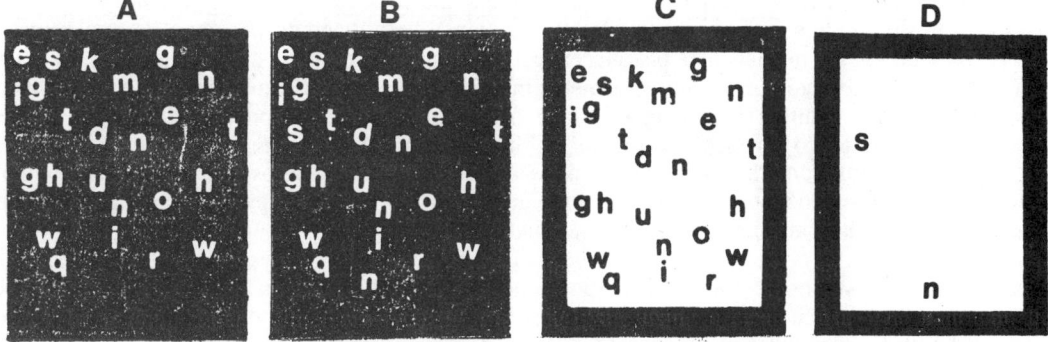

Fig. 9-10 : Principle of substraction technique, diagrammatic representation. A and B radiographs are similar except that B contains two additional letters. To identify these letters, a negative (C) of A is made. When C is superimposed on B, black density of letters in C merges with black density of B except for two additional letters. Print then shows added letters in D.

not possible. Because of high density (blackness) of the mask, special viewer with bright light is essential to view the superimposed films, or print should be made. It is also essential that the mask has the same density of all structures as is on the survey radiograph. Therefore, a special type of film is used to make a negative. This film has a gamma of one i.e. it does not exaggerate or decrease the density difference of the survey radiograph. Both the negatives and prints in substraction technique require exposure to white light source.

The principle of substraction tecnhique can be used to locate a change that occurred between the first and second radiograph provided 'no motion existed in between. Digital equipment has been used to apply intravenous digital substraction angiography in human patients to evaluate various arterial disorders. The principal advantage of such techniques is that these are less invasive. However, such an advantage is being superceded by magnetic resonance imaging which also is free from hazards of radiation exposure and contrast overload.

SECTION-E

NUCLEAR SCINTIGRAPHY

RISHI TAYAL

Nuclear scintigraphic imaging is a highly sensitive advanced procedure in which radioisotopes (see chapter 11) are used to detect functional abnormalities of the body systems. Two basic requirements of the technique are:

    i)   A gamma ray emitting pharmaceutical that concentrates in the area of interest.

    ii) A detctor or a gamma camera to provide information or image of the pharmaceutical's distribution.

The principle of nuclear scintigraphy is based on the use of a pharmaceutical which after entry into the blood stream gets localised in a particular tissue or organ. Before the pharmaceutical is injected it is labelled with a radioisotope. Thus localisation of the isotope can then be detected by using dectetor or camera due to emission of gamma rays from the area of interest. For bone imaging, pyrophosphate or methylene diphosphate is generally used as the pharmaceutical while for lung perfusion imaging in horses, macroaggregated albumen is used. So a variety of pharmaceuticals are available for different organs. Most commonly used radioisotope is Technetium-99m. This isotope has the advantage of a short half life of 6 hours and thus animal can be discharged next day after the scan. In addition, radiation exposure to the animal and attending personnel is minimum.

The distribution of the labelled pharmaceutical in a particular organ can be detected by using either a hand held detector or a gamma camera. In both cases sodium iodide crystal is used which absorbs gamma rays emitted by the radioisotope from the patient and converts it to light flashes. The light is converted to an electrical impulse. This is recorded on a meter in case of hand-held detector. In a gamma camera, electrical impulses are shown on a oscilloscope or converted to an image. Image can be produced in colour or in a grey scale. A computer system is usually attached to the camera for gathering data and its display and quantification.

A scan appears as a image formed of dots. The interpretation is based on the appearance of increased (**hot spots**) or decreased (**cold spots**) radioactivity regions. An active process produces hot spots. In lung perfusion studies, lack of perfusion in an area produces cold spots. Similarly cold spots will be observed in case of an abscess.

Nuclear scintigraphy has been used in both small and large animals, in experimental studies as well as in clinical cases. Mild sedation of the animal is necessary when camera is used as images are recorded over an interval of 2-3 minutes. Sedation is not required if hand-held detector is used as the device is placed at one site only for a few seconds.

Nuclear scintigraphy has been used to detect functional disorders of the kidney, liver, gastrointestinal tract, lungs, thyroid gland and many ohter organs. However, in small and large animals, scanning of the bone has most widely been investigated and applied in clinical practice. Using this technique, it is easier to detect localised increase or decrease in bone turnover as a result of trauma or disease. This is so because while plain radiographs reflect changes in the bone density and structure, which take a long time to appear on a radiograph after injury, bone scans reflect canges in skeletal metabolism and physiological conditions. **The high diagnostic sensitivity of scintigraphy can be related to the following factors:**

i) The multidirectional emission of gamma rays from a lesion.

ii) High sensitivity of the detector to the gamma rays.

iii) Fact that changes in radiotracer distribution are observed earlier than the tissue density changes can be appreciated on a radiograph, e.g., stress fractures can be scintigraphically imaged within 24-48 hours of their occurrence while it may take many days before these are detected on a radiograph. In horses it has been reported that it is possible to predict anatomical changes in the bone associated with relatively advanced stages of degenerative joint disease as much as a year before such changes are detected on plain radiographs.

Any inflammatory or pathological process that causes increased bone activity can be diagnosed by using nuclear scintigraphy. The technique is thus very useful to diagnose occult lameness. There are numerous other applications of scintigraphy e.g. to detect brain lesions. It has also been used to study renal and cardiac function, lung ventilation and pefusion and ureter patency in both small and large animals. It has been used to obtain useful images of the vertebral column of small animals and horses. Scans can be used to monitor the progress of fracture healing as uptake of methylene diphophate, pharmaceutical used for bones, diminishes and becomes more diffuse in older fractures.

The main problems associated with scintigraphy are: the cost of the gamma camera involved, precise and strict safety precautions required, non-specificity to the aetiology and difficulty encountered sometimes in interpreting the scans, especially in the case of skeletal system. There are many processes which can alter bone blood supply or metabolism and changes will be reflected in a scan. Thus a bone scan may not be diagnostically specific if other clinical information is not available or considered. However, these limitations do not decrease the diagnostic value of scintigraphy.

Apart from organ scanning, scintigraphy is also useful in detecting neoplasms. There are number of radioisotopes or radiolabelled agents with some affinity for tumours. These isotopes, after injection into the circulation of the patient, get concentrated more selectively in certain types of neoplasms to produce an area of increased radioactivity which can be detected by the detector or gamma camera. However, there are certain limitations e.g., some of the isotopes may also get concentrated in normal tissue or in inflammatory lesions. Some of the neoplasms may not concentrate the agents at all. As more specificd agents are developed, the technique will find wider applicability in this area too.

SECTION-F

DIAGNOSTIC ULTRASOUND

K.K. MIRAKHUR
TRILOK NANDA

GENERAL PRINCIPLES

Ultrasound is defined as sound waves of frequencies greater than audible to the human ear i.e. greater than 20,000 Hz. Frequencies between 1 and 10 MHz are mainly used for the purpose of diagnostic ultrasound. **A sound wave travels in a pulse and when it is reflected back it becomes an echo. It is this pulse-echo principle which is used for ultrasound imaging.** A pulse is generated by one or more piezoelectric crystals in an ultrasound transducer. When this crystal is stimulated electrically it changes its shape and produces sound waves of a particular frequency. The frequency of a transducer is determined by the times the crystal expands and contracts per second. In an ultrasound transducer, the crystal is shocked by a single extremely short pulse of electricity to vibrate at a frequency determined by its thickness.

As the transducer is placed in close contact with the body surface through a coupling medium, it undergoes continuous modification which occurs through three processes viz absorption, reflection and scattering.

Absorption: It occurs when the energy in the sound beam is absorbed by the tissues thereby converting it into heat. **Absorption process forms the basis of therapeutic ultrasound.**

Reflection: It is the redirection of a portion of the ultrasound beam back towards the source. **The reflection gives rise to echo and forms the basis of ultrasound scanning.** Interfaces between tissues of different accoustic impedence give rise to different echoes. These echoes are converted by piezoelectric effect into electrical signals and displayed onto an oscilloscope screen.

Scattering: It occurs when the beam encounters an interface that is irregular and smaller than the sound beam. The portion of the beam that interacts with this interface is scattered in all directions. Since the scattering interfaces are small, only a small portion of the beam is involved. There are two other closely related phenomena, refraction and diffraction of which **refraction is a common cause of artifacts.** When the sound beam crosses an interface, only a small percentage of it is reflected and the rest continues on through the tissues.The amount of sound that is reflected at an interface determines the amplitude of the echo. This depends on the difference between the interfaces of two different acoustic impedences. The smaller the difference, lesser is the reflection of the sound. The beam should be reflected just enough to be detected by the scanner. **Lage echoes leave very little of the beam to produce echoes from another interface deeper in the tissues. This is the reason that bowel gas or bone does not allow the scanning of tissues deeper to these.**

Once the echoes are converted into electrical signals, these are processed and transformed into a visual display of the measure of the amplitude of the echo. This is known as **echo quantification.** For displaying this echo amplitude information, different modes are used. **There are three modes of display in diagnostic ultrasound: A, B and M.**

A-Mode (A stands for amplitude): This is the simplest form of display. It displays two parameters of the echoes in the form of spikes i.e. distance from the transducer and the amplitude. The horizontal line shows the distance and the amplitude is depicted on the vertical line. It is rarely used these days.

B-Mode (B stands for brightness) : The brightness of the dot is made proportional to the amplitude of the echo. It yields a two-dimensional image of the area covered by the transducer. The picture represents a slice through the body in the plane of the beam. The information of the amplitude is maintained in the brightness of the dots on the screen. **Any machine in which the transducer is moved in the scanning plane by hand is a static B-mode whereas machine in which the sound beam automatically and rapidly moves in the scan plane is a real time B-mode scanner. In real time, the image is continuously updated to allow movement to be seen and thus the popularity of this mode.**

M-mode (M stands for motion): This is an adaptation of real time scanning. It records the position and motion of the echo and resembles A-mode except that A-mode also records the amplitude in addition to the position. Each spike on the display is replaced with a dot. These images are moved along a horizontal axis showing the movement of structures along that line. Interfaces that are stationary will produce straight line while moving interfaces will produce wiggly lines. A wiggly line represents the motion of echo with time.

An ultrasound machine is shown in fig. 9-11.

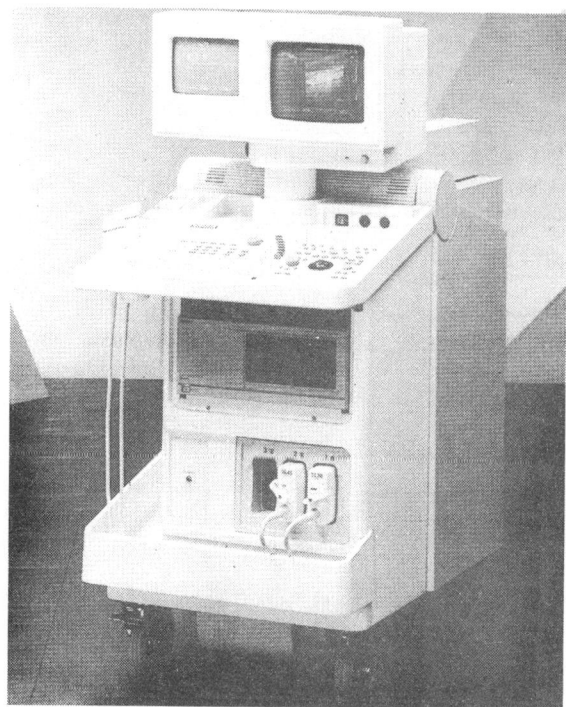

Fig. 9-11 : Quantum 2000 - an ultrasound imaging system (Courtesy Siemens India Pvt Ltd).

TRANSDUCERS

The frequency of a transducer is determined by the times the crystal expands and contracts in one second. The axial resolution of the ultrasound system is limited by the pulse length and so to get high resolution one must have a short pulse. A higher frequency penetrates less far but provides better resolution. Ideally, a transducer should emit sound waves at only a single frequency determined by its crystal thickness, however, this is not possible. Even though a transducer is designed to work at a single frequency, it actually emits ultrasound waves of many freqencies on either side of the main frequency. This range is known as band width of transducer. Over the years, transducers have been developed with quite narrow band widths.

There are three main types of transducers available: linear array, convex and sector.

Linear array: These transducers are composed of thin rectangular clips lined up side by side, each producing sound waves. The beam thus produced is rectangular in shape and permits a good visualisation of superficial structures with an easy analysis of the anatomical relationship.

Convex: The composition of this transducer is similar to that of linear except that the crystals are placed in a curvilinear fashion. Thus with the same contact area, imaging of a greater area can be effected.

Sector: Such transducers contain a single crystal which oscillates or rotates to produce a fan shaped beam. The small size gives it more manoeuvrability and access to more of the thoracic and abdominal organs through a small contact area. The superficial structures, however, are not well visualised.

SCANNING PROCEDURE

Ultrasound scanning involves considerable cooperation between the patient and sonographer. A certain amount of expertise is required to produce clinically useful images.

The transducer must be in close contact with the patinet's skin to have minimum attenuation of the sound beam. In veterinary patients, it requires clipping of hair and cleaning of the skin. There should not be any air between the transducer and skin as it will completely block the beam. A good contact is provided by application of mineral oil or an aqueous gel between the transducer and contact surface. Mineral oil is better in all respects except that it is messy. Any coupling medium used should be applied liberally.

Both gas and bone act as a barrier to the ultrasound beam as these relfect the beam. Therefore, a proper acoustic window should be found to produce a useful image of organs.

Real time ultrasound

When the images displayed in B-mode scan are formed rapidly and presented in sequence, the movement of organs will be viewed in real time. In order to form these sequential images, it is necessary to sweep the ultrasound beam over the tissues by either mechanical or electronic means. **The sound beam can be swept either in an arc or in a linear fashion. When it is in the form of an arc, it is called a sector scan.**

Mechanical sector scanners provide a very good patient contact. All mechanical scanners have two traits in common i.e. they are annular transducers and have a fixed focus. Electronic sector scanners provide the best images. Nothing moves in this but the beam is swept through the patient by changing the firing sequence of the elements in the transducer. The major disadvantage of this system is its high cost.

Linear array refers to the alignment of single small transducers that are electronically fixed in a given sequence so rapidly that no evidence of image replacement is noticed on the screen. The resultant image is rectangular.

Scanning controls

There are two important controls that require continuous monitoring and adjustment: the TCG (Time compensated gain) and the gain.Sound beam is diminished in intensity by approximately one decibel per cm per MHz. Thus unequal echoes will be returned by equal acoustic interfaces located at different distances from the scan heads. TCG controls allow equal acoustic interfaces to return equal echoes regardless of the depth. The purpose is to get a balanced picture. The correct way is to adjust this control visually. If the far-field echoes are weak, TCG should be turned up and if the near field is washed out, it should be turned down. The gain detemines the overall density of the picture. The gain should be adjusted to receive the echoes from all organs and tissues.

IMAGE INTERPRETATION

Images are usually displayed as white against a black background. Various terms used to describe the image are as follows:

Hyperechoic or echogenic: These present the bright echoes which appear as white on conventional scans. Such images are given by highly reflective interfaces such as bone and air.

Hypoechoic: These appear as grey images or dark screens and are given by interfaces of moderate reflection such as soft tissues.

Anechoic or echolucent: In the absence of any echo, the image is seen as black. It is represented by complete transmission of sound such as through fluids.The image formed on the scan screen is actually a mixture of the images of different echoes depending on the area scanned. Fluids will give an anechoic image as the sound beam passes uninterrupted. **There is often a normal bright area immediately deep to fluid and this phenomenon is called acoustic enhancement.** Similarly, bone or gas or mineral deposits reflect the sound waves totally and the image seen is bright with no visible structure beneath it. This phenomenon is called **acoustic shadowing** and helps in detection of urinary or biliary stones. Soft tissues present an image of mixed shades of grey depending upon their proportion of fat, fibrous tissue and fluid.

Artifacts: A sonologist should be aware of the common artifacts to avoid errors in image interpretation. These artifects are:

i) Acoustic shadows: These are caused by attenuation or reflection of the sound beam at an acoustic interface. When this happens, the pulse is unable to reach the deepe interfaces to produce any echo. In order to cast a shadow, the interface must reflect a large percentage of the sound beam. **The most common acoustic shadows of clinical significance are those caused by cystic, renal and biliary calculi. Gas causes near total reflection of the beam.**

ii) Reverberation: Reverberations are the largest source of positive artifact echoes that are not real. When sound beam arrives back at the transducer, a portion of the sound beam is absorbed by the transducer crystal to produce a small electrical pulse that records the echo. The remainder, however, is reflected back into the patient. So the echo bounces back to the transducer and is again reflected through the patient and back to the transducer. **This process of echo bouncing back and forth between the two interfaces is known as reverberation.**

iii) Mirror image: This effect occurs at highly reflected interfaces. Returning echoes reach the transducer under a time delay and are registered on the image as being a highly echogenic interface that is in the path of the beam.

iv) Comet-tail: This is caused by a highly reflective interface most commonly the air fluid interface. Comet-tail occurs most commonly in partially consolidated lung at the interface between the diaphragm and lung, and at the interface between the bowel wall and bowel gas.

Applications of ultrasound in ruminants have not been fully exploited, but for pregnancy diagnosis. There could be numerous organs which can be scanned using an ultrasound scanner. The organs which are usefully scanned are the liver, kidney, urinary bladder, spleen, uterus, ovaries, teat and udder. In ruminants, the use of a rectal linear scan head could be quite versatile for the diagnosis of ovarian and uterine disorders. In small animal and equne practice, ultrasound is now routinely used as a diagnostic aid. Figs. 9-12 to 9-21 show sonograms (ultrasound scans) of mare, sow and bitch.

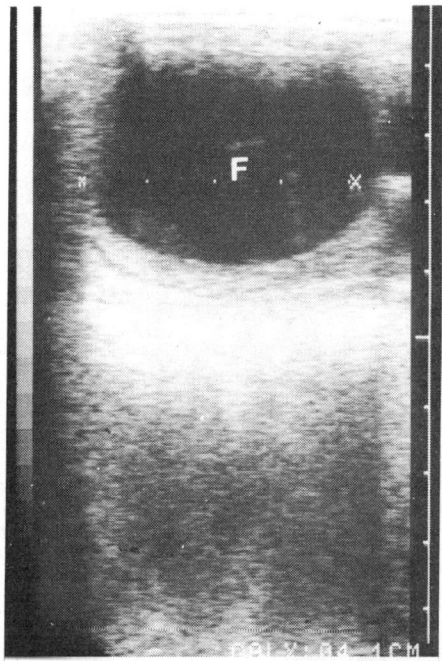

Fig. 9-12 : Sonogram of a mare ovary showing ovulatory follicle (F) of 4.1 cm.

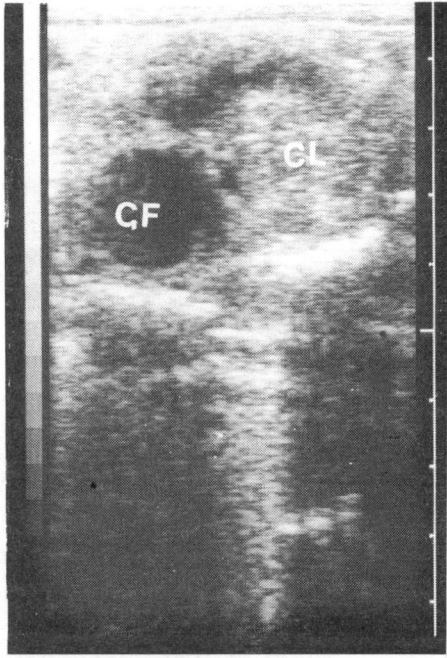

Fig. 9-13 : Sonogram of a mare ovary showing fully grown corpus luteum (CL) and a growing follicle (GF).

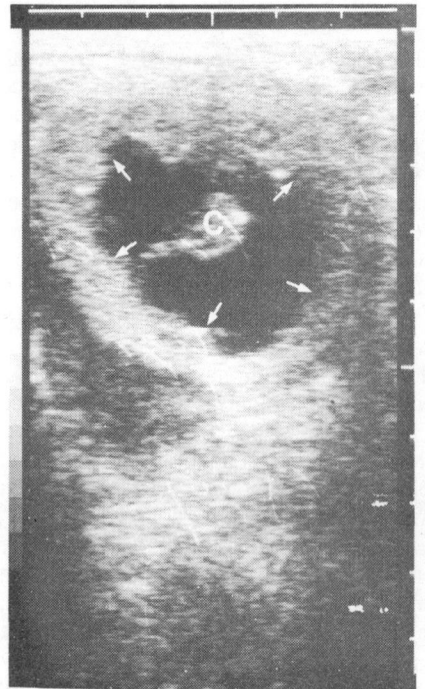

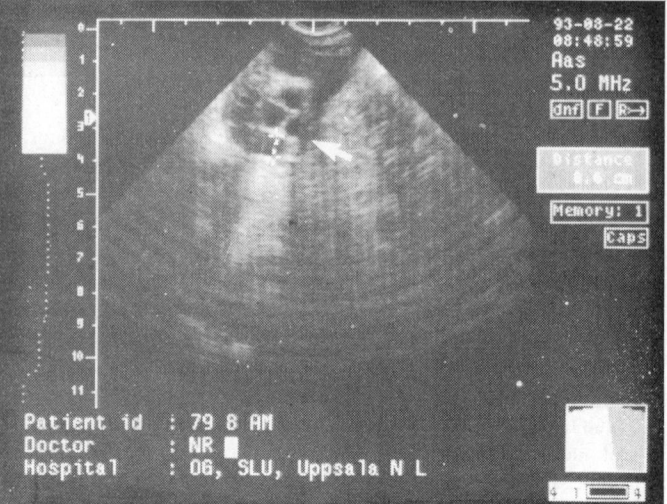

Fig. 9-14 : Sonogram of gravid uterus of a mare. C = conceptus. Arrows indicate inner uterine wall.

Fig. 9-15 : Sonogram of porcine ovary, day of standing oestrus. Arrow points follicles.

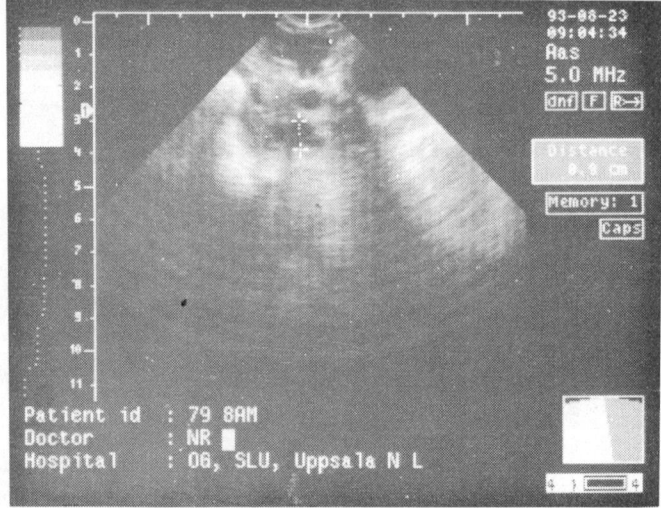

Fig. 9-16 : Sonogram of porcine ovary shown in fig. 9-15, Day 2. Ovulation has started and only 3 follicles have remained.

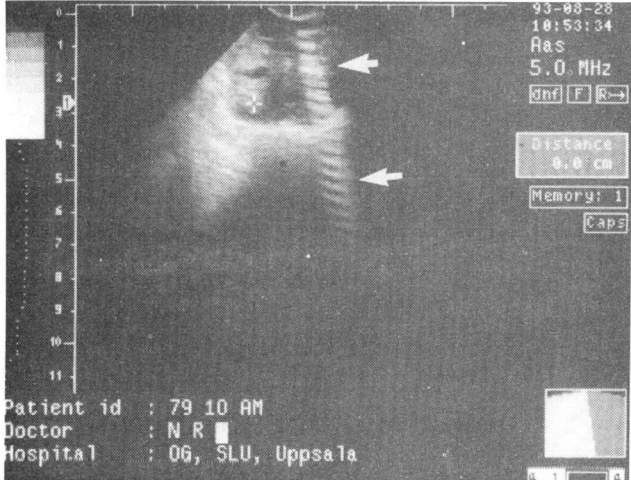

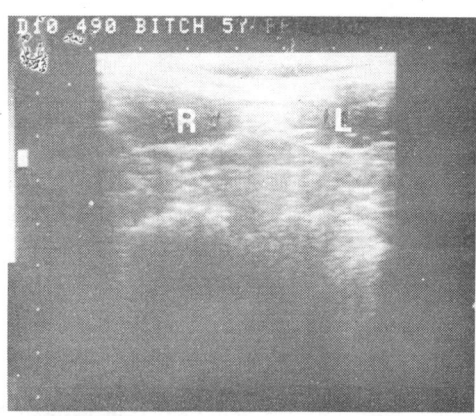

Fig. 9-19 : Sonogram of uterus of bitch showing anechoic right (R) and left (L) horns suggesting pyometra.

Fig. 9-17 : Sonogram of porcine ovary shown in fig. 9-15. Day 7. White cross indicates growing corpus luteum. Arrows indicate artifact due to air between rectum and probe.

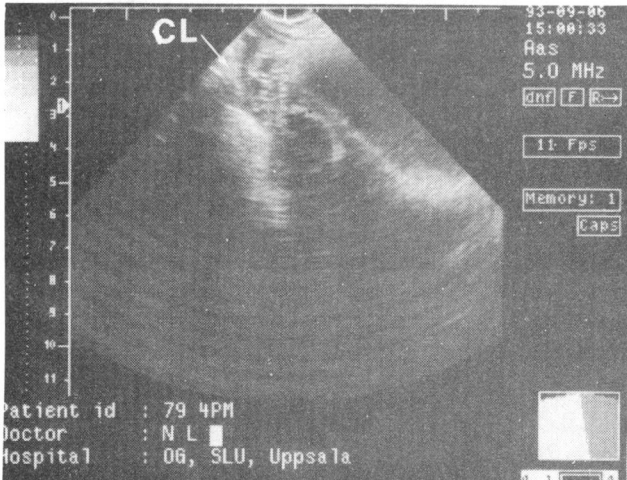

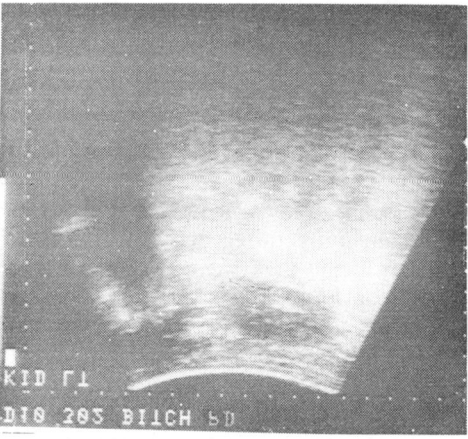

Fig. 9-20 : Sonogram of left kidney of bitch showing normal echotexture and corticomedullary junction.

Fig. 9-18 : Sonogram of porcine ovary shown in fig. 9-15, Day 14. Fully grown corpus lutea (CL).

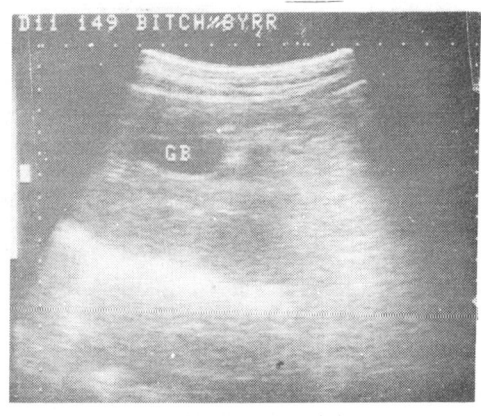

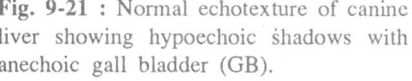

Fig. 9-21 : Normal echotexture of canine liver showing hypoechoic shadows with anechoic gall bladder (GB).

SECTION-G

COMPUTED TOMOGRAPHY

K.K. MIRAKHUR

Various X-ray techniques record the shadow of the object to be viewed on a photographic film. In conventional radiographic examination, entry of the X-ray beam from one side and exit from the other superimposes pattern of one structure over the other. The images thus formed of the internal body structures cause confusion on a radiograph. It has been realised since long that this is not the most efficient method of imaging. It is of great importance that method of examination ensures maximum utilisation of information obtained with maximum efficiency. Methods have thus been developed which essentially can image a single plane at one time and one of these is computed tomography or CT imaging.

The classic form of tomography may have originated in 1921 when the Dutch radiologist Ziedes des Plantes moved the X-ray tube and film during exposure to blur all planes but one of interest. Structues in planes other than the plane of interest will have their image fall at different places on the firm during the motion while those near the pivot will stay fixed and relatively sharp in the developed image. Lots of modifications and innovations have taken place since then to improve upon the limitations of one type over the other.

A tomogram is a picture of a slice. **Computed tomography is a diagnostic modality that is fundamentally different X-ray method in which an organ is scanned in successive layers by a narrow beam of X-rays, in such a way that the transmission of X-ray photons across a particular layer can be measured and, by means of a computer, used to construct a picture of the internal structure.** It is a fast, non-interventional computer assisted radiographic technique and has high and detailed resolution power, co-efficient of tissues being imaged into standard relative units called CT numbers. The CT number scale is calibrated in Hounsfield units (HU) with water (the reference material) given a volume of O HU, the most dense body material (bone) corresponding to a value of + 1000, and the least dense material (air) corresponding to a value of -1000.

PRINCIPLE

The credit for the development of a practical X-ray system for CT with functional apparatus that gave excellent performance goes to Hounsfield. The principle of digital computed tomography was based on the facts given by a mathematician J. Radon who proved that a two or three dimensional object can be reconstructed uniquely from the infinite set of all of its projections. The aim of the system is to produce a series of images by a tomographic method. In the actual equipment, the patient is scanned by a narrow beam of X-rays. The X-ray tube, detectors and collimators are fixed to a common frame while those rays which pass through the head are detected by two collimated sensing devices (scintillation detectors) which always point towards the X-ray source. Both the X-ray source and detectors scan across the patient's part linearly and taking 160 readings of transmission through the part being examined. At the end of the scan, the scanning system is repeated. This continues for 180° when 28,800 (180 x 160) readings of transmission will have been taken by each detector. These are stored in a disc file for processing by a mini computer. A picture is reconstructed from the data. For this purpose, a separate detector measures the intensity of the X-ray source and the readings taken from this can be used to calculate absorption by the material along the X-ray beam path:

$$\text{Absorption} = \log \frac{\text{Intensity of X-rays at source}}{\text{Intensity of X-rays at detector}}$$

If the body is divided into a series of small cubes, each with a calculable value of absorption, then the sum of the absorption values of the cubes which are contained within the X-ray beam will equal the total absorption of the beam path.Each beam path, therefore, forms one of a series of 28,800 simultaneous equations in which there are more equations than variables, and so the values of each cube in the slice can be solved. In short, there must be more X-ray readings than picture points.

The picture is built up in the form of an 80 x 80 matrix of picture points to each of which a numerical value is ascribed. Each of these points indicate the value of the abosrption coefficient of the corresponding volume of material in the slice. After appropriate scaling, the absolute values of absorption coefficient of various tissues are calculated to an accuracy of half per cent. These values are printed out on a line printer or viewed on a cathode ray tube.

The image of a cross section can be formed by projecting many beams at many angles into the edge of a desired slice; the more and finer the beams or rays, the finer the detail. What is stored in the projection toward the edge of a slice is the amount of material across the slice perpendicular to the line along which the recording is made. From these projections must be formed an image. The necessity of this extra step has yielded the name of CAT scan for computerised axial tomogrraphy or computer aided tomography, which is popularly called CT scan meaning computed tomography.

CT SCANNER UNIT (Fig. 9-22)

There are three components of CT scanner: a unit for data acquisition (operating console, a patient handling table and a scannign gantry), a unit for data analysis (computer) and a unit for image display and manipulation (image display console).

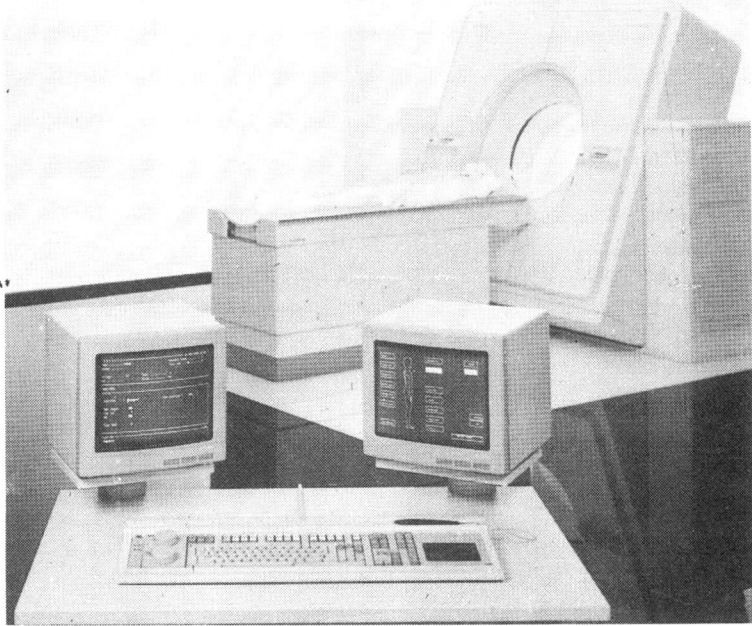

Fig. 9-22 : CT scan equipment (Courtesy Siemens India Pvt Ltd).

Portion of scanner that houses the X-ray source, collimators and detectors, and provides a mechanism to move these around the patient is known as gantry. Detectors are sensitive to ionising radiations and are able to quantify the energy deposited in them. For examination of the head, it is placed within a rubber cap in a circular orifice around which the X-ray source and detéctors rotate. The rubber cap forms the front face of a box containing water, and when water is pumped out the rubber cap exapands and the patient's head is inserted at the correct angle. The water is then made to flow back allowing the cap to collapse on the patient's head. The variations of transmission through the box during the scan will be considerably less than if the head was to be scanned in air. This reduces the range of the readings from the machine to make easier the calculations required to be made by the computer.

In the first X-ray CT units, a single pencil of rays moved back and forth parallel to itself while the entire assembly slowly circled the subject. Scatter radiation in the X-ray case was largely rejected because apertures before the source and detector minimised acceptance of photons arriving along other paths. Overall speed was increased by the simultaneous use of several rays and detectors.

A single standard X-ray tube can supply several beams or diverging rays. The apparatus is contained in a large shield sahped like a doughnut through which the subject is positioned. The source of X-ray spins round the subject and in the process emits many rays diverging in a fan pattern. Each ray is shown terminating upon a detector to measure the emerging intensity. These detectors record momentary total flux and do not count individual photons for the sake of rapidity. They may be some form of solid state detector or can be ionisation chambers. The detectors can be mounted at one end of a rigid support or gantry to whose end is attached the source so that they can rotate together around the subject. This is so called 'third generation' scanner. Alternatively, in 'fourth generation' scanners a thousand or more detectors totally encircle the subject and stand still while only the source moves. Fourth generation scanners do not necessarily produce better images. If the source is shifted towards the pivot, magnification increases while the amount of the subject that can be viewed decreases. With a fan beam, spacing is not quite constant across the subject. A full image requires rotation of 180°C plus the fan angle.

In fourth generation scanners, each fixed detector sees a sequence of rays converging on it from a changing direction as the source rotates giving the impression of a reverse fan beam. Scatter rejection is difficult in fourth generation scanners because rays enter any one detector at angles changing over 60°. In some cases, one detector monitors the direct X-ray beam and all others are compared to it, thus compensating for changes in X-ray output. The quantum efficiency of a detector is the fraction of incident photons that contribute to the output, that is, any one incident photon will be detected. A detector should not add noise of its own. If it does, it can give the same average output as a quiet detector but with greater fluctuations.

A computer is used to store data collected during the scanning and to calculate the data to produce the image. In the computer, a set of numbers is stored that corresponds to the different denisities of the subject at different spots. The spots or picture elements are often individually referred to as a 'pixel'. To commence the estimation, one assumes some distribution of densities among the points. One picks one row (or column), adds up the absorptions at the various points along its length, substracts this from the actually observed value at the end of that ray and distributes this difference equally among all the points along the ray. That line then matches reality. This is then repeated for each of the other lines that give rise to the projections and each computation partially undoes the matching of the previous computations. One can improve performance by imposing constraints such as no vlaue under zero for any pixel, or a known value for bone absorption when a value in that general range is encountered. Readings from the scanning unit for one cut are stored on a removable disc pack and processed during the scanning of the cut. The processor is time scaled with the scanner unit, its speed being such that it is able to keep pace with the flow of patient's data through the scanner unit. The removable disc pack can store more than 60 pictures any of which can be selected and viewed on the viewing unit. A key board terminal and video monitor, to allow the operator to contact the computer, manipulate the image parameters and view the image.

CONTRAST ENHANCEMENT

Effective X-ray CT visualistion of various structures is possible because of the availability of density resolution. Density resolution of most CT scanners approaches 0.3% to 0.5% (3 to 5 HU, respectively). Therefore, the difference between soft tissue and fat can be resolved readily. Hyperdense soft tissue structures can be identified readily by their high CT numbers relating to those of fat and, to a lesser extent, relative to those of the surrounding normal soft tissue.

It is possible to use the machine for determining approximately the atomic number of the material within the slice. If two different pictures are taken of the same slice at different kV and the scale of one picture is adjusted so that the values of normal tissue are the same on both pictures, the picture containing material with a high atomic number will have higher values at the corresponding place on the low kV picture. One picture can then be substracted from the other by the computer so that areas containing high atomic number can be enhanced. A contrast medium containing 420 mg of atomic iodine per ml can be readily detected at a concentration of one part in 1000 by the machine.

Tissue density may be artificially enhanced by an intravenous injection of substance containing large atoms. Sodium iothalamate has been found to be ideal and the tissue density of a variety of tumours can be enhanced. Use of contrast agents may be advantageous to facilitate visual distinction of one type of tissue from the other type of soft tissue. The need for this distinction is most apparent when there is atomic distortion of regional structures. It can also help provide information regarding the vascularity of structures. The administration of an iodinated contrast medium is essential for the elucidation of intracranial lesions (Fig. 9-23). Pathological conditions can lead to leakage of contrast medium into extravascular spaces in and around the lesion. Extrvasated contrast medium would be evident as a region of high density which clearly defines the presence and extent of a lesion on CT scan.

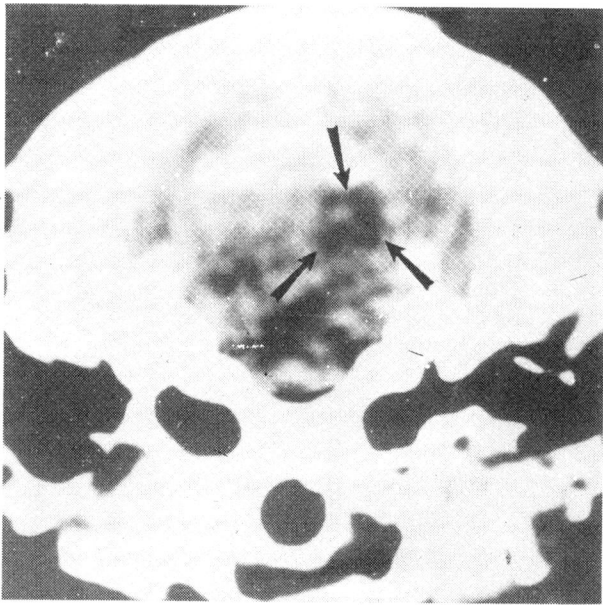

Fig. 9-23 : Contrast enhanced computed tomographic image of the brain of a 10 years old male toy poodle. The brain within caudal fossa is very mottled due to artifacts caused by surrounding dense bone. Arrows point a lucent area on the right side of the cerebellum. Cerebellar infraction was suspected (From Thomson, C.E. et al, Vet. Rad. Ultrasound 34, 2, 1993. Used with permission)

INTERPRETATION

Each section or slice is scanned by the X-ray beam in a transverse axial plane. The pictures taken from the cathode ray tube must be viewed in such a way as if the cranial contents at a particular level are being looked at from above. Decreasing the number of projections or angle of view does not decrease to a greater extent the spatial resolution but decreasing the number of rays does decrease spatial resolution. It means that if limited by dose or computation time, more fine lines per view are desirable than directions of view. Generally, no finer detail than about the width of a ray will be seen.

The computer printout accompanying each section is the numerical record of the absorption coefficients of small volumes of tissue at each picture point. The values of the absorption coefficients are calculated on a standard scale, adopting zero as the value for water. Since the absorption coefficients are measured to an accuracy of 0.5 per cent, the method is able to detect and register small differences in tissue density. In normal operation the density variation of tissue is confined to values within a 4-6 per cent range above zero. The picture brightness and contrast can be adjusted within this small range from peak white down to full black. Examination of CT scan pictures is conducted in much the same way as a radiograph. Structures are identified and their shape, size and position defined. Changes in tissue density are then searched. For a lesion to be precisely located in the picture, the density of the abnormal tissues must be either lower or higher than that of the surrounding normal tissue.

There are several sources of artifacts in CT X-ray images. An opaque object at one point obliterates each ray through that point and thus eliminates information about points beyond. This manifests as dark radiating strips that have nothing to do with body structures. There can be other distortions of brightness at any sudden extreme change in opacity which may be due to a mathematical limitation called Gibbs phenomenon. These effects can occur when processing of an image is being done and need not occur in ordinary images. There may be other sources of artifact too. The granularity associated with quantum noise can be considered as an artifact of random or irregular form. If there is a subject movement then streaking artifacts result. If the detectors are not well matched then a ring appears which can change with time, intensity, and spectral input or voltage. Fourth generation scanners eliminate this by making each detector contribute to each image element. If the number of intervals in a scanning system is too small in comparison to the desired spatial resolving power then a star pattern can result. However, this is usually not a problem with modern commercial scanners.

Computing procedures have become so sufficiently rapid that one can change the orientation of the section being observed while watching it in real time, once the individual points have been computed and stored. When about five sections per second can be alternated, one can achieve an excellent three dimensional impression of the inside of a subject by running back and forth in any direction. Interslicing smoothening makes viewing less jumpy and more staisfactory.

CLINICAL APPLICATION

Computerised tomography has been gaining importance in veterinary practice since last few years. The unique cross sectional imaging ability of CT makes it possible to determine the location of an orbital lesion, involvement of intraorbital structures and extension into priorbital regions such as paranasal sinuses and intracranial component. Tumours have been clearly defined on the transverse CT scans by their inherent density and gross distrortions of normal orbital anatomy. CT has proved successful for the visualisation of pituitary tumours in dogs. It would accurately differentiate between symmetrical bilateral adrenocortical hyperplasia and a unilateral adrenal mass. Anatomical structures and pathological processes in the central nervous system have been examined usefully in small animals. The method is useful in the diagnosis of tumours, malformations, inflammations, degenerative and vascular diseases and trauma. Multiplaner quantitative CT has been used for bone mineral analysis in dogs. Computed tomographic anatomy of the normal canine ethmoid regions using both transverse and dorsal imaging planes has been investigated. Dorsal imaging plane allows a more accurate assessment of cribiform plate involvement than the transverse imaging plane.

Computed tomography with X-rays has been an extremely significant development. Developments in electronic technology have conributed to practicality of alternatives. Digital storage of information has become more routine due to developments in solid state circuit technology, thus providing stable arbitrary resolution and storage time with fast access. Improvements in integrated circuits have allowed processing of many detector channels rather easily to reduce imaging time. The output of a computer into a television display itself is a real advantage since it readily supplies contrast enhancement that can be adjusted for best appearance while viewing.

Rather large and relatively expensive imaging systems, such as a CT scanner, perform a cost-effective and sometimes unique functions but do raise financial questions that are debatable. It is understood that computerised tomography and procedures similar to it are not readily availble to majority of veterinarians. However, if human hospitals and laboratories can be receptive to the needs of veterinarians, CT shall be able to provide a fairly good amount of diagnostic precision for various animal diseases.

SECTION-H

MAGNETIC RESONANCE IMAGING

R.R. PARSANIA
D.B. PATIL

Magnetic resonace imaging (MRI) is a highly sensitive and non-invasive technique which provides accurate and detailed anatomic images with good contrast and spatial resolution. It has multiplaner capability and no radiation burden, and that makes MRI very attractive.

In veterinary medicine, MRI is still in its infancy and its use is infrequent. The high purchase and maintenance cost of the equipment has limited its use. The impact on veterinary medicine will be through access to the equipment in human facilities. Moreover, with refinement in technology, the cost of MRI may decrease. In developed countries, MRI has become increasingly available as some veterinary institutions have obtained units and veterinarians have also taken advantage of fixed and mobile human facilities. Todate, MRI has been used primarily as a research tool, especially for CNS diseases in small animals. All body regions in small animals can be imaged, but animals much larger than human beings (maximal transverse diameter of 60 cm) cannot be imaged without significant alterations in the equipment. In large animals, extremities and head can be imaged.

The spectrum of MRI application is very wide. It has numerous musculoskeletal applications. It has been an effective tool in the evaluation of internal derangement of human knee joint. MRI can be used to answer many questions about musculoskeletal diseases in animals such as understanding the pathogenesis of navicular disease, traumatic arthritis and osteochondrosis in equines, and wobbler syndrome in dogs. Amongst the newer applications of MRI are magnetic resonance angiography and magnetic resonance spectroscopy. It is especially useful for differentiating an inflammatory process from a neoplastic mass. It is able to differentiate tumour from peritumoural oedema. MRI helps to diagnose soft tissue tumours such as lipomas, haemanginomas and cysts but its general lack of specificity has to be taken into account when interpreting results. It is more specific and sensitive in detecting, localising and differentiating osteomyelitis, cellulitis and abscess. It is insensitive to small calcifications. In the future, MRI will become a powerful research tool in veterinary medicine. Further, with improvement in technology and subsequent decrease in the expense involved, MRI will be used for diagnosis of complex clinical cases in referal centres for small and large animals. There are certain contraindications to the use of MRI. One is the presence of ferromagnetic materials in the body (aneurysm clips), as the magnetic force may cause them to move cardiac pacemker's malfunction in the magnetic field. Depending on the bore size and magnetic field strength, the anaesthetic circuit has to be extended to provide a safe distance from the magnet. Pregnancy is also a contraindication. This section discusses only the basics of MRI.

Certain atomic nuclei will absorb and reemit radio waves when placed in a strong magnetic field. This phenomenon called nuclear magnetic resonance (NMR) was first discovered in 1946 by Bloch and Purcell and was first used in live animals by Jackson in 1967. **Radio waves reemitted can be used to construct a diagnostic anatomic image through a computer-assisted technique termed magnetic resonance imaging (MRI).**

Positive charged atomic nuclei act like small magnetic dipoles and are randomly arranged in the absence of any orienting force. A strong and homogenous magnetic field is required to align these protons into a parallel or low energy state. The summed magnetic force extended by dipoles is known as net magnetisation vector. The magnetic field used is one to two Tesla, about 30,000 times stronger than the earth's magnetic field. Hydrogen proton is of greatest overall importance because of its magnetic properties and relative abundance in the living tissue.

In the magnetic resonance phenomenon, transmission of radio frequency (RF) pulse perpendicular to static magnetic field to the aligned nuclei causes rapid nuclear oscillation between different energy states. Some previously aligned nuclei will be excited into a higher energy, antiparallel state. The net magnetisation vector changes its direction, which was parallel to static magnetic field, to one rotating in synchrony with the RF pulse at an angle with static magnetic field. The frequency at which the nuclei will resonate is directly proportional to the strength of the magnetic field and is called the **Larmor frequency.**

When the RF pulse is discontinued, the excited nuclei emit the absorbed energy and thus return to their ground energy state. This energy is in the form of a radio signal temed the **free induction signal.** The decrease of the free induction signal with time is known as **free induction decay.** The released energy can induce a voltage in an electrical reciever coil surrounding the specimen. The information is converted into a diagnostic image through a process of computer assisted reconstruction. **The frequency of the radiowaves is the same from absorption to reemission of energy by any single nucleus, while the rate of emission of decay of the energy is different from nucleus to nucleus, depending on the local environment. This effect is exploited to differentiate proteins in one tissue from those in another. Change in local environment due to pathological processes can also be descriminated.**

The process through which excited nuclei return to equilibrium by giving up their energy to the environment is known as relaxation. It occurs by both spin-lattice and spin-spin relaxation which are defined by two exponential time constants T_1 and T_2, respectively. The absorption of RF energy by protons causes them to change their orientation with respect to the main magnetic field. The duration of the RF pulse will affect thechange in alignment. A single RF pulse, normally, tips the protons from O degree (complete alignment) to 90 degrees (right angle to the main magnetic force). As there is 90 degree change in orientation, the pulse is referred to as a 90 degree RF pulse. When the protons have been flipped by a 90 degree pulse, these are said to be located in the transverse plane. When they realign with the main magnetic field, these are said to be located in the longitudinal plane. The rate of realignment from the transverse plane to the longitudinal plane is called the **T1 relaxation time.** As it involves energy exchange between the excited nuclear spins and the surrounding non-resonating molecular lattice, it is also referred to as the **spin-lattice relaxation time.** The physical and chemical nature of the environment surrounding the excited nuclei affects the values of T1. Broadly, smaller molecules including pure water (T1 value-3 sec) relax much more slowly than medium sized molecules such as lipids (T1 value - a few hundred milliseconds). T2 relaxation occurs through the interaction of protons with the magnetic fields of other nuclei and because of inherent inhomogenities of static magnetic field. As soon as a group of protons are tipped into the transverse plane they begin to process at the Larmour frequency. Some of the protons process faster or slower than the others. The rate at which the protons become desynchronised with one another is called the **T2 relaxation time.** As T_2 relaxation occurs in transverse plane, it is known as **transverse**

relaxation. The T2 relaxation times for biological tissues are much shorter than T1 relaxation times and vary from 25 to 120 milliseconds.

Relaxation characteristics of tissues are more important to signal strength and contrast in the image. **Tissues with shorter T1 or longer T2 values are associated with NMR signals of relatively higher intensity (whiteness) on MR image.** As cerebrospinal fluid has a longer T1 relaxation time than brain tissue, it appears dark on a T1 weighted image but white on a T2 weighted image. The soft tissue contrast obtained in MRI can be enhanced with the use of contrast media. The main advantage of contrast enhancement is the differentiation of isomagnetic tissues and evaluation of organ function. Contrast media for T1-weighted imaging include paramagnetic ions such as gadolinium, ferric agents and other compounds including organic molecules and transitional elements. The most commonly used contrast agent is gadolinium chelated to diethylene triaminepentaacetic acid (GD-DTPA). This complex is highly useful in differentiating vascular meningiomas (Fig. 9-24). The agent may cross damaged or abnormal blood brain barriers within or surrounding lesions in a way similar to iodinated contrast media used in computerised tomography.

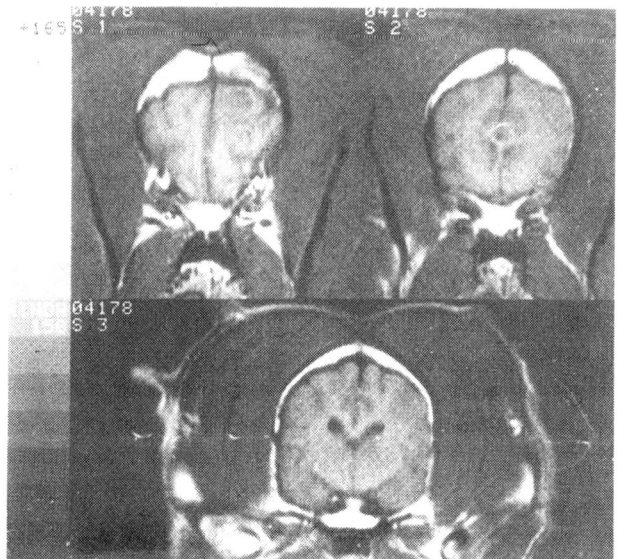

Fig. 9-24A : Transverse T1-weighted magnetic resonance image at the level of frontal cortex (top-left), base nuclei (top-right) and thalmus (bottom) in a male castrated boxer 9 years old. Prominent white matter tracts have greater signal intensity than overlying cerebral cortex.

The differences between T1 and T2 times of distinct tissues can be exploited to change the appearance of the MR images. In order to emphasise the T1 and T2 characteristics of tissue, a special spin echo pulse sequence is used. For this sequence two variables are altered, the pulse repetition time (TR) and the echotime (TE). If TR and TE are short, a T1 weighted image is generated. On these images, tissues with short T1 relaxation times will produce a much stronger signal than tissues with a long T1 relaxation time. The various signal intensities are assigned different gray values on an image so that short T1 tissues are white and long T1 tissues are black. If TR and TE are long, a T2 weighted image is generated. On these images, long T2 tissues will be white and short T2 tissues will be black. In another commonly used imaging sequence called gradient echo sequence, three variables are altered: TR, TE and flip angle. Flip angle is more significant than others. These images are desirable in bone and joint imaging, as the cartilage is emphasised on a small flip angle image.

The actual MR image is composed of an array of pixels (picture elements) that correspond to volume of tissues termed voxels. Brightness of the individual pixel varies with the excitatory properties of the tissue in the corresponding voxel. The use of large voxel size reduces spatial resolution and may limit image quality.

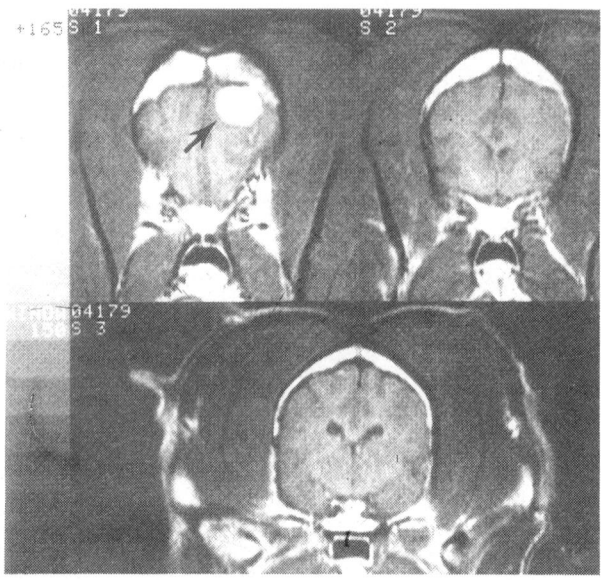

Fig. 9-24B : Area of fig 9-24A after infusion of gadolinium-diethylenetriaminepentaacetic acid (Gd-DTPA). A roughly circular area of increased signal intensity (arrow) is visible on the right side of the cerebral cortex. This area in fig 9-24A was poorly demarcated as a low intensity signal. A meningioma was resected (Figs. 9-24A and B are from Thomson, C.E. et al, Vet. Rad., ultrasound 34, 2, 1993. Used with permission).

Spatial orientation in MRI is partly achieved through the application of magnetic gradients accross the imaging volume. Suprimposed on the main magnetic field are three smaller magnetic fields known as gradient fields, which are perpendicular to one another. These three fields enable the formation of cross-sectional images from the NMR signal. As the resonance frequency of the proton is detemined by the magnetic field, a small gradient field superimposed on the main field will cause the protons on one end of the field to resonate at a slightly different frequency than those at the other end of the field. This feature is exploited to extract spatial information from the NMR signal. Only proton in a slice with a particular resonant frequency rate will be excited when the corresponding RF is applied. If the gradient of magnetic fields applied across the imaging volume is steep, the slice of tissue that can be distinguished will be thin (1-2 mm). The orientation of the slice can be in any plane, depending on which combination of coils is activated. Selective activation of the gradients within each transverse slice yields information on individual voxels of the imaging volume. Hence, the signal from individual voxels within a thin slice of tissue can be collected and converted into an actual anatomical image using computer assisted reconstruction based on Fourier techniques. A low field magnetic resonance scanner is also used for the CNS imaging in dogs. Compared to a high field magnetic resonance scanner which provides a better signal-to-noise ratio, low field magnetic scanners are cheap and do not require a special shielding. However, low signal intensity results in longer scanning time. Still, due to shorter T1 relaxation, spatial and contrast resolution is acceptable.

REFERENCES

Ambrose, J. 1973. Computerised transverse axial scanning (tomography). Part 2-Clinical application. Brit. J. Radiol, **46**, 1023.

Attenburrow, B.P. and Vennart, W. 1989. The application of radioisotope scanning and imaging in general veterinary practice. Vet. Annual **29**, 15.

Bartum, B.J. and Crow, C.C. 1983. Real Time Ultrasound-A Manual For. Physicians and Technical Personnel. W.B. Saunders Co., Philadephia.

Bloch, F, Hansen, W.W. and Packard, M. 1946. The nuclear induction experiment. Phys. Rev. **70**, 474.

Burk, R.L. 1992. Computed tomographic imaging of nasal disease in 100 dogs. Vet. Rad. Ultrasound **33**, 177.

Bushong, S.C. 1975. Radiologic Sciences For Technologists. The C.V. Mosby Co., Saint Louis.

Cartee, R.E., Ibrahim, A.K. and Mcleary, D. 1986. B-mode ultrasonography of the bovine udder and teat. J. Am. Vet. Med. Assoc. **188**, 1284.

Cartee, R.E. 1985. An intoduction to ultrasound. Anburn Vet. **40**, 7.

Curry, T.S., Dowdey, J.E. and Murry, R.C. 1984. Christensen's Introduction to the Physics of Diagnostic Radiology. 3rd edn., Lea and Febiger, Philadelphia.

Devous, M.D.S and Twardock, A.R. 1984. Techniques and application of nuclear medicine in the diagnosis of equine lameness. J. Am. Vet. Med. Assoc. **184**, 318.

Forster, E. 1985. Equipment for Diagnostic Radiography, M.T.P. Press Ltd., Kluwer Academic Publishers, Netherland.

Hornof, W.J., Koblik, P.D., Stombeck, D.R., Morgan, J.P.S and Hansen, G. 1989. Scintigraphic evaluation of solid phase gastric emptying in the dog. Vet. Rad. **30**. 242.

Jene, M.T. 1986. Computed tomography characteristics of primary brain tumors. J. Am Vet. Med. Assoc. **188**, 851.

Karkkainen M.,k Mero, M., Nummi P. and Punto L. 1991. Low field magnetic resonance imaging of the canine central nervous system. Vet. Rad **32**, 71.

Kevin, A.H. and Lantz. G.C. 1990. Comparision of survey radiography with ultrasonography and X-ray computed tomography for clinical staging of subcutaneous neoplasms in the dog. J. Am. Vet. Med. Assoc. **196**, 1705.

Koblic, P.D. and Berry, C.R. 1990. Dorsal plane computed tomographic imaging of the ethamoid region to evaluate chronic nasal disease in the dog. Vet. Rad. **31**, 92.

Mackay, R.S. 1984. Medical Images and Displays. A Wiley Interscience publication. John Willey and Sons, New York.

Markel, M.D., Morin, R.L., Wikenheiser, M.A., Robb, R.A. and Chao, E.Y.S. 1991. Multiplanar quantitiative computed tomography for bone mineral analysis in dogs. Am J. Vet. Res. **52**, 1479.

McCarthy, R.J., Feeney, D.A., and Lipowitz, A.J. 1993. Preoperative diagnosis of tumors of the brachial plexus by use of computed tomography in three dogs. J. Am. Vet. Med. Assoc. **202**, 291.

Metcalf, M.R., Tata, L.P. and Sellett. L.C. 1989. Clinical use of $^{99}M_{Tc}$-MADP scintigraphy in the equine mandible and maxilla. Vet. Rad. **30**, 80.

Park, R.D, Beck, E.R. and Lecouteur, R.A. 1992. Comparision of computed tomography and radiography for detecting changes induced by malignant nasal neoplasms in dogs. J. Am. Vet. Med. Assoc. **201**, 1720.

Powis R.L. and Powis, W.J. 1984. A Thinker's Guide to Ultrasound Imaging. urban Schwarzonberg.

Purcell E.M., Torrey H.C., and Pound R.V., 1948. Resonance absorption by nuclear magnetic moments in a solid. Phys. Rev. **69**, 37.

Richard, A.L. 1982. Computed tomography of orbital tumors in dogs. J. Am. Vet. Med. Assoc. **180**, 910.

Selman, J. 1977. The Fundamental of X-ray and Radium Physics, 6th edn., Charles C. Thomas-Publisher, Illionis.

Stewart W.A., Parent J.M.L., Towner R.A., and Dobson H., 1992. The use of magnetic resonance imaging in the diagnosis of neurological disease. Can Vet. J. **33,** 585.

Siwengel, J.R. 1982. Computerized tomography for diagnosis of brain tumors in dog. J. am. Vet. Med. Assoc. **181,** 605.

Theodorakis, M.C.,, Bermudes, A.J., Manning, J.P. Kortiz, G.D. and Hilldige, C.J. 1982. Liver scintigraphy in ponies. Am. J. Vet. Res. **43,** 1561.

Thomson C.E., Kornegay J.N., Burn R.A., Drayer B.P., Hedley D.M., Levesque D.C., Gainsburg L.A., Lane S.B., Sharp N.J.H. and Wheeler S.J. 1993. Magnetic resonance imaging-A general overview of principles and examples in veterinary neurodiagnosis. Vet. Rad. **34,** 2.

Twrdak, A.R., Allhands, R.V. Boero, M.J., Baker, G.I. And Kneller, S.K. 1987 Nuclear scintigraphy of the equine skeletal and pulmonary systems: Overview of the technique, its capabilities and limitations. Procdings 32 Annual Convention Am. Assoc. Equine Pract. pp. 495-504.

Voorhout, G., Stolp, R., Luuberink, A.A. M.E. and VanWaes, P.F.G.M. 1988. Computed tomography in the diagnosis of canine hyperadrenocorticism not suppressible by dexamethasone. J. Am. Vet. Med. Assoc. **192,** 641.

Voorhout, G., Stolp, R., Rijnberk, A. And VanWaes,.P.F.G.M. 1990. Assessment of survey radiography and comparison with X-ray computed tomography for detection of hyperfunctioning adrenocortical tumors in dogs. J. Am. Vet. Med. Assoc. **196,** 1799.

Widmer, W.R., K.A. Buckwalter, E.M. Braunstein, D.M. Visco and B.L. O' Connor 1991. Principles of magnetic resonance imaging and application to the stifle joint in dogs. J. Am. Vet. Med. Assoc. **198,** 1914-1992.

SELECTED QUESTIONS

1. Name different layers of a fluoroscopic screen.
2. What are the advantages of using an image intensifier with a fluoroscopy unit?
3. What is a spot film device?
4. What are the main disadvantages of using fluoroscopy?
5. What devices can be used for viewing image formed by an image intensifier?
6. What is xeroradiography?
7. What advantage is offered by substraction radiography?
8. What is the principle of nuclear scintigraphy? What are 'cold' and 'hot' spots?
9. What are the main problems associated with scintigraphy?
10. What is pulse-echo principle that is used in diagnostic ultrasound?
11. Why bowel gas or bone does not allow the ultrasound scanning of tissues deeper to these?
12. What is a real time B-mode ultrasound scanner?
13. What is a sector scan?
14. What do you understand by computed tomography?
15. What is the role of contrast agents in computed tomography and magnetic resonance imaging?
16. What is the scope of following alternate imaging tehniques in veterinary practice?
    i) Diagnostic ultrasound, ii) Computed tomography iii) Magnetic resonance imaging.

CHAPTER
10 COMPUTERS AND RADIOLOGY

G.C. GEORGIE
S.M. BEHL

Computer is fast becoming a common item in human society. There is an unending list of jobs being done these days by these machines. Present day computer is being used as an automated electronic device which stores information, does calculations and provides answers very fast. It is not surprising that medical science has so easily accepted the role of computers in its various activities, this being more true with radiology. There is hardly any alternate imaging technique, discussed in chapter-9, which does not use computers. There are expectations that these electronic devices will become automated diagnosticians in the future to quickly provide answers to complex disease problems. Human mind, however, can not be replaced as computer's ability and efficiency is dependent upon the information fed by the man. Moreover, a computer is unlikely to equal human mind's ability to tackle unexpected situations, to handle visual and auditory data that unravel finer problems of a patient and to deal with social and ethical problems. It is also difficult to visualise a computer-patient or a computer-client relationship. Above all, a computer lacks commonsense, a most essential commodity required while dealing with a patient, be it a man or an animal.

Before the role of computers in the radiology unit is discussed, it would be worthwhile to briefly explain what a computer is. These electronic units are friendly but challenging and at times frustrating. Scientists, professionals and students of veterinary science of third world countries have, so far, very little exposure to computers. The aim of the information provided in this chapter is to make them familiar with the basics of a computer and to understand the terms used by individuals working with computers. **It is not possible to explain all computer jargon in such a brief chapter. The terms given in bold prints in this text are glossarised at the end of the chapter to meet this deficiency to some extent.**

WHAT IS A COMPUTER ?

Unless one is a computer designer or a repairman there is very little need to know the internal circuits of a computer. However, one must know what it does. Very few people though can accurately define or describe a computer. From the smallest **microcomputer** to the largest **mainframe,** computers are machines that have the ability to : (i) be programmed (a programme being a series of instructions that tell the computer which commands are to be executed), (ii) store programme (and once a programme is stored, it is executed without human intervention through a process loop), (iii) read or input data, (iv) access data stored in **online** files, (v) perform arithmetic functions on data items, (vi) perform logical operations and (vii) output data. These seven functions are essential and achieved by the integration of **hardware** and **software** of the computer.

TYPES OF COMPUTERS

Based on the developmental epochs in computeronics, computers are known as belonging to first generation to fourth generation. The fifth generation computers with capacity to work like human brains using reasoning capacity are under development from the beginning of this decade. Integrated circuits, instead of vacuum tubes (1st generation) and transistors (2nd generation), were used from the third generation computers (1965-1977). These have solid state or magnetic core **memories** often exceeding one **Mb** and are capable of concurrent operation of **multiprogramming.** Some machines have more than one central processing unit making it possible to do **multiprocessing.** The microcomputers were born with the third generation. The fourth generation computers have large scale integrated circuits, more internal memory and capacity of using a variety of input and output e.g., voice. A wide variety of high level programming languages is supported by the latter two generations.

In the common parlance of a computer vendor or manufacturer and a computer user, the computers are also classified as microcomputer (often a **personal computer** or **PC**), minicomputer and mainframe. The microcomputers usually have limitations of memory size and word length (**bits**) they are able to use, speed and **peripherals** they are capable of supporting as well as the high level languages they are able to understand. They also support only limited software. In comparison, minicomputers are superior in all these aspects and are able to handle multiprogamming and **multitasking.** The mainframes have enormous capacities in all these areas. Their speed of processing is around 1 to 0.1 microseconds compared to 4 to 0.4 microseconds for a mini and 30 to 3 microseconds for a microcomputer.

Many computers make use of mathematical expressions represented as **binary** digits and all operations are done using these digits at very high rates. Accuracy achieved with binary system is very high, almost unlimited. However, the speed of operation is relatively less and as processing is done in series, the output is available after completion of computation. These computers are called digital computers. On the other hand, the **analog** computers operate by establishing continuous current or voltage signals. They operate by measuring rather than by counting. These types of computers have very high speed and are capable of parallel processing of all compliments. They output continuously, but precision and accuracy are limited. Hybrid computers have the capacity to assess continuous analog signals in some parts and binary signals in other parts and the performance is in between to an analog and a digital computer.

A COMPUTER SYSTEM

When working with computers, one has to make use of a whole computer system. A complete computer system has five components (Fig. 10-1). There are people who work with computers using them and controlling them. The users can enter data or obtain information. The people involved with computer system are : (a) programmers who write step-by-step instructions (programmes) to tell the computer how to process the data (a computer does not possess a brain and has to get detailed instructions as a necessity) and (b) operators who communicate with the computer using **system commands.** The communication from the computer to the user (and operator) is through a peripheral device, usually, by way of disply on a **CRT** (cathode-ray-tube) or a printer (**hard copy**).

Analysts (system analysts) design application for the computers and set up procedures for entering data, obtaining desired information and checking the validity of the output. **Input** is the data to be processed, the processing taking place within the computer and **output** is the result of processing the input. The processing of the data in the computer is a continuous loop (Fig. 10-2) and often the output becomes the input for the next processing cycle. If the procedures are not well defined, the desired result will not be obtained. A procedure is, therefore, a precise step-by-step method of solving a problem.

The third component, hardware, consists of the central processing unit (CPU) and **input/output (I/O) devices.** The CPU contains the **arithmetic/logic Unit (ALU)** which causes instructions to be executed and data to be processed. It also contains storage, called the memory which is used to store : (a) the programmes that process the data, (b) the data being processed and (ċ) the **control software** i.e., the instructions to coordinate the CPU and I/O devices.

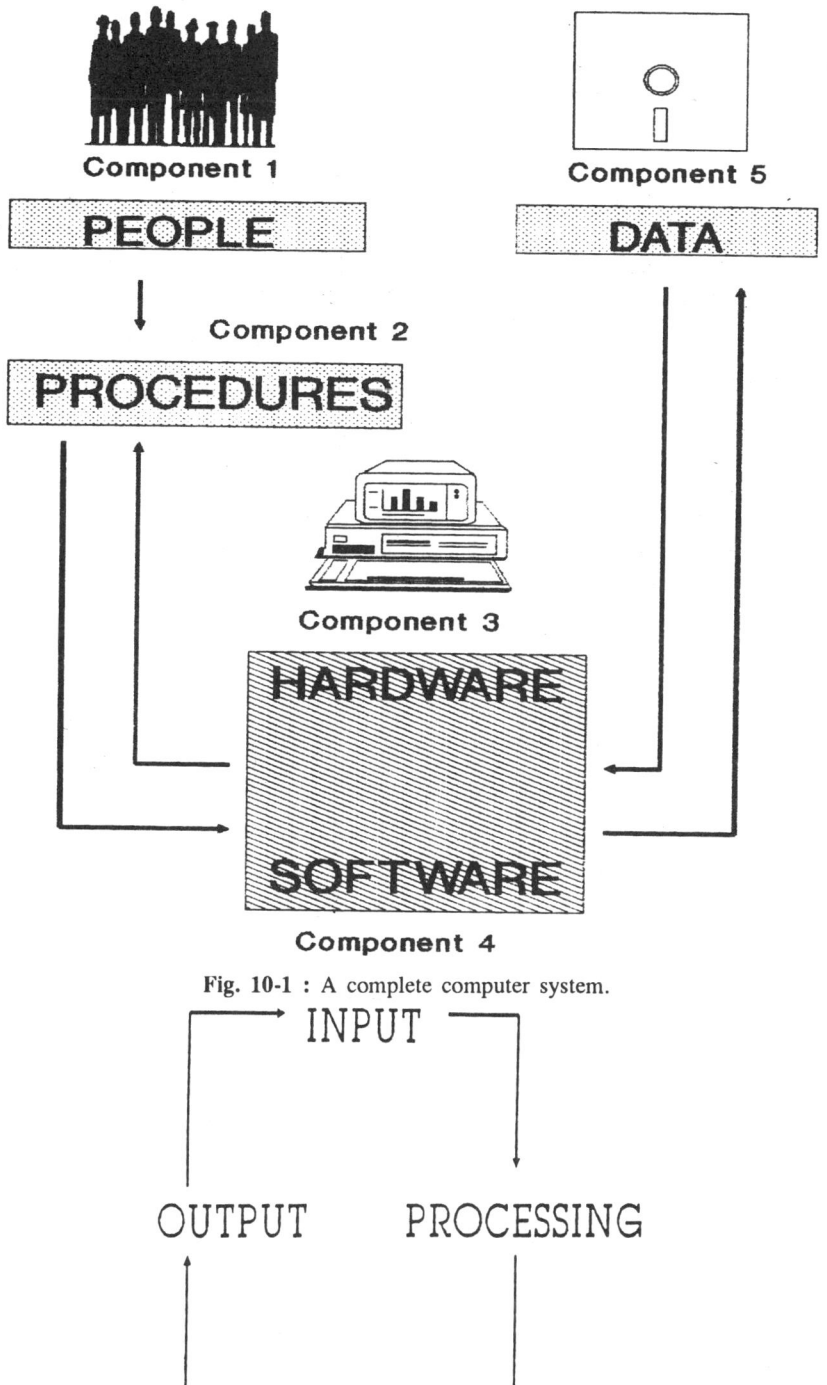

Fig. 10-1 : A complete computer system.

Fig. 10-2 : Data processing cycle in the form of a continuous loop.

Softwares consisting of **application programmes** and control software form the fourth component. The control software is also referred to as operating system. The **operating system** provides better communication between the operator or the programmer and the computer. Present day operating system softwares are very user-friendly and once the commands have been learnt by the user even a person with little knowledge of computer systems can enter, manipulate and retreive data. The functions that can be performed by a computer are often determined by the operating system. Application programmes vary depending on the end use for which they are intended. Custom made application programmes are available in the market for many uses such as **word processing,** statistical analysis, financial analysis, accounting, computer games, preparation of graphics etc. These are written in one of the many high level programming languages. An application programme for an exclusive and specific purpose can be written in one of the computer languages either by a programmer or by any one with knowledge of programming language.

The fifth component of the system, the data, is the unprocessed facts that are to be processed to obtain information. These are made available for use by the computer from the data storage media, either online or offline (i.e. separately).

THE HARDWARE

The basic hardware requirements for most of the applications to which a computer is put to (viz, inputting, processing, outputting and storing data) are illustrated in Fig 10-3. The four general types of equipment shown (input device, CPU, output device and disk subsystem) correspond to the four data processing functions : inputting, processing, outputting and storing data.

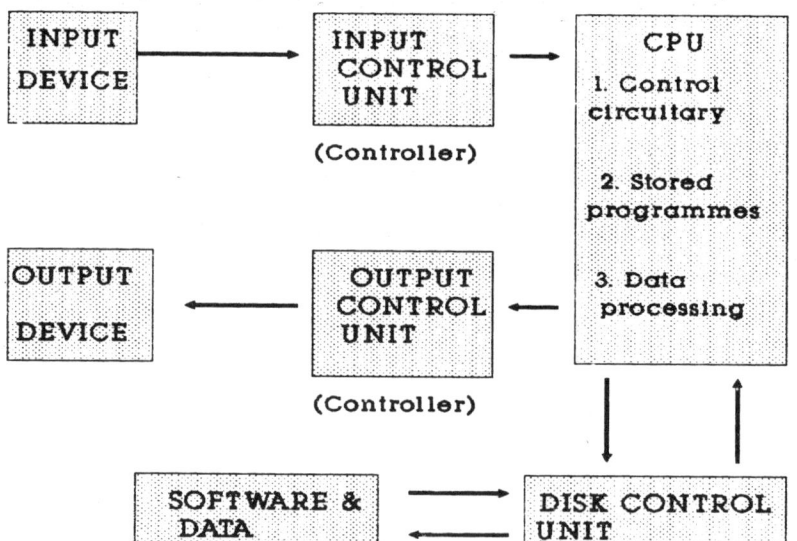

Fig. 10-3 : Hardware requirements for four basic functions of a computer.

(i) Input media : Media frequently used to input data are : (i) Punched cards and tapes, (ii) magnetic tape-reels and cassettes, (iii) magnetic hard disks and diskettes, (iv) encoded documents that can be read by scanners, and (v) data terminals or data collection devices that enter data directly into the computer system.

(ii) Central processing unit : The CPU is both the heart and brain of a computer system and within it resides memory, one or more **central control units** (CCU) and one or more arithmetic/logic units (ALU). The CCU has several functions :

(i) It fetches an instruction from a location in memory which is stored and placed within a **register** within the CCU.

(ii) It breaks the instruction into its components.

(iii) It gets the data from memory to make it available for processing.

(iv) It makes sure that the data is processed in the ALU or that a logical decision is made.

(v) It directs the other components of the system, making sure that the data is read from the input device or recorded on an output medium as required.

The ALU executes the arithemtic and logic operations for the data items brought from the memory for processing. After the ALU performs the required operations on the data, the updated or new data is stored in the memory.

(iii) Memory : The memory within a computer, called **main memory,** can be of different types. A memory called **ROM (read only memory)** contains instructions or data that can only be read by the computer and cannot be altered or changed by a programme. The **ramdom access memory (RAM)** can be used to store data and the data in this memory can be changed and updated by a programme. Another type of memory is called **cache** (pronounced *cash*). The data stored in this is assessed faster than in ROM or RAM. The capacity for performance by a particular computer is greatly influenced by the availability of ROM, RAM and cache. In selecting a computer system, another very important consideration is the kind of **secondary storage** devices that are supported by the machine. The secondary storage devices are used to store data outside the main memory. There are many alternatives available on online secondary storage devices. The time that it takes to retrieve data from secondary storage devices and its storage capactiy are important considerations telling on the performance of computers. The most widely used direct-access storage medium is a magnetic disk. Depending on the device drive, disks can either be hard or flexible (floppies). Hard disks are rigid, metal disks stacked one over other (depending on the device, there may be 1 to 11 disks stacked). The flexible floppy disks are made from the same soft material as used in manufacturing audiotapes for tape recorders. Other storage systems are videodisks, magnetic tapes, computer output microfilms and punched cards.

(iv) Output media : Output units and devices record data in either human-readable or machine-readable formats. Frequently used output media include : (i) punched cards and tapes, (ii) magnetic tape-reels and cassettes, (iii) magnetic hard disks and diskettes, (iv) cathode-ray tubes, (v) microfilms, (vi) printed documents, and (vii) audiodevices.

DISKS AND DISKETTES

It is not possible to exclude from this chapter a little more information about disk and diskettes which are the most common input and output media that are used with microcomputers and minicomputers. It is a vast subject, however, we shall pick up only just those details that a computer user finds most useful.

The basic concepts behind a disk/diskette are very simple and depend on the same phenomenon as in audio and video tape recorders. A recording head magnetises microscopic particles embedded on a surface; moving the particles past the magnetised head causing particles to become magnetised. By applying a digital signal to the head, we can record information on the magnetic disk/diskette. This recorded information will be in binary singnals unlike on a audiotape or a videotape where continuous variable analog signals are recorded. Less space is required to store data in binary (base 2) system than in a decimal (base 10) system. More the data becomes condensed, faster it can be assessed and processed. Mathematical functions can also be done faster in a binary system. The amount of data that can be jammed on to a single circular **track** on the disk depends on the data encoding scheme. The disk controllers are designed to read and write only a segment **(sector)** of a track at time. The division of the disk/diskettes into cylinders (two tracks lying one over the other on the both sides of a double sided disk) and tracks **(formating)** and the particular number of **bytes** of data in each sector depends on the controller hardware and the operating system. Typically, however, operating system developers choose sector sizes of 128, 256, 512 or 1024 bytes (1024 bytes = 1 Kb i.e. kilobyte).

CONTROL SOFTWARE AND OPERATING SYSTEMS

The capability of a system is derived as much or more from the control software as from its hardware. Although fast working hardware and a CPU with enough memory are critical features yet appropriate control software are equally or more important than the hardwares. Two computer system's having identical hardware may be totally different from each other in performance and task execution if the control software and operating systems are dissimilar. Control software consists of system commands, a monitor or executive programme, initial programme load routines and other routines stored online. They are all designed to allocate the resources of the computer system. The nucleus of the system's control software is usually called the operating system. Other control software are provided for telecommunication, computer networks etc. The manufacturer usually provides a computer's operating system. Several operating systems are available. The most common are **disk operating systems (DOS)** such as PC-DOS, MS-DOS, O/S2, Unix, Window etc. These are proprietary items of different computer companies. These operating systems were developed : (i) to increase throughput, (ii) to improve communication between computer and user, (iii) to decrease downtime by giving a diagnostic message when an error occurs in operating the system, (iv) to decrease the job set up time, and (v) to increase performance so that multiprogramming and time sharing and **realtime system** operation are optimised.

COMPUTER LANGUAGES

Although computers understand and operate by executing commands only in **machine language,** most computers are now programmed by using a high level language. When a high level language is used, a statement written in a programme is translated by a **compiler** or **interpreter** into instructions that the computer can execute i.e. machine language. The programmer using high level languages need not understand how the commands generated by the programming language are made executable. The compiler or interpreter undertakes this function. On the other hand, when machine language is used, the programmer must know how each command is executed and where the data and commands are stored in the memory. **Assembler languages** are somewhat easier to use than machine language. Again, in its use, the programmer must know exactly how the computer will execute a command. Perhaps, the best of the two worlds – assembler and high level language – is being combined in the **C-language.** This language is now being used to programme microcomputers. For an experienced programmer, C is very powerful even though the programmer must still know how a command will be exectued by the computer. Complex programmes can be written with minimum effort and executed efficiently and rapidly through C. If one is going to learn to programme, one should select as first language **BASIC.** It is english-like, relatively easy to learn and is used extensively to write various application programmes. There are many other languages from which to choose. **PASCAL** is available to many microcomputers. **COBOL** and **PL/1** (programming language1), and **FORTRAN** are also available on many microcomputers. While selecting a computer system, it is important to know what languages are available to it.

HARDWARE AND SOFTWARE EVALUATION

In selecting a computer system for use, some steps are considered important: (i) Determine the applications for which the computer is to be used and the requirements needed for each application, (ii) determine the software needed for each application, (iii) determine the hardware needed to suppport the software, and (iv) consider the amount of real memory, the amount of online storage and the kind of output device (the printer and video monitor).

Once the hardware and software requirements are determined, it is good to involve one or more individuals who understand computers (i) to study the literature that is available, (ii) to vistit other installation, (iii) have a demonstration of the hardware and software that is being considered, and (iv) to perform a benchmark i.e. a typical mix of jobs should be run and results evaluated.

COMPUTERS IN GENERAL MEDICAL PRACTICE

The role of computers in general medical practice can be considered in four forms :
  (i)   For information management
  (ii) For focusing attention
  (iii) For patient-specific consultation.
  (iv) For general management.
In other words, computers are used as support systems in making clinical decisions and management of hospitals or departments.

For information management

Computers provide facilitated access to patient data (Fig. 10-4) and current literature search on a specific problem. It is the clinician who has to decide the requirements.

For focusing attention

Some computers project diagnoses or problems related to a disease or some abnormal clinical data which a clinician might otherwise overlook from his considerations. For example, clinical laboratory systems which provide possible explanations for abnormal values. The computer helps the clinician to consider all possibilities but it is the clinician who has to arrive at a decision based on his knowledge. Once a therapeutic protocol has been decided by the clinician, the computer helps in projecting drug interactions.

For patient-specific consultation

Some computers have been designed to directly assist or advise the clinician in handling a particular patient. It means a computer can be used as a consultant while attending a specific patient. All data related to the patient is fed to the computer which then projects an advice. The advice can be in the form of differential diagnosis, additional information to narrow down the possibilities of the cause of the disease, a single best explanation for the clinical data fed or possible therapy protocols that could be followed.

For general management

In addition to earlier described specific assistance to a clinician, computers are now routinely used for general management of hospitals e.g., filling of hospital schedules, patient registration, patient clinical record, patient scheduling and tracking and financial management. Computers also aid in tracing past history of the patient, disease problems and treatment received etc. Computers are also increasingly being put into use as a teaching aid in classes and laboratories.

COMPUTERS IN RADIOLOGY SECTION

In radiology, computers are now firmly placed. Over and above their use in designing (CAD-computer aided designing) and manufacture (CAM-computer aided manufacture) of X-ray systems, they are also used in improving the performance of imaging devices (Fig. 10-5).

Computers can be used to store X-ray images. During storage certain characteristics of the image can be altered to the advantage of the radiologist. For example, in digital substraction angiography the superimposed images are removed to clearly visualise blood vessels. It is also possible to enhance contrast of the image and to remove fogging effect as a result of scatter radiation. If a normal radiograph of an area is fed to the computer, it becomes possible for the computer to collate and indicate if any other radiograph of that area is normal or abnormal. It is, ofcourse, the radiologist who has to ultimately find what the abnormality is. Computers also help in reconstructing a three dimensional image (a plain radiograph is a two dimensional image) for better diagnosis.

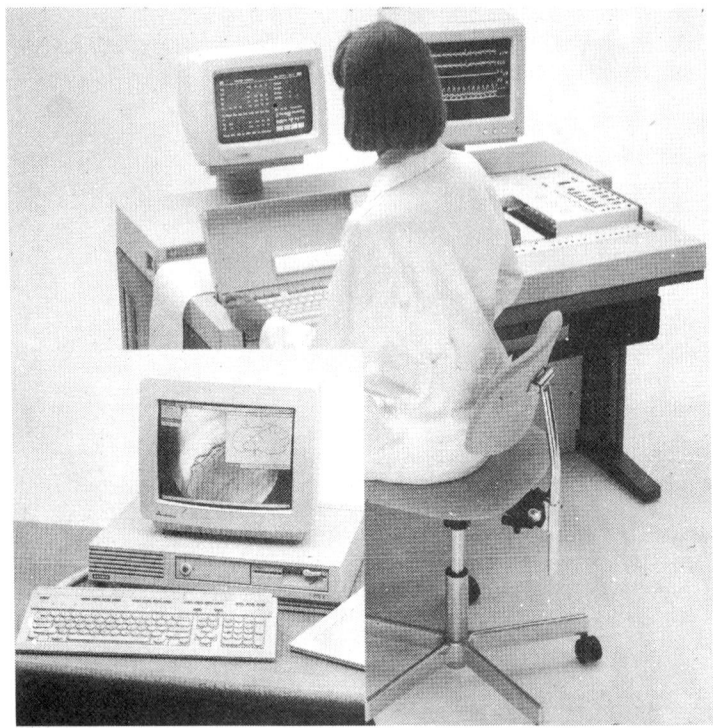

Fig. 10-4 : A computer aided lab data management system (Courtesy Siemens India Pvt Ltd).

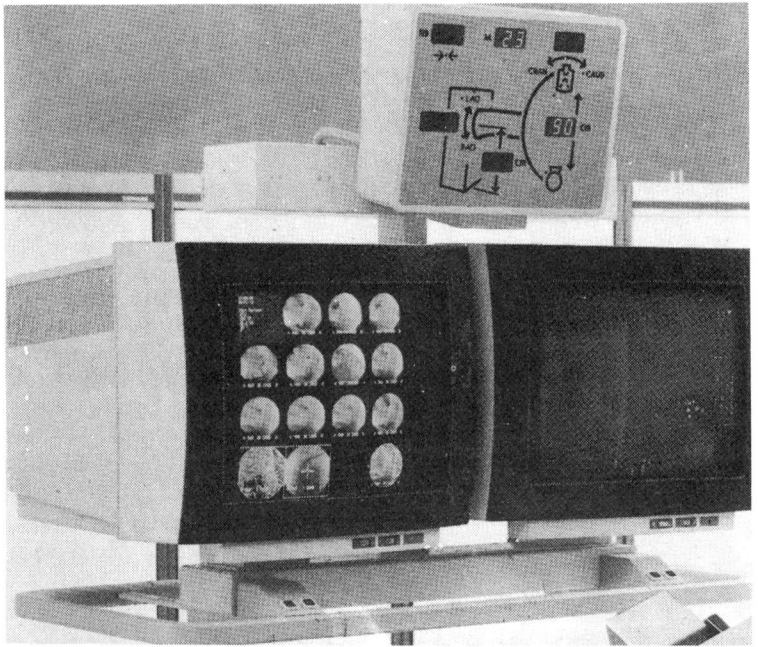

Fig. 10-5 : A digital cardiac imaging system (Courtesy Siemens India Pvt Ltd).

Digital images can be electronically stored, transmitted, processed and displayed under computer control. Thus an integrated information service for diagnostic purposes can be created. Digital radiography based on storage phosphor technology is an imaging system in which X-ray images are correctly exposed due to storage screen's extremely wide exposure range. It offers virtually endless opportunities for qualitative and quantitative image assessment. Bone and soft tissues are clearly visible from a single exposure. Computers thus act as electronic view boxes for interpretation of digital images. It is possible that a day will come when X-ray films will no longer be used in the radiology section.

Computers are also used in computed tomography, diagnostic ultrasound, image intensifiers, magnetic resonance imaging etc. In addition, they are used in other activities of the radiology section e.g., to maintain radioactive decay charts of radiopharmaceuticals, to maintain record of X-ray film library, to file statistical reports etc. The role of computers discussed here in radiology is not exhaustive but should be sufficient to provide an idea about the extent of their usefulness.

To sum up, in future computers are likely to occupy an important place in routine medical and veterinary practice though clinicians will retain the ultimate responsibility for determining diagnostic and therapeutic strategies. It is expected that in third world countries leading veterinary institutes will now focus their attention on the use of computers for providing efficient clinical services. However, it might take longer before these electronic devices are used for clinical decisions support on the same lines as in medical practice.

In the short time microcomputers have been available, each generation of computers have been physically smaller but many times faster and more efficient. Faster and more efficient I/O devices have also been developed. It has become possible to store in computer's memory more programmes and data. The availability of many useful and user-friendly application programme softwares has made it possible for even non-computer personnel to use them, though there will always be some applications that require large mainframes. Instances of applications that can be run most cost effectively on microcomputers and minicomputers are increasing day by day. Whenever large volume of data and complex processing steps on the data are required with high precision and accuracy, then in todays world, the computers provide the solution.

GLOSSARY OF TERMS

Analog: Numerical data in the form of continously variable physical quantities.

Analyst: An individual who investigates a problem and then designs a solution for the problem.

Application programme: A programme used to process data for a particular use e.g., graphics, accounts, payrolls etc.

Arithmetic/logic unit (ALU): The part of the CPU in which the arithmetic and logic operations are executed.

Assembler language: A low level language used to write application programmes and system software and can be used only when the basic architecture of the computer that will run the programme is understood.

Basic: An acronym for beginners all purpose scientific instruction code.

Binary: A numbering system based on two possibilities.

Bit: The smallest storage unit within a computer, acronym for Binary Digit.

Byte: An addressable location consisting of 8 bits within the memory or an external storage medium.

Cache: A high speed buffer memory used to hold programmes and data needed for the programme currently being executed, data and programmes can be swapped back and forth between real and cache memory.

Cathode-ray-tube (CRT): A computer output device that displays instructions and information.

Central control unit (CCU): The circuitry within central processing unit responsible for fetching instructions, breaking instructions to parts, seeing that arithmetic and logic operations are done and storing the processed data in real memory.

Cobol: An acronym for Common Business Oriented Language.

Compiler: A programme that translates a high level language into machine language of the computer.

C-Language: A high level programming language developed by Bell Laboratories and often associated with UNIX operating system.

Control software: Software that is part of, or controlled by, the operating system whose function is to coordinate the use of components and providing communication between computer and the user.

Disk opeating system: See operating system.

Formating: A program which is the part of an operating system software which enables the physical and logical division of disk/diskettes into cylinders, tracks and sectors for use as data storage media.

Fortran: An acronym for Formula Translation, a high level language very useful for writing mathematically oriented programmes.

Hard copy: Printed output from a computer.

Hardware: Computers and peripheral equipments such as tape drives, disk drivers, terminals and printers.

Input: Data to be processed.

Input/Output (I/O) device: A device used to capture data, enter it into the system and to record the output.

Interpreter: A programme that checks the source code (a code that computer cannot understand) and translates the code into machine language.

Machine language: The commands the computer understands and can execute.

Mainframe: A large-scale computer. Sometimes the term is also used to describe the computer's CPU.

Main memory: Memory located within CPU.

Mb : Abbreviation for megabytes. One Mb = 1000 kilobytes (kb). See also byte.

Memory: Storage within the computer or on external medium used to store data and programmes. Before data can be processed, it must be stored in real memory.

Microcomputer: A computer system that has its entire CPU on a single microprocessor chip. Some advanced microcomputers have coprocessors also.

Minicomputer: A digital computer that usually has more memory, a more powerful instruction set and a more comprehensive operating system than a microcomputer.

Multiprocessing: Two or more CPUs are linked to form a computer system, one of them acting as the 'boss' or 'host' and other(s) as 'employee' or 'server', or the CPUs can share resources equally.

Multiprogramming: See multitasking.

Multitasking: Also known as multiprogramming. Two or more programmes are executed at the same time by a single CPU.

Online: Under direct control of a computer, a device receives all of its instructions from the computer.

Operating system: Software that controls the execution of application programmes and allocates the system's resources.

Output: The results of processing data.

Pascal: A high level programming language suited for the development of structured programming.

Personal computer(PC) : A microcomputer.

Peripherals: Devices like monitors, disk drives, printers etc. attached to the computer and forming part of the computer system.

PL/1: Abbreviation for Programming Language one. A high level language incorporating some of the features of Fortran and Cobol.

Random access memory (RAM): Real memory within the computer that can be accessed randomly.

Realtime system: Data entered into the system is processed in time to influence the transaction. Usually multiple users can have access to the system.

Read only memory (ROM): Memory that contains data that cannot be changed or accidently destroyed. The contents of the ROM can be used only as input.

Register: Temporary storage area assigned specific function such as those which are used to perform mathematical operations.

Secondary storage: Storage outside the main memory of the computer, on magnetic disks, tapes etc.

Sector: An area on a disk. The address where data is stored is made up of sector and track number.

Software: Control and application programmes that contain instructions for the computer.

System command: An instruction used for telling the computer what task is to be performed.

Track: A recording surface on a disk/diskette. Each track is addressable by its number, and the operating system software usually retains information on which track a record is stored.

Word processing: Text rather than numerical data is stored and utilised in the preparation of letters, memos and other types of printed manuscripts.

REFERENCES

Barnett, G.C., Cimino, J.J., Hup, J.A. and Hoffer, E.P. 1987. Dxplain - an evolving diagnostic decision support system. J. Am. Med. Assoc. **258,** 67.

Dologite, D.G. and Mockler, R.J. 1988. Using Microcomputers. Prentice Hall, New Jersey.

Henry, R.E. 1986. Personal computers: Their use in a low-volume imaging department. Am. J. Rad. **146,** 1302.

Jost, R.G., Rodewald, S.S., Hill, R.L. and Evens, R.G. 1982. A computer system to monitor radiology department activity : a management tool to improve patient care. Radiology, **145,** 347.

Leeson, Marjorie M. 1985. Computer Information-a modular system. Science Research Associates Inc. Chicago.

Ridgeway, A. and Thum, W. 1968. The Physics of Medical Radiography. Addison-Wesley Publishing Company, Reading, Massachussets.

Shortliffe, E.H. 1987. Computers programmes to support clinical decision making. J. Am. Med. Assoc. **258,** 61.

Spackman, T. and Bensman, K.W. 1987. Development of useful picture archiving and communication system. Am.J. Red. **148,** 1025.

SELECTED QUESTIONS

1. What are the main advantages of using a computer?
2. What is the role of computers in :
    i) General medical practice; ii) Radiology.

CHAPTER

11 PRINCIPLES OF RADIATION THERAPY

V.K. SOBTI
PREM SINGH
K.I. SINGH

Radiation therapy for the treatment of neoplasms of domestic animals has been used since the discovery of X-rays. Dr. R. Eberlin of Berlin Veterinary School was the first to report (1906-1912) on the use of radiotherapy in veterinary practice. Although an useful adjunct to veterinary medicine in treating neoplasms, the technique has yet to be introduced in clinical practice in India for treating neoplasms of domestic animals. Chemotherapy is widely used but radiation therapy introduction has been stalled due to expenses involved.

The basis of radiation therapy revolves around the principle that ionising radiation kills cells, the exact mechanism, however, is not understood. Before fundamentals of radiation therapy are discussed, four terms should be understood: radiosensitivity, radioresponsive, radiocurable and radioisotopes. **Radiosensitivity** refers to the susceptibility of the cells or tissue to the killing effect of absorbed radiation. **Radioresponsive** is the degree to which a normal or neoplastic tissue visibility changes during or after radiotherapy. **Radiocurability in veterinary medicine** is a two year patient survival after radiotherapy without further progress of the neoplasm and subsequent metastases. To understand the term **radioisotope** or **radionuclide** some details are required and are discussed here in brief.

What are radioisotopes?: The nucleus of an atom is composed of protons and neutrons. The proton carries a single positive charge while a neutron carries no charge. However, mass of neutron is nearly the same as that of a proton. The protons and neutrons together are referred to as nucleon. The atomic number (Z) of an element refers to the number of protons in the nucleus while total number of neucleons is called atomic mass number. A nuclide refers to an atom with a particular atomic number and mass number. A nuclide is generally represented by its chemical symbol with mass number as a superscript and atomic number as a subscript e.g., $^{14}_{6}C$. Because both chemical symbol and atomic number identify the chemical species, subscript is usually omitted. Thus the carbon nuclide can now be written as ^{14}C.

The isotopes are those nuclides which have the same atomic number but different mass number i.e. **isotopes of a given atom have same number of protons but different number of neutrons.** Most natural elements are mixture of isotopes e.g., natural carbon is predominantly ^{12}C with a mixture of ^{13}C and ^{14}C. Isotopes of some elements, e.g., carbon and iron are stable i.e. neutrons and protons ratio in such atoms is one. However, some or all isotopes of some heavy elements such as cobalt, cesium etc. are unstable. **Unstable nuclides undergo the process of spontaneous decay to form stable nuclides by the process of radioactive decay. During the process of this decay, there is emission of radiation energy from the isotopes. Such isotopes are called radioisotopes**

or radionuclides. The most naturally occuring radioactive elements emit either alpha or beta rays though some may emit both or gamma rays. All elements with atomic number greater than 83 and/or atomic mass number greater than 209, and also some others, show the property of being radioactive. Why some elements are radioactive and others not is not fully understood.

The naturally occuring radioisotopes, e.g., uranium, radium, thorium, were either formed at the time of earth's formation and are decaying very slowly or are being continuously formed in upper atmosphere by the action of cosmic radiation. Artificially produced radioisotopes have been identified for nearly all elements. These are produced by irradiation of stable nuclides by subatomic particles such as neutrons in a nuclear reactor. In this process, nucleus of bombarded atom captures a neutron and thus becoming unstable and exhibiting the property of spontaneous breakdown known as **radioactivity** which follows an exponential decay. The length of time required for one-half of the atoms to be decayed in a given amount of radionuclide is called **radioactive half life** (T 1/2) and may range from 10^{-9} seconds to 10^{10} years. It is constant for any given radionuclide. For example, half life of uranium is 4.5 billion years and of radium is 1622 years.

Application of radionuclides in medicine: The field of medical science which employs radioactive materials for diagnosis and treatment of cases is called **nuclear medicine.** Radionuclides are also used in biological research and sterilisation of surgical instruments, syringes and dressings, especially of disposable types. Radionuclides for diagnostic application (e.g. in scintigraphy) have a short physical half life as use of such materials reduces radiation dose to patients in wide variety of diagnostic examinations. A number of radionuclides are used for treatment of neoplasms as will be discussed later.

MECHANISM OF ACTION OF RADIATION

The mechanism by which cells are killed by ionisation is not fully understood. Two theories postulated are : direct or target theory and indirect theory.

Direct or target theory

The theory proposes that radiant energy acts by a direct hit on the target molecules within the cell. To ionise the molecules, either a single or multiple hits are required. During this process, energy gets deposited within the molecule which is greater than the binding energy of the electrons. This results in ejection of the orbital electrons, a change in the chemical configuration of the molecule and thus damage to the cell. **The DNA molecule is the most important target of radiation in the cell, especially linkages and bonds within the DNA molecule.**

Depending upon the radiosensitivity of the tissue, and dose and duration of radiation, there are three principal effects on the DNA molecule:

i) Genetic damage: If damage occurs in the germ cells, response is observed in the next generation.

ii) Production of cancer: If proper dose is not used upto a particular period, there will be derangement of the DNA resulting in abnormal metabolic activity causing production of malignant disease.

iii) Cell death: DNA plays an important role in cell division and is also important for maintaining life of the cell. When radiation damages DNA, cell division is interfered. **This explains the death of cancerous cells by ionising radiation.**

These principal effects explain how ionising radiation is not only an important causative agent for cancer production but can also be used to treat cancerous tissue.

Indirect theory

This theory proposes that radiant energy exerts its effects by producing free 'hot' radicals, such as peroxides, within the cell that damage the specific target. Water molecule is a major constituent of the cell and gets ionised into H^+, OH^- and other unstable particles such as HO_2 and H_2O_2. Since these radicals are highly unstable, they react rapidly among themselves and other solutes within the solution producing a crucial biological change in the cell which leads to cell death.

EFFECTS ON BIOLOGICAL TISSUES

Following fundamentals explain how energy gets absorbed from its source to the tissue and what are the factors which make the tissue more susceptible to this energy:

Linear energy transfer (LET)

It is a measure of the rate at which energy is transferred from ionising radiation to the exposed tissue. The biological damage increases as LET increases. ·

Oxygen effect

Due to proliferation potential of a tumour, tissue amount is unable to receive required circulation. Thus many cancerous cells are hypoxic and also radioresistant. **Radiation therapy is more effective in oxygenated cells.** As these cells die during radiation therapy, hypoxic cells receive more oxygen and become radiosensitive. Hyperbaric oxygen and hyperthermia are thus often recommended to increase oxygenation of tissue.

Metabolic effects

In 1906 two French scientists observed that radiosensitivity is a function of the metabolic state of the tissue. This is of interest in radiotherapy as **radiosensitivity is directly proportional to the mitotic activity of the cell and indirectly proportional to their level of specialisation.** This explains why permanent cells such as neurons, and skeletal and cardiac muscles are relatively radioresistant and labile or dividing cells such as germ cells, marrow cells, interstitial epithelial cells, lymphoid cells and respiratory cells are more radiosensitive. That is also the reason why foetus is considered more sensitive than a child or an adult.

TISSUE TOLERANCE TO RADIOTHERAPY

The normal tissue response to radiotherapy is a limiting factor in planning radiotherapy and the reason why every malignancy can not be cured. Any neoplam can be controlled by using tumoricidal doses of radiation but because it has to be done with minimal normal tissue complications, a bar is imposed. Thus risk must be evaluated for each patient.

Neoplams's radiosensitivity is based on three factors:
 i) Neoplasm lethal dose.
 ii) Normal tissue tolerance dose.
 (iii) Therapeutic ratio.

Neoplasm lethal dose

It is that dose of radiation which **in vivo** produces lethal effects on the neoplasm i.e. 80-90% regression of the neoplasm in the treated area. It varies for neoplasms of different histological types but shows a narrow range of variation for neoplasms of the same type.

Normal tissue tolerance dose

It is that dose of radiation which normal tissue can absorb without any pathological effects. This also varies for different types of tissues e.g., 200 rads for the eye lens, 2000 rads per two weeks for the kidneys and lungs, and 4000 rads per four weeks for the brain.

Therapeutic ratio

It is the ratio of normal tissue tolerance dose to the neoplasm lethal dose. Neoplasms can be classified into threee categories on the basis of therapeutic ratio: Sensitive, moderately sensitive and resistant. Sensitive neoplasms are those where therapeutic ratio is higher e.g., squamous cells carcinoma. These neoplasms require much less dose for treatment than normal tissue tolerance dose. In such neoplasms, about 74% regression occurs after radiotherapy. Reverse is true for resistant neoplams (e.g. fibrosarcoma) and only about 34% regression occurs. In moderately sensitive neoplasms, the therapeutic ratio is one and regression is about 54%, mast cell neoplasm is an example of such neoplasms.

Besides these three factors discussed earlier, quantity and quality of radiation, dose rate of radiation and time factor, and volume of the neoplasm to be irradiated must also be considered while planning radiotherapy.

INDICATIONS FOR RADIOTHERAPY

Radiotherapy is **usually indicated for localised solid neoplasms that can not be excised completely.** It is **usually not indicated if neoplasm has the potential of high incidence of metastases.** Apart from these factors, **other indications are:**

  i)   When surgery is expected to or has already failed.
  ii) When regional or distant metastases has not occurred.
  iii) When radical surgery is unable to remove whole of the neoplasm.
  iv) When bulk of the neoplasm needs reduction in size so that it can subsequently be removed surgically.

It is, therefore, essential that thorough clinical examination of the patient is done before radiation therapy is planned. Because of potential side effects, debilitated patients should not be subjected to radiotherapy. Complete blood count and urine analysis should be carried out. Radiographs of the primary site of the neoplasm, of the thorax and of the abdomen should be obtained to check evidence of metastases.

METHODS OF RADIOTHERAPY

A patient can acquire life threatening complications, briefed later, during radiation therapy due to destruction of healthy cells. Therefore, present research in this field is aimed to make neoplastic cells more oxygeneted than normal cells before radiotherapy is done. The injuries which normal cells undergo are compensated for by the cells by shortening the cell division time and by increasing their relative number. This is done through **4 R's of radiotherapy** i.e. reoxygenation, repopulation, redistribution and repair. Radiotherapy is not done by a single dose, rather multiple treatments are given over a period of time, termed **fractioned therapy.** In animals, it is usually in 10-12 fractions of a radiation dose of 4-5 Gy each time, usually three times per week. In man it is usually 1.8 to 2.0 Gy with a total dose of 60-70 Gy over a period of 6-7 weeks. During fractioned therapy, as oxygenated cells are killed earlier, more oxygen becomes available to hypoxic radioresistant cells to make them radiosensitive. Since repair of most normal cells occurs much faster than that of neoplastic cells, fractioned therapy provides less chance for repair of neoplastic tissue. The ability of neoplastic cells to repopulate is believed to be slower than the normal tissue after radiation. Moreover, due to fractioned radiotherapy, redistribution of cells within the cell cycle occurs with increasing the possibility of radiation therapy at a phase when cells are more radiosensitive than if a single dose is given.

Methods or techniques used in radiation therapy are briefly outlined here:

A. Teletherapy

The radiation source is kept at a distance from the lesion. It is of four types:

i) Superficial X-ray therapy: It is given through X-ray machine with energy range of 60-100 keV.

ii) Deep X-ray therapy: It is given through X-ray machines with energy range of 200-300 keV.

iii) Supervoltage therapy: It can be provided through (i) X-ray machines having linear accelerator (1 MeV to 20 MeV) or Betatron (20 MeV to 100 MeV) or cyclotron, (ii) through isotopic X-ray machine with cobalt or cesium in a sealed form.

iv) Particulate beam therapy: Electron, neutron or proton beam can also be used as a mode of teletherapy.

The radiation output of various teletherapy machines is the primary factor in determining treatment time. For most machines, radiation exposure ranges from 50-300 centi Gy/min. A dose of 400 C Gy requires a treatment time of 1-8 minutes. Other factors which determine treatment time include; neoplasm and patient thickness, number of radiation fields, field size and treatment dose.

B. Brachytherapy

It is the therapeutic use of radioisotopes either within the interstitium or on the surface of a neoplasm. Brachytherapy sources are usually in the form of surface applicators, needles, seeds or grains etc. Permanently implanted isotopes are : ^{198}Au, ^{222}Rn and ^{125}K. Removable isotopes include ^{192}Ir, ^{60}CO and ^{137}CS. Specific methods of brachytherapy include:

i) Interstitial brachytherapy: When the sources of radiation are within the interstitium of the neoplasm e.g., ^{198}Au, ^{60}CO etc. Its advantages are: a) continuous low-dose irradiation of the neoplasm and high total doses obtained within the neoplasm, b) the dose to the surrounding normal tissue falls off quickly because of the inverse square effect, and c) implantation requires a single anaesthetic procedure and hospitalisation time is short than in teletherapy. The disadvantages are: a) implantation is invasive, b) often a difficult procedure and c) special training and facilities are required.

ii) Pliesotherapy (surface brachytherapy): e.g., use of ^{90}Sr for superficial lesions.

iii) Systemic brachytherapy: e.g., ^{131}I and ^{32}P can be administered systemically.

Choice of technique

In general, the use of different radiation techniques for various types of lesions can be summarised as under:

a) Small superficial lesions of the skin: superficial X-ray therapy.
b) Small superficial/shallow lesions: radium implant, deep therapy or particulate beam therapy.
c) Deep small lesions: supervoltage or deep therapy.
d) Shallow or deep, moderate sized lesions: supervoltage or particulate beam therapy
e) Substantial lesions: supervoltage therapy
f) Extensive lesions and specific malignant conditions (leukaemia, thyroid cancer etc.): systemic fluid isotope therapy.

Success of these above methods depends upon many factors, such as energy of the radiation used, type of the neoplasm (whether benign or malignant), location of the neoplasm (superficial or deep), size of the neoplasm, involvement of various organs (liver or bone) and clinical stage of the patient.

COMPLICATIONS OF RADIOTHERAPY

Generally two types of complications are observed: Immediate and latent. Immediate complications are those which are observed within minutes or days after irradiation e.g. epilation (hair loss), moist desquamation of skin, skin erythema, chromosome aberration, haematological depression etc. When complications are not observable within months or years and occur after a long gap of time then these are called latent complications e.g., leukaemia, cancer, life span shortening and lethal genes in coming generations.

Complications also depend upon the area to be irradiated. For example, in case of ophthalmic neoplasm irradiation, the effects can be in the form of conjunctivitis, keratitis, cataract etc. These complications are not life threatening but may impair vision. In case of radiotherapy of the bone, complications may include fracture, septic osteoradionecrosis and sarcoma formation.

REFERENCES

Fenny. D.A. and Jhonston, G.R. 1983. Radiation Therapy; application and availability. In: Current Veterinary Therapy VIII: Small animal practice. Ed., Kirk, R.W., W.B. Saunders Company, Philadelphia.

Gillette, E.L. 1979. Radiotherapy. In: Veterinary Cancer Medicine. Eds. Theilen, G.H. and Madewell, B.R., Lea and Febiger, Philadelphia.

Robbin, S.L. 1974. Pathologic Basis of Disease, W.B. Saunders Company, Philadelphia.

Thrall, D.E and Dewhirst, M.W. 1986. Application of radiotherapy in the control of neoplasia. In: Oncology, Ed. Gorman, N.T., Churchil Livingstone, New York.

Thrall, D.E. and Dewhirst, M.W. 1989. Radiation therapy. In: Clinical Veterinary Oncology. Eds., Withrow, S.J. and MacEven, E.G., J.B. Lippincott Company, Philadelphia.

SELECTED QUESTIONS

1. What do you understand by the following terms?
   i) Radiocurability ii) Radionuclides iii) Radioactivity iv) Fractioned therapy v) Brachytherapy
2. Explain indirect theory of mechanism of action of radiation.
3. List the factors which make the tissue more susceptible to radiation energy.
4. Explain the factors which affect radiosensitivity of neoplasms?
5. What is the role of oxygenation of tissues in radiation therapy?

-CAT SCANNING-
DONOT BE AFRAID
THERE IS NO CAT HERE

CHAPTER

12 CONTRAST MATERIALS AND TECHNIQUES

SECTION - A

CONTRAST MATERIALS

P.K. PESHIN

When a body part is interposed between the path of X-rays and an X-ray film, its shadow is formed. This shadow on the film is processed to obtain a permanent record in the form of a plain radiograph. In plain radiography, no deliberate attempt is made to alter the density of the body tissues and at times, the demarcation of adjacent tissue is not clear due to lack of contrast. **When the radiodensity of the tissue itself or its surrounding structures is deliberately altered to obtain a radiograph with enhanced visualisation and demarcation, it is called contrast radiography.** The substance used for the purpose is called **contrast medium.** Those materials which increase radiodensity of the structure or tissue in relation to surrounding tissue are called **positive contrast media** and those which relatively decrease the radiodensity are **negative contrast media.** In **double contrast radiography** both contrast media are used together. Contrast radiography offers following advantages:

    i)   Structures or organs can be evaluated more effectively for their size, shape and position.

    ii)   Valuable information can be gained regarding serosal and mucosal surfaces of hollow organs or their contents which otherwise are not apparent on plain radiographs.

    iii) In some instances some idea of the function of the organ can be formed.

POSITIVE CONTRAST MEDIA

As has already been discussed in chapter 2, atomic number (Z) is generally positively correlated with the radiodensity of a compound. The highest Z in the body is that of the bone i.e. 138. That is the reason why shadows of the ribs against air filled lungs (Z = 4.0) provides an excellent contrast. **For compounds to be used as positive contrast agents, Z of the element has to be above 50 e.g. Z of barium is 56 and that of iodine 53.**

Ideal positive contrast medium

Any contrast medium used should have following qualities:

    i)   It should have desired Z number.

    ii)   It should be inert or its metabolic byproducts should not be toxic. Uranium (Z = 92) could be the best positive contrast medium but for its radioactivity and being an element of actinide series. Similarly lead (Z= 82) compounds produce excellent contrast but are toxic to living beings. Emulsion of lead oxide and soap in water is a cheap contrast agent used for angiography in experimental animals to be euthanised. In general, high osmolar contrast media are more toxic than low osmolar contrast media.

iii) It should be retained in the area of interest only for a desired period.

iv) Agent used for outlining excretory organs must specifically be excreted through that route in sufficient concentration so as to produce radiodensity of diagnostic value, e.g. barium sulphate for alimentary canal, maglumine iodipamide for biliary system.

Adverse reactions

Adverse reactions after injection of positive contrast agents can be described under two categories:

i) Chemotoxic or local reactions.

ii) Systemic or hypersensitivity reactions.

It should, however, be mentioned that incidence of adverse reactions is low and that systemic reaction's incidence with non-ionic media is lower than with ionic (which dissociate in solution). Most currently used media in veterinary practice are ionic and available non-ionic media are expensive.

i) Chemotoxic or local reactions: These reactions include all types of allergic responses, urticaria, head jerks, muscle fasciculations etc. Local reactions are attributed to some chemical property of the injected agent and usually occur when high volume or concentration of the agent is injected. Hyperosmolarity of triiodonated compounds is the major contributing factor.

ii) Systemic or hypersensitive reactions: These reactions may occur even with a small dose of the contrast agent but increasing the dose increases the incidence. Reaction may occur due to a mediator release (histamine), antigen-antibody reaction or due to involvement of acute activation systems (complement system, coagulation system, kinin system etc). One or all of these factors may be involved. Reasons for side effects with the use of conventional ionic contrast agents are their hypertonicity, chemical toxicity, ionic charge and occasional anaphylactic reactions. Most important of these is hypertonicity which is capable of inducing endothelial lesions in blood vessel, damage to blood brain barrier, damage to red blood cells which become rigid and deformed and cause blockage of capillaries, localised or generalised vasodilation, osmotic hypervolaemia, reflex cardiac changes, diuresis, acute renal failure etc. Patients with preexisting cardiac and renal abnormalities, diabetes mellitus and dehydration are at higher risk. Premedication with steroids reduces adverse effects.

Low osmolarity contrast media

The first low osmolarity contrast medium was produced by Nyegaard and was named metrizamide (non-ionic) which has now been in clinical use in both humans and animals for over a decade. It is, however, expensive and used only for myelography. Subsequently much cheaper agents iopamidol and iohexol (both non-ionic) were produced. These agents have fewer side effects than metrizamide, are stable in solution unlike metrizamide and can also be autoclaved. Ioxaglate is another low osmalarity agent but is ionic. It is neurotoxic and should not be used in subarachnoid space. These new contrast agents produce good radiographic contrast and in some cases even better than conventional agents. Present research is aimed at producing third generation of non-ionic low osmalarity contrast media which are expected to have far fewer side effects. Although these low osmalarity contrast media have numerous advantages over conventional contrast media, especially in patients with specific problems, yet their use in veterinary practice is restricted because of their high cost. It is expected that as soon as their prices are reduced, they will find a common place in veterinary radiology.

Broad classification of positive contrast media

Broadly, positive contrast agents can be classified into following categories:

i) Barium sulphate preparations.

    ii)   Water soluble iodine preparations.

    iii) Cholycystapaques.

    iv) Viscous and oily agents.

    i) Barium sulphate preparations: Barium sulphate is exclusively used for outlining alimentary tract. It is insoluble and is not absorbed in the body. Therefore, its use should be avoided if perforations are suspected. If it enters peritoneal or pleural cavity, it remains permanently *in situ* and may provoke granulamatous reaction. It is available as poweder, suspension or paste.

    ii) Water soluble iodine preparations: These form the largest single group of contrast media. All conventional and low osmolarity contrast media fall in this category. Their adverse reactions, ideal properties and information about low osmolarity agents has already been briefed in this section. Commonest conventional agents are the sodium and megluimine salts of iothalamic, diatrizoic or metrizoic acids. Since these are ionic, they dissociate in solution.

    iii) Cholycystapaques: These are water soluble organic iodine preparations but exclusively excreted through the biliary system. These agents are thus exclusively used for outlining biliary system and gall bladder. Intravenous preparations (Meglumine iodoxamate, ioglycamate or iotroxate) are better than oral preparations (Sodium iopodate, Iopanoic acid) as absorption through oral route is variable.

    iv) Viscous and oily preparations: Since water soluble contrast media are quickly eliminated, viscous and oily preparations are indicated when this is not a desirable feature. These agents are less irritant. Because of their immiscibility with water, these preparations are not suitable for intravascular use. Their use in veterinary practice is limited to lymphangiography, dacrocystorhinography and hysterosalpingography. These can be used for myelography if non-ionic agents are not available. Propyliodone is a viscous agent while iodised oil and iophendylate are oily agents.

    Some information about positive contrast media and their uses are listed in Table 12-1.

NEGATIVE CONTRAST MEDIA

    Room air, carbon dioxide and oxygen are most commonly used negative contrast media. Nitrogen can also be used. **An ideal negative contrast media should be inert, quickly dissolved in the body fluids and quickly eliminated from the body.** Room air is cheapest, is safe but less readily absorbed. After a gas has been infused into the tissues or a body cavity, its elimination is according to the pressure gradient. Figs 12-1 to 12-3 show movement of air, oxygen and carbon dioxide from the gas cavity to extracellular fluid. **Since solubility of carbon dioxide in water is very high, it is eliminated quickly.** Gases also provide poor contrast and are best used in double contrast studies. Indications include arthrography, fasciagraphy, pneumoperitoneography, pnenmocystography etc.

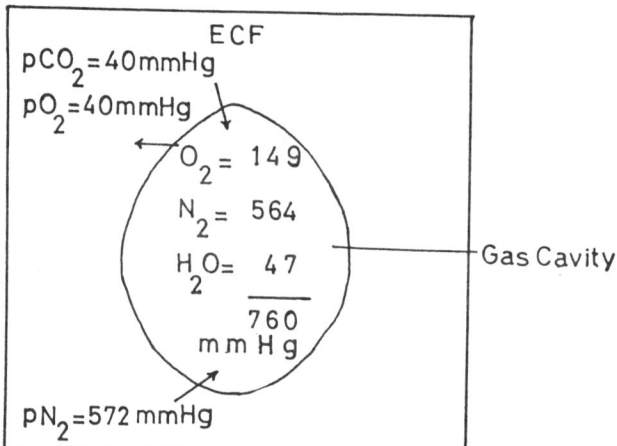

Fig. 12-1 : When air is infused in the peritoneal cavity, the movement of gases (arrows) is according to the pressure gradient, and there is accumulation of carbon dioxide in the cavity after equilibrium has reached.

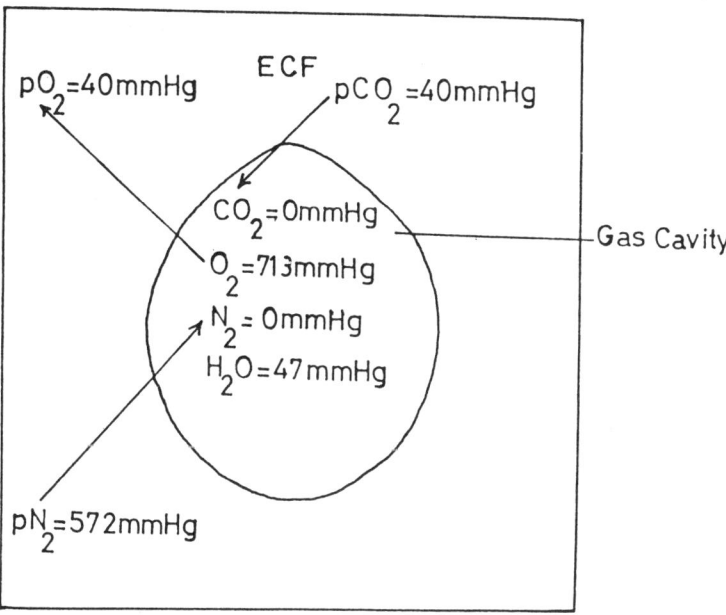

Fig. 12-2 : Since solubility of oxygen in water is lesser than carbon dioxide, the time taken to reach the equilibrium would be more if oxygen is infused in the peritoneal cavity.

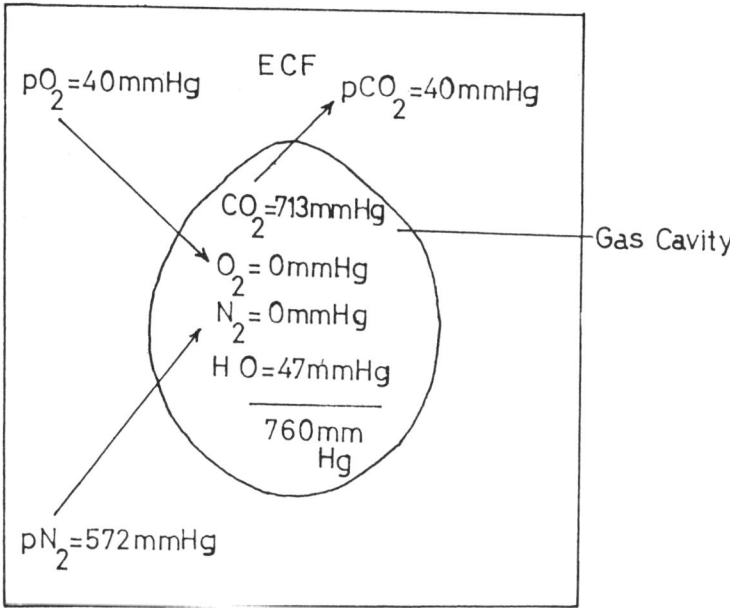

Fig. 12-3 : Since solubility of carbon dioxide in water is very high, gas will be eliminated quickly after its infusion in the peritoneal cavity.

TABLE 12-1 : Positive contrast media and their uses

S.No.	Generic name	Trade name	Iodine content mg/ml	Osmolarity	Ionic/ nonionic	Uses
		A. WATER SOLUBLE IODINE PREPARATIONS				
1.	Meglumine iothalamate	Conray 280	280	1500	ionic	Angiography, Phlebography, Osteomedullography, Sialography, Arthrography, Cystography, Urethrography, Intravenous pyelography, Dacrocystorhinography, Double contrast peritoneography, Vertebral venography.
2.	Sodium iothalamate	Conray 420	420	2000	ionic	
3.	Maglumine diatrizoate	Hypaque 65%	390	-	ionic	
4.	Sodium diatrizoate	Hypaque 85%	440	-	ionic	
		Renographin-60	60	-	ionic	
		Renographin-76	76	-	ionic	
		Urographin-150	150	-	ionic	
		Urographin-290	290	-	ionic	
		Urographin-370	370	2070	ionic	
5.	Sodium . metrizoate	Isopaque-350	350	-	ionic	
		Isopaque-440	440	-	ionic	
6.	Sodium iodamide	Uromiro-300	300	-	ionic	
7.	Meglumine iodamide	Uromiro-380	380	-	ionic	
		Uromiro-420	420	-	ionic	
8.	Metrizamide	Amipaque	300	470	non-ionic	Myelography, Angiography, Intravenous pyelography, Arthrography.
9.	Iopamidol	Niopam-200	200	-	non-ionic	As for agents Sr. No. 1-7
		Niopam-300	300	61C	non-ionic	
		Niopam-370	370	800	non-ionic	
10.	Iohexol	Omnipaque-140	140	-	non-ionic	
		Omnipaque-180	-	-	non-ionic	
		Omnipaque-300	300	690	non-ionic	
		Omnipaque-350	350	880	non-ionic	
11.	Meglumine ioxglate	Hexabrix	320	-	ionic	
12.	Sodium ioxglate	Hexabrix	320	-	ionic	

Contd.......

(*Table 12-1 Contd.........*)

B. VISCOUS AND OILY PREPARATIONS

No.	Preparation			Uses
13.	Propyliodone	Dionosil-aqueous	280	Lymphangiography, Dacrocystorhinography, Hysterosalpingography, Myelography (replaced by new low osmolar non-ionic water soluble media), Bronchography, Sialography.
		Dionosil oily	340	
14.	Icdized poppy seed oil ·	Lipiodol-ultra fluid	480	
15.	Iophhendylate	Myodil	300	

C. BARIUM SULPHATE PREPARATIONS

No.	Preparation		Uses
16.	a) Barium sulphate suspension	Micropaque standard	Oesophagraphy, Gastrography, Reticulography, Barium meal for GIT, Barium enema.
		Baritop-100	
		Microbar suspension	
		Micropaque-HD	
	b) Barium sulphate powder	Baritop-G	
	c) Barium sulphate paste	Microtrast	
		Microtrast	

D. PREPARATIONS EXCRETED THROUGH BILIARY SYSTEM

No.	Preparation		Uses
17.	Iopanoic acid	Telepaque	Cholecystography (Oral)
18.	Sodium iopodate	Biloptin	-do-
19.	Iocitamic acid	Cistobil	-do-
20.	Calcium Iopodate	Solubiloptin	Cholecystography (IV)
21.	Meglumine Iodipamide	Biligrafin forte	-do-
22.	Meglumine iodoxamate	Endobil	-do-
23.	Meglumine ioglycamate	Biligram	-do-
24.	Meglumine iotroxate	Biliscapin	-do-
25.	Meglumine iodipamide	Chlografin	-do-
		Biligrafin	-do-

SECTION - B

CONTRAST TECHNIQUES

D. KRISHNAMURTHY
A.P. SINGH

DACROCYSTORHINOGRAPHY

Dacrocystorhinography is the contrast radiographic study of the nasolacrimal duct. The procedure is indicated in cases suspected of partial or complete obstruction, atresia, inflammation, deviation or distortion of the nasolacrimal system. Quick radiographic exposures are required if water soluble agents are used because of their rapid drainage.

Technique

i)   Sedate the animal and restrain in lateral recumbency with eye to be examined uppermost.
ii) Instil topical anaesthetic into the eye.
iii) Cannulate superior punctum with 18 to 20 gauge bevelled polyethylene catheter and secure it with a stay suture to the side of the head.
iv) Flush the eye and nasolacrimal system, as far as possible, with normal saline.
v)   Repostition the animal so that eye to be examined can be positioned near a cassette. Insert a cassette at proper position.
vi) Inject 0.5 to 1 ml of the contrast agent into the catheter. While injecting, use digital pressure at the point of enterance of the catheter into the punctum to avoid any leakage of contrast material.
vii) Make lateral and ventrodorsal projections.
viii) Flush the eye and nasolacrimal system with normal saline and remove the catheter.

Normal nasolacrimal duct in cattle is visualised on a radiograph as a smooth well defined structure extending from the lacrimal sac and passing rostrally and slightly medially to open against the medial wall of the nostril at the level of alar fold. In sheep and goat, the duct opens in the vestibule caudal to the nostril. In the dog, the duct opens caudal to the external nares on the lateral floor of the nasal cavity. Figs 12-4 and 12-5 show the normal course of the nasolacrimal duct in the calf and sheep.

SIALOGRAPHY

Contrast radiographic study of the salivary glands and ducts is called sialography. In ruminants, main indications are:

i)   to diagnose space occupying lesions of the parotid gland,
ii) to locate the site of obstruction in the stenson's duct, and
iii) to locate the site of leakage of saliva in cases of sialocele.

Technique: A. Retrograde: i) Sedate the animal in lateral recumbency.
ii) Cannulate the salivary duct through its opening in the oral cavity or exteriorise it under local anaesthesia and cannulate.
iii) Secure the cannula tightly by a ligature to prevent backflow of the contrast agent.

iv) Infuse 3-6 ml of water soluble contrast agent and obtain a lateral projection at the end of injection.

A normal parotid sialogram of a calf is shown in Fig. 12-6.

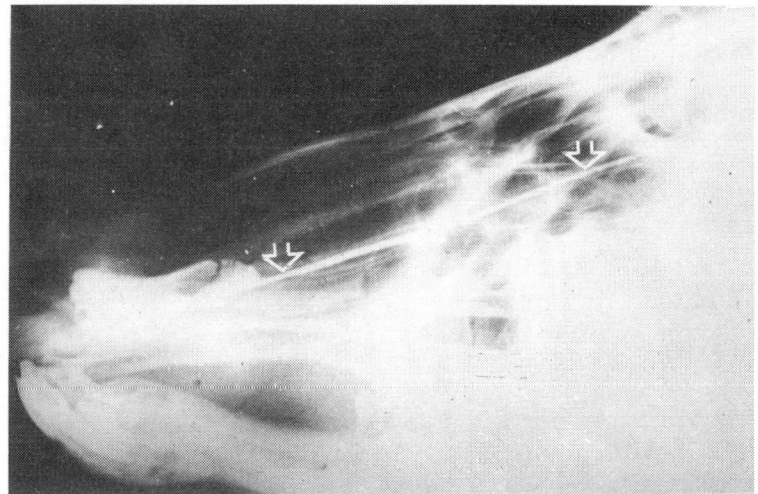

Fig. 12-4 : The course of nasolacrimal duct (open arrows) after dacrocystorhinography in a calf.

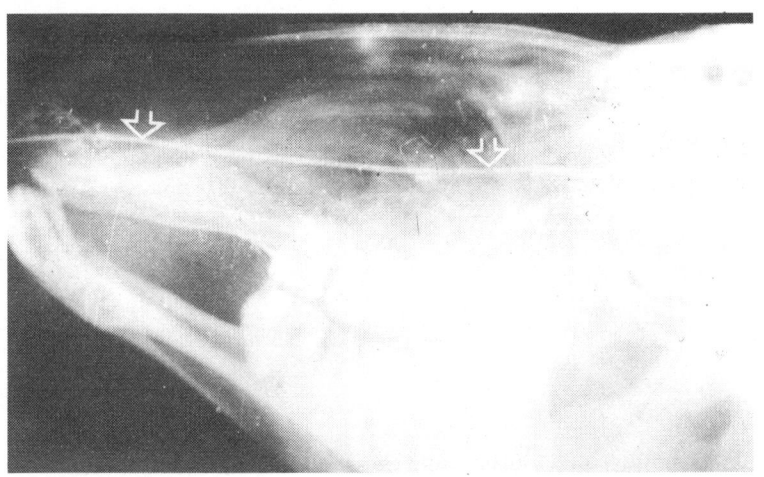

Fig. 12-5 : Dacrocystorhingram of a sheep. Note the course of nasolacrimal duct (open arrows).

B. Antegrade - Using all aseptic precautions, a water soluble contrast agent is directly infused into the glandular tissue and serial radiographs are taken.

Any agent which decreases the flow of saliva e.g., atropine, should not be administered before sialography.

BRONCHOGRAPHY

Bronchography is the radiographic visualisation of the bronchial tree after infusing a contrast material into the airways. Oily contrast media are less irritant. Water soluble preparations are irritant but are more uniformly distributed and eliminated rapidly. Bronchography should be done cautiously in patients with cardiopulmonary disease. Only one lung should be investigated at a time.

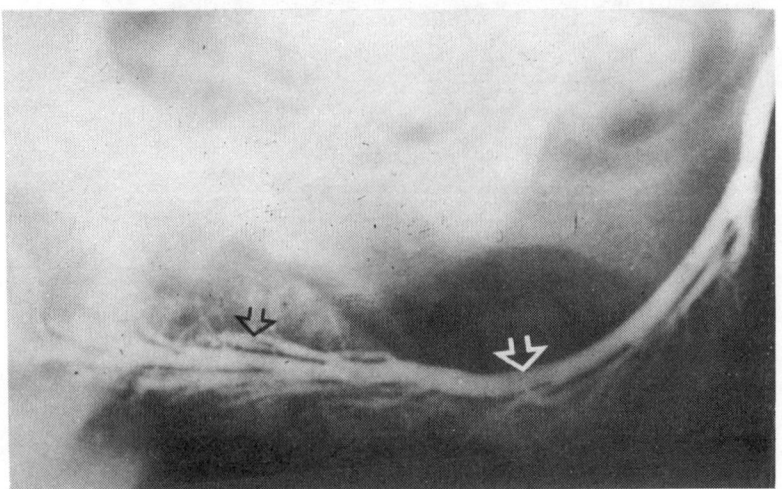

Fig. 12-6 : Parotid sialogram of a calf.

Technique

    i)   Fast the animal for 24-36 hours.

    ii) After anaesthetising the animal, restrain in lateral recumbency with the lung to be investigated lowermost.

    iii) Intubate the animal and slide a polyethylene tube into the endotracheal tube upto the level of fourth rib for the apical lobe and for the rest upto the level of sixth rib.

    iv) Infuse contrast material into the tube.

    v) Tilt the animal or roll in different directions so that contrast material is properly dispersed.

    vi) Obtain lateral projections.

    vii) If other lung is also to be investigated, repeat the procedure on the other side after 3-5 days as otherwise details of both sides get superimposed on each other.

Fig. 12-7 shows normal bronchogram of a sheep.

BARIUM SWALLOW (OESOPHAGRAPHY)

The technique is used to evaluate both structural and functional status of the oesophagus after introduction of a positive contrast agent. Oesophagraphy is indicated to diagnose cases of oesophageal obstruction, stenosis, diverticulum, perforations and mucosal diseases. It should not be used if rupture of the thoracic part of the oesophagus is suspected. Barium sulphate suspension in water, with its viscosity similar to that of light cream, is usually used as a contrast agent. Water soluble iodine based agents are recommended if cervical oesophageal rupture is suspected.

Technique

    i)   Obtain a survey radiograph of the area.

    ii) Administer slowly orally the contrast agent. In case of barium swallow, approximate dose is 1-2 ml/kg.

    iii) As the last swallow is being administered, a lateral radiograph of the oesophagus is made.

The mucosal folds, in the absence of an obstruction, appear as linear streaks (Fig. 12-8) and barium meal is quickly cleared into the stomach. In case of an obstruction, barium gets accumulated cranial to the obstruction site (Fig. 12-9). Diverticulum is identified as an oesophageal outpouch filled with contrast agent (Fig. 12-10A). Oesophageal dilatation is shown in fig. 12-10B.

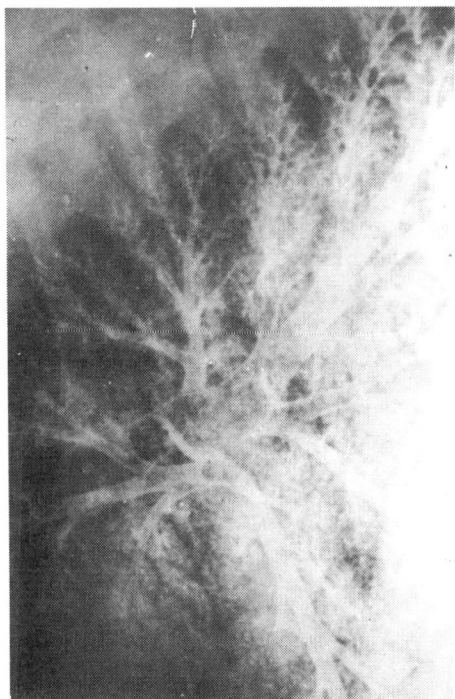

Fig. 12-7 : Bronchogram of a sheep.

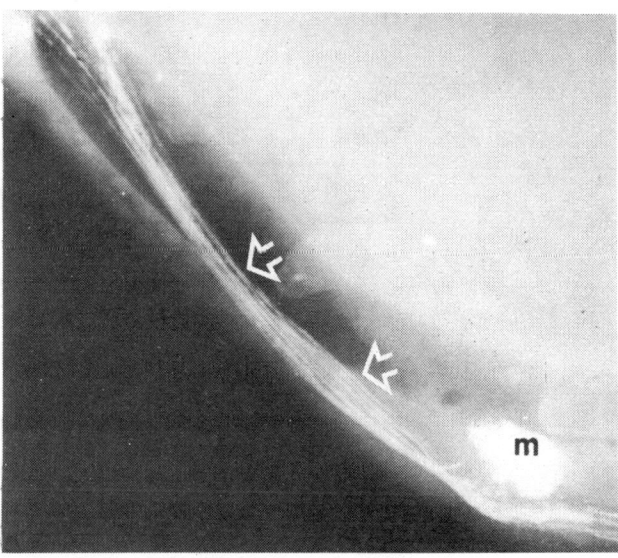

Fig. 12-8 : Oesophargam of a cow. Linear streaks of contrast material (open arrows) indicate mucosal folds of oesophagus. m = A small mass causing partial obstruction of oesophagus.

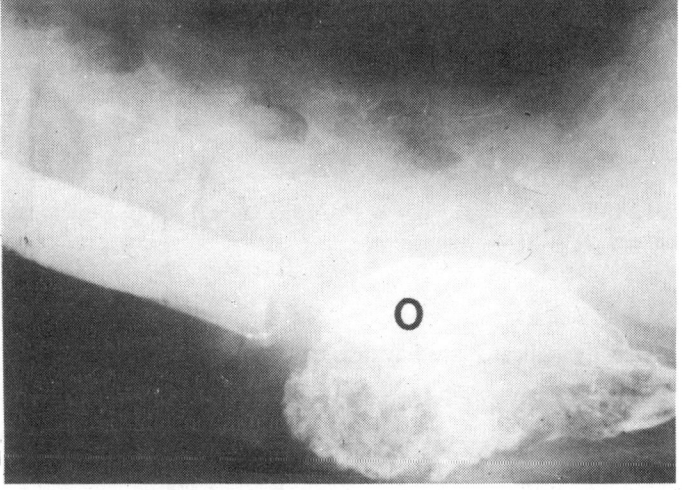

Fig. 12-9 : Oesophagram of a buffalo showing cervical oesophageal obstruction (O).

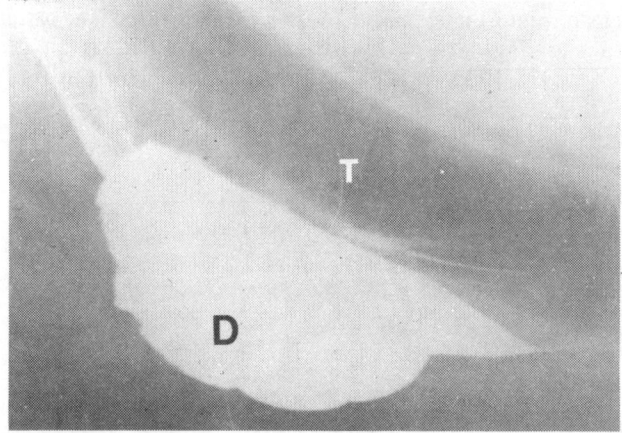

Fig. 12-10A : Oesopha'gram in a buffalo showing cervical oesophageal diverticulum (D). T = Trachea (Figs. 12.9 and 12-10 are From Singh and Nigam, Mod. Vet. Pract. 61, 867, 1980. Used with permission).

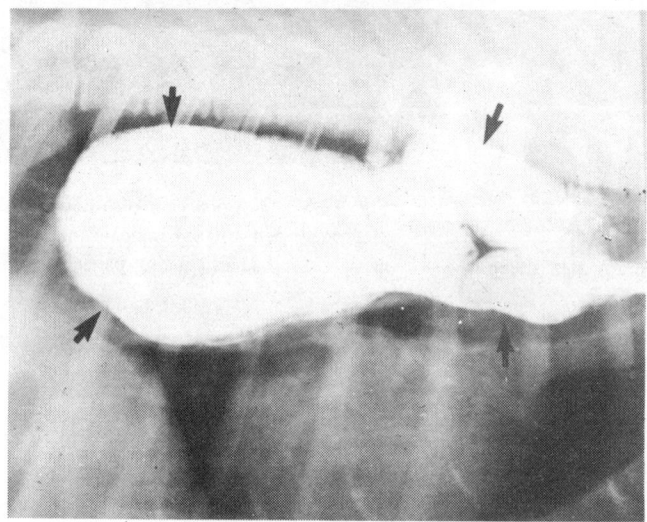

Fig. 12-10B : Oesophagram of a dog showing (arrow) oesophageal dilatation (Courtesy N.N. Balasubramanian)

RETICULOGRAPHY

The technique is usually indicated to diagnose cases of reticular hernia in buffaloes and cattle.

Technique

i)   Keep the animal off feed for 24 hours and off water for 12 hours.
ii) Administer thick barium suspension (about 1-2 kg) orally.
iii) Restrain the animal either in lateral recumbency or in supine position.
iv) Take a lateral radiograph about 40 minutes after administration of barium.

A lateral radiograph with the animal in supine position (Fig. 12-11) may provide better diagnostic information than a lateral radiograph with the animal in lateral recumbency (Fig. 12-12). The technique also helps to differentially diagnose diaphrgmatic hernia form pleurisy, phrenic abscesses or pericardiophrenic adhesions which may simulate diaphragmatic hernia in survey radiograph.

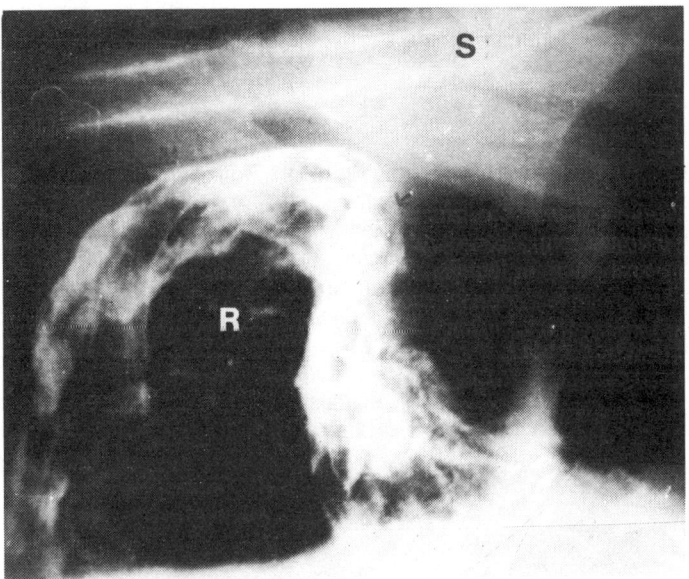

Fig. 12-11 : Reticulogram of a buffalo. Lateral projection with the animal in supine position. R = Reticulum, S = Sternum.

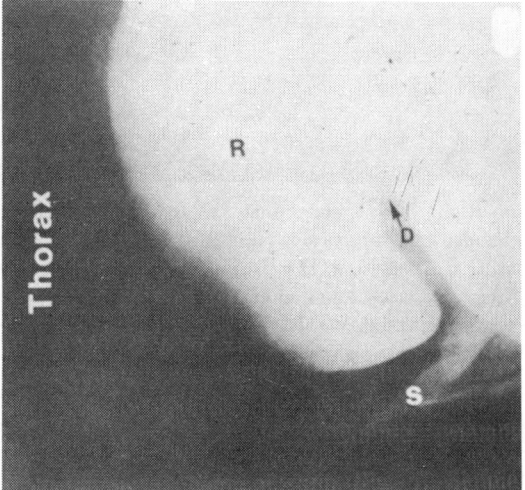

Fig. 12-12 : Reticulogram of a buffalo. Lateral projection with the animal in lateral recumbency. R = Reticulum herniated in the thoracic cavity, D = Diaphragm, S = Sternum. (From Kumar et al, Vet. Med. Small Anim. Clin, 75, 305, 1980. Used with permission).

BARIUM SERIES

The technique is used to examine radiographically the gastrointestinal tract. It is routinely used in small animals but is of limited value in ruminants. However, it can be used in sheep, goats and calves to visualise major part of the tract. The procedure is indicated to evaluate structural and functional status of the gastrointestinal tract. Since barium is reported to cause granuloma formation in the peritoneal cavity, the technique should be avoided if rupture of the stomach or intestines is suspected.

Technique - (Calves, sheep & goats)

   i) Keep the animal off feed for 36 hours and off water for 12 hours. Administer 0.5 kg of magnesium sulphate 24 hours before the study.
   ii) Administer about 100 gms of activated charcoal orally 12 hours before the study in an attempt to clear the gastrointestinal tract of gas.
  iii) Do warm soap-water solution enema about 3 hours before study.
   iv) Administer orally 70% W/V solution of barium sulphate @ 25-30 ml/kg.
    v) Obtain lateral and ventrodorsal projections at various intervals. Most organs upto caecum are visualised at 4 hours while colon gets demarcated between 6-8 hours after administration of the contrast agent. Entire course of intestines is not visualised at one time. Figs 12-13 to 12-18A show the normal radiographic anatomy of the gastrointestinal tract in sheep. Fig. 12-18B is a case of dog. **In case of dog,** use 15-20% barium sulphate suspension, 6-12 ml/kg. Laxative is not used.

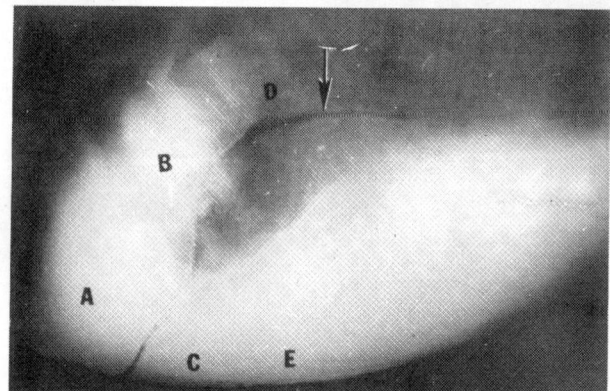

Fig. 12-13 : Right lateral abdominal radiograph obtained 10 minutes after oral administration of barium sulphate in a sheep. A = Reticulum, B = Omasum, C = Abomasum, D = Dorsal sac of rumen, E = Ventral sac of rumen. Arrow indicates longitudinal groove of rumen (Figs. 12-13 to 12-17 are from Sharma et al, Vet. Rad. 25, 17, 1984. Used with permission).

BARIUM ENEMA

Barium enema is indicated to outline the colon and rectum in suspected cases of intraluminal or extraluminal obstructions. The technique should not be used if perforations are suspected.

Technique (Ruminants)

   i) Keep the animal off feed for about 36 hours and off water for 12 hours before the study.
   ii) Administer orally a laxative dose of magnesium sulphate 12 hours before the study.

iii) Warm soap-water enema is indicated about 2 hours before the study, till returning fluid is clear.
iv) Sedate the animal deeply after controlling in lateral recumbency.
v) Raise the hind quarters. It helps in retention of contrast agent.
vi) Insert a cuffed lubricated catheter into the rectum. Inflate the cuff.
vii) Attach the catheter to the enema container filled with micropulverised barium sulphate suspension (15-20% W/V, approximate dose 15-20 ml/kg) to infuse the contrast agent to fill the colon
viii) Obtain right lateral and ventrodorsal projections.
ix) Evacuate contrast material from the colon and rectum as much as possible by elevating the cranial part of the abdomen.

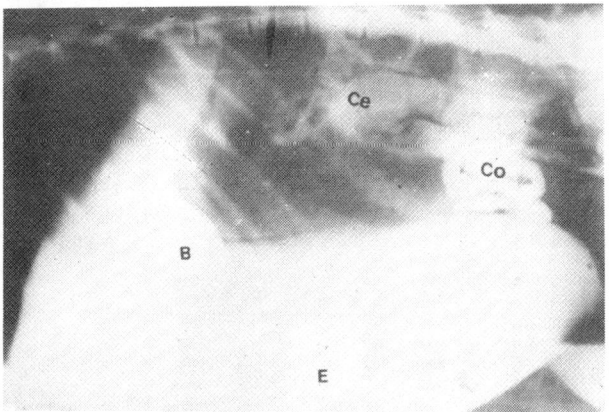

Fig. 12-14 : Right lateral abdominal radiograph of sheep 4 hours after oral administration of barium sulphate. B = Omasum, E = Ventral sac of rumen, Ce = Caecum, Co = Colon.

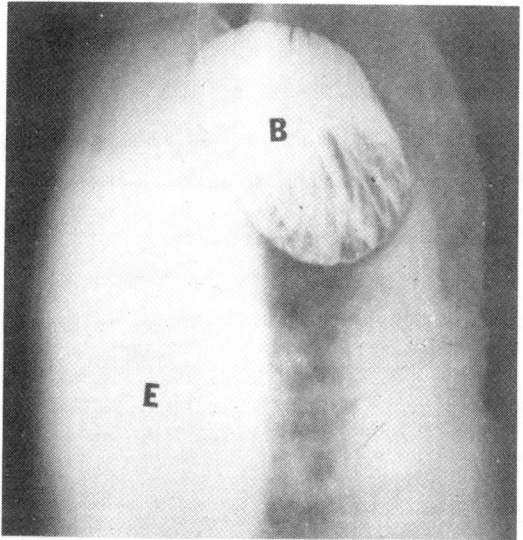

Fig. 12-15 : Ventrodorsal abdominal radiograph of sheep obtained 6 hours after oral administration of barium sulphate B = Omasum E = Rumen. Note the typical laminar mucosal pattern of omasum.

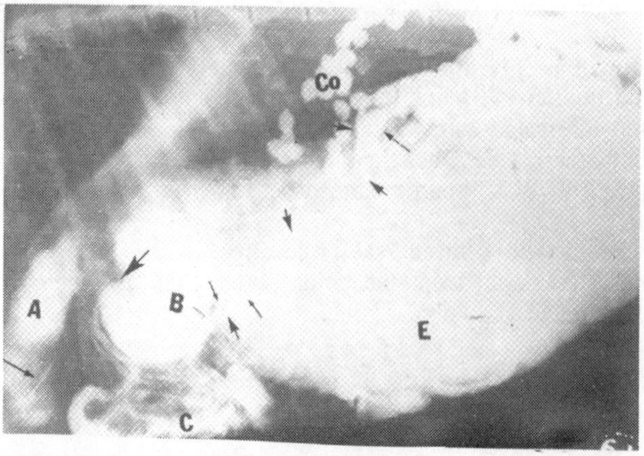

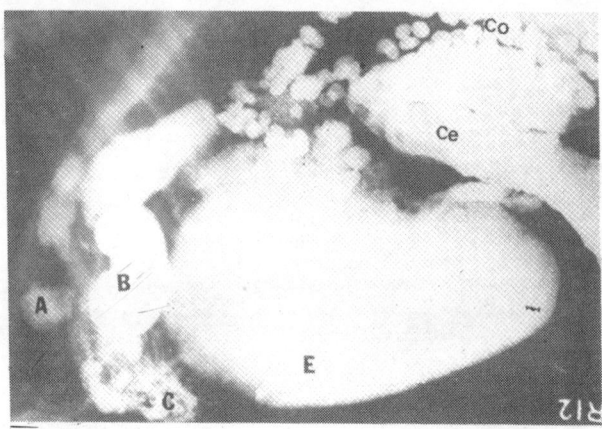

Fig. 12-16 : Right lateral abdominal radiograph of a sheep 9 hours after oral administration of barium sulphate. A = Reticulum, B = Omasum, C = Abomasum, E = Rumen, Co = Colon. Thicker arrow indicates omasoabomasal orifice, thinner small arrows indicate course of the duodenum, Long thin arrow indicates the presence of a nail in reticulum.

Fig. 12-17 : Right lateral abdominal radiograph of a sheep 12 hours after administration of barium sulphate. A = Reticulum, B = Omasum, C = Abomasum, E = Ventral sac of rumen, Ce = Caecum, Co = Colon.

For a **double contrast study,** infuse air in the rectum and colon through the same catheter after positive contrast has been evacuated. Obtain lateral and ventrodorsal projections. **Triple contrast** is obtined if pneumoperitoneum is also created. Double contrast enhances visualisation of barium coated mucosal surface whereas triple contrast helps in clear demarcation of the colonic wall. Fig. 12-19 shows normal appearance of the colon in sheep after contrast infusion. **In case of dog,** fasting is recommended for 24 hours. A mild laxative a night before, and general anaesthesia during the study are indicated.

PERITONEOGRAPHY

It is radiographic study of the peritoneal cavity and its contents after introduction of a negative contrast agent (**pneumoperitoneography**) or a combination of a negative and a positive contrast agent (**double contrast peritoneography**). The technique is indicated to visualise outlines of various abdominal organs and to locate a suspected abdominal mass. Pneumoperitoneography helps in better visualisation of vertebral column. It should not be used if diaphragmatic hernia is suspected because of risk of pneumothorax.

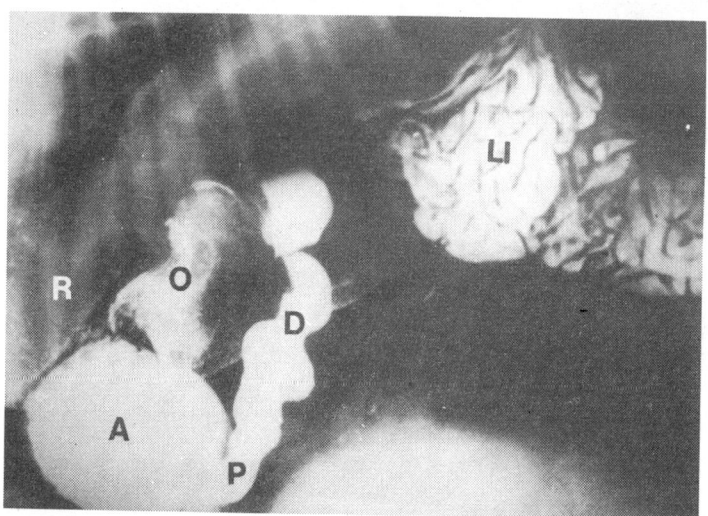

Fig. 12-18A : This radiograph of sheep shows how creation of pneumoperitoneum enhances visualisation of various organs when used in conjunction with barium series. R = Reticulum, O = Omasum, A = Abomasum, P = Pylorus, D = Duodenum, LI = Loops of intentines.

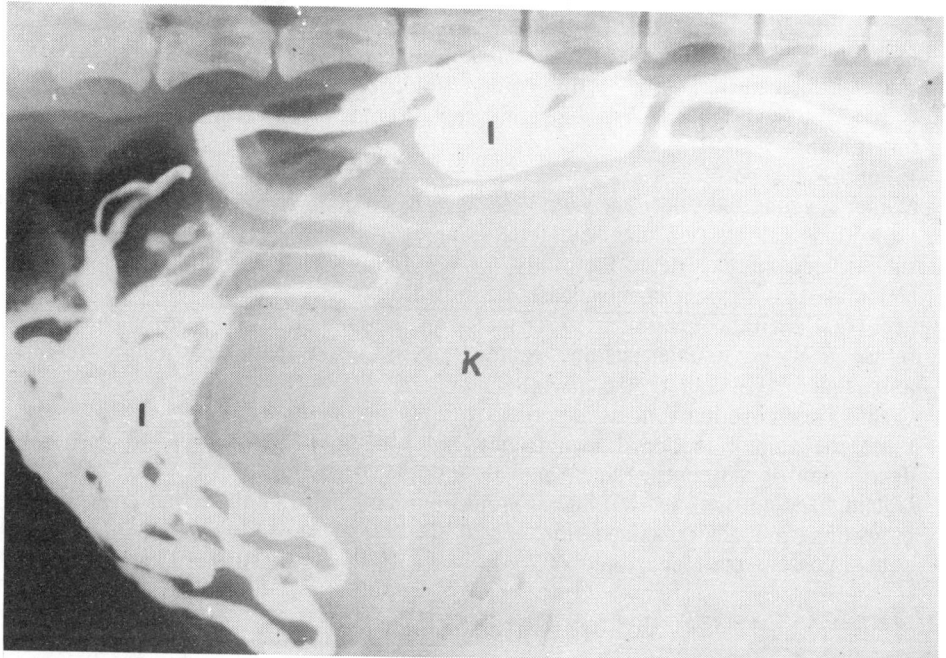

Fig. 12-18B : Lateral abdominal radiograph of a dog 2 hours after barium meal. The intentines (I) are displaced by unusually large hydronephrotic left kidney (K) as a results of congenital stenosis of left ureter. (Courtesy N.N. Balasubramanian).

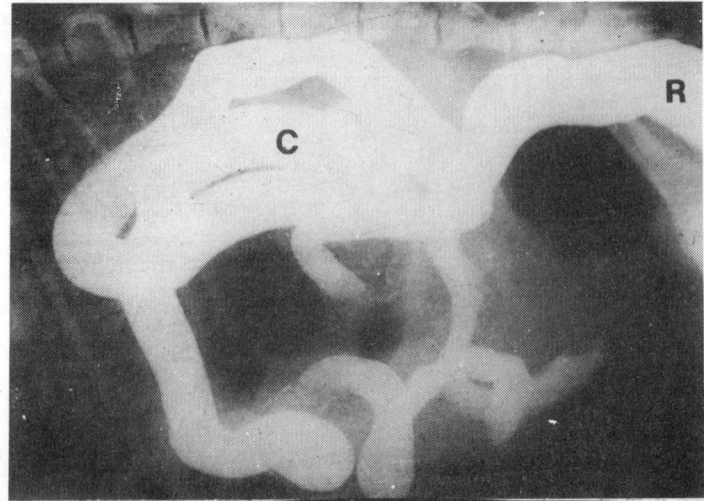

Fig. 12-19 : Normal appearance of colon in the sheep after barium enema.
C = Colon, R = Rectum.

Technique

Pneumoperitoneography : i) Prepare the animal as described for barium series.
  ii) Insert a 5 cm long 14-16 G needle (18 G in a dog) attached to a catheter into the peritoneal cavity through left paralumbar fossa using aseptic precautions.
  iii) Make sure by aspiration that needle has not punctured any abdominal organ.
  iv) Through the catheter, slowly infuse carbon dioxide, oxygen or compressed air till abdomen is moderately distended. Avoid overinflation. Room air can also be infused by using a foot pump.
  v) Withdraw the needle.
  vi) Obtain standing left and right lateral radiographs. In small ruminants and dogs, ventrodorsal projections can also be obtained. If large amount of gas gets accidently infused into the subcutaneous tissue, defer the study till absorption of gas occurs. Gas present in the abdominal cavity gets spontaneously absorbed.

Double contrast peritoneography : i) Preparation of the animal is same as for barium series.
  ii) Infuse a water soluble positive contrast agent in the peritoneal cavity (e.g. sodium iothalamate 70% W/V @ 0.5 ml/kg).
  iii) Create pneumoperitoneum as described for pneumoperitoneography.
  iv) Place the animal in dorsal recumbency and roll from side to side to ensure uniform distribution of the contrast agent into the peritoneal cavity.
  v) Obtain standing right and left lateral projections. In small ruminants and dogs, ventrodorsal projections are also obtained. Fig. 12-20 to 12-22A show pneumoperitoneographs and double contrast peritoneographs of sheep and calf. Figs 12-22B and C are of dog.

ORBITAL AND CAVERNOUS SINUS VENOGRAPHY

It is contrast radiographic study of the orbital veins and carvernous sinuses. The technique can be used to study normal radiographic anatomy of the area and as an aid to delineate intracranial lesions particularly those involving the floor of the cranial vault and orbit.

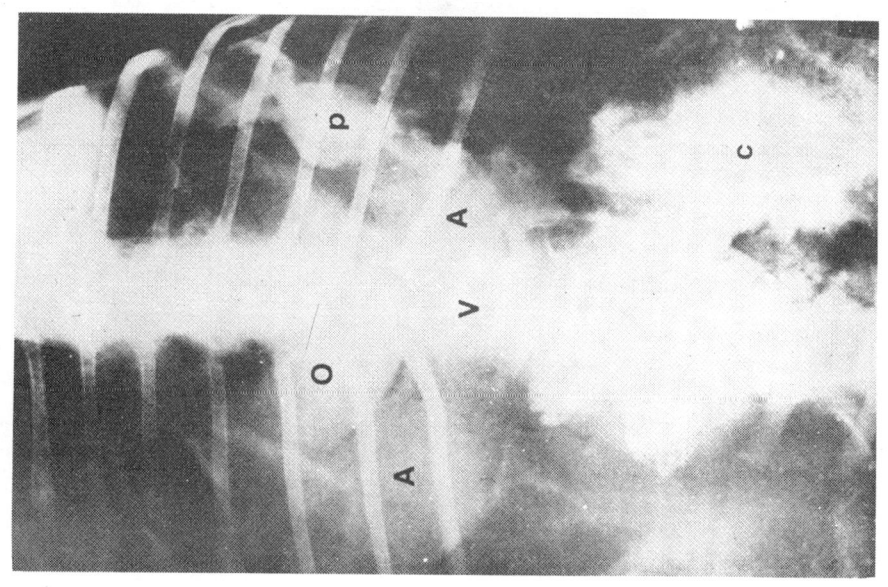

Fig. 12-21 : Ventrodorsal projection of abdomen of a sheep after double contrast peritoneography. O = Omasum, A = Abomasum, P = Pylorus, V = Vertebral column, C = Deposited contrast material.

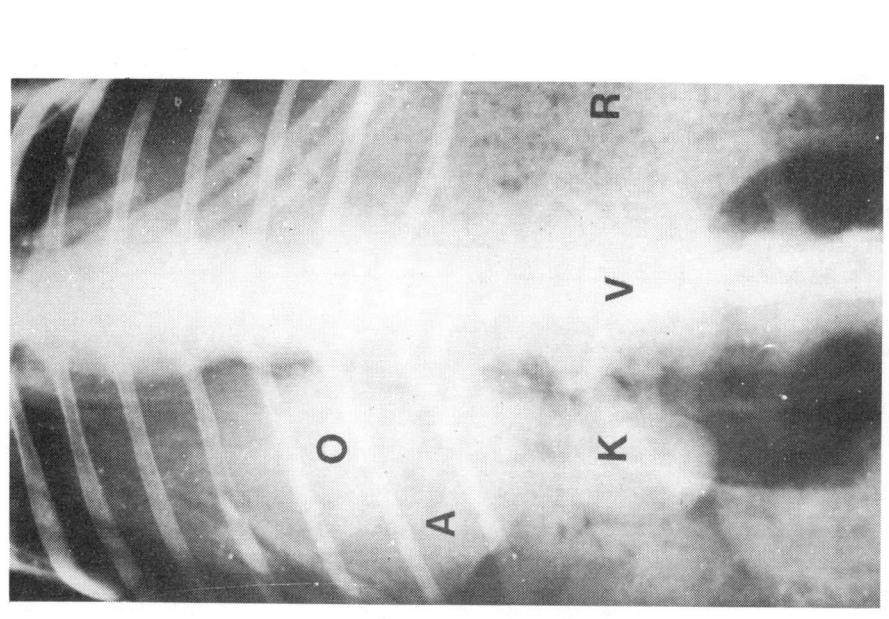

Fig. 12-20 : Ventrodorsal radiograph of abdomen of a sheep after creation of pneumoperitoneum. R = Rumen, O = Omasum, A = Abomasum, K = Kidney, V = Vertebral column (Fig. 12-20 to 12-22 are from Singh et al Zbl. Vet. Med 28A, 720, 1981. Used with permission).

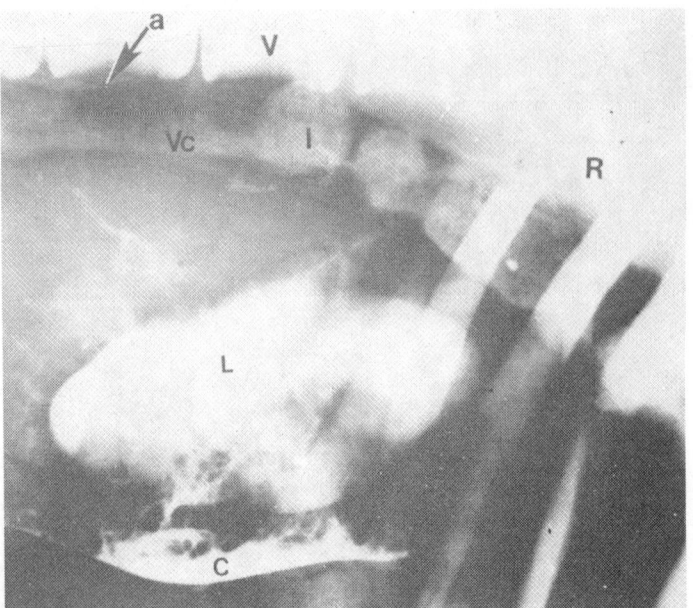

Fig. 12-22A : Lateral projection of abdomen of a calf after double contrast peritoneography. a = Aorta, V = Vertebra, Vc = Vena cava, I = Intestine, R = Right kidney L = Left kidney, C = Deposited contrast material.

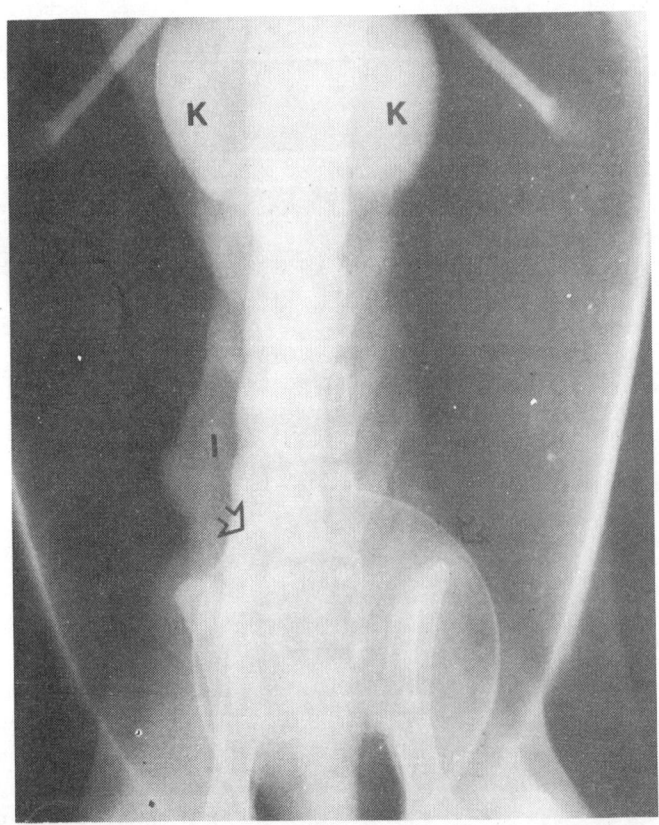

Fig. 12-22B : Pneumoperitoneogram of a dog. K = Kidneys, I = Intestines. Arrow indicates urine filled bladder.

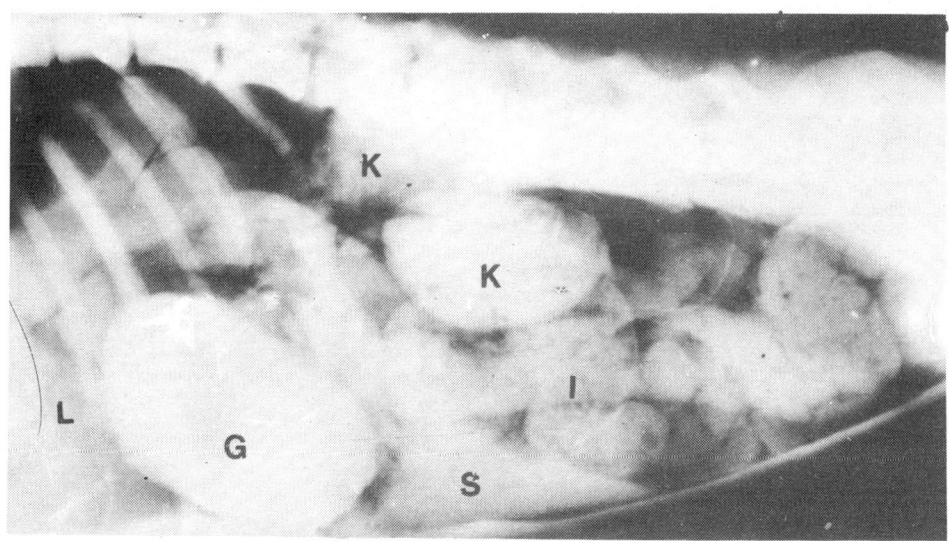

Fig. 12-22C : Double contrast peritoneogram of a dog. K = Kidneys, L = Liver, G = stomach, S = Spleen, I = Intestine.

Technique (Ruminants)

i) Control the animal in lateral recumbency with the site to be examined uppermost. Anaesthetise the animal.

ii) Using aseptic precautions, insert an 18G needle into the facial or angularis oculi vein, or catheterise it.

iii) Change the side of the animal and insert a cassette against the site under examination.

iv) Inject rapidly water soluble contrast material (10-15 ml sodium iothalamate) through the needle or catheter.

v) Obtain a radiograph when last 2-3 ml of the contrast agent is being injected. Press the external jugular vein with fingers during the injection and exposure.

Figs. 12-23 and 12-24 show the normal orbital and cavernous sinus venograms of the goat.

RENAL ANGIOGRAPHY

The technique is used to visualise renal vascular architecture and also helps to assess renal cortex-to-medulla ratio.

Technique

i) Anaesthetise the animal.

ii) Using aseptic precautions expose and catheterise the femoral artery with a radiopaque catheter.

iii) Under fluoroscopic guidance, advance the catheter into the renal artery through abdominal aorta. Infuse small test dose of water soluble iodine based contrast agent to confirm placement of the catheter.

iv) Infuse contrast agent (e.g. sodium iothalamate 70% W/V @ 0.5 to 1.0 ml/5 kg). Obtain radiographs during last phase of injection and immediately after completion of infusion.

v) If facilities exist, serial radiographs are better.

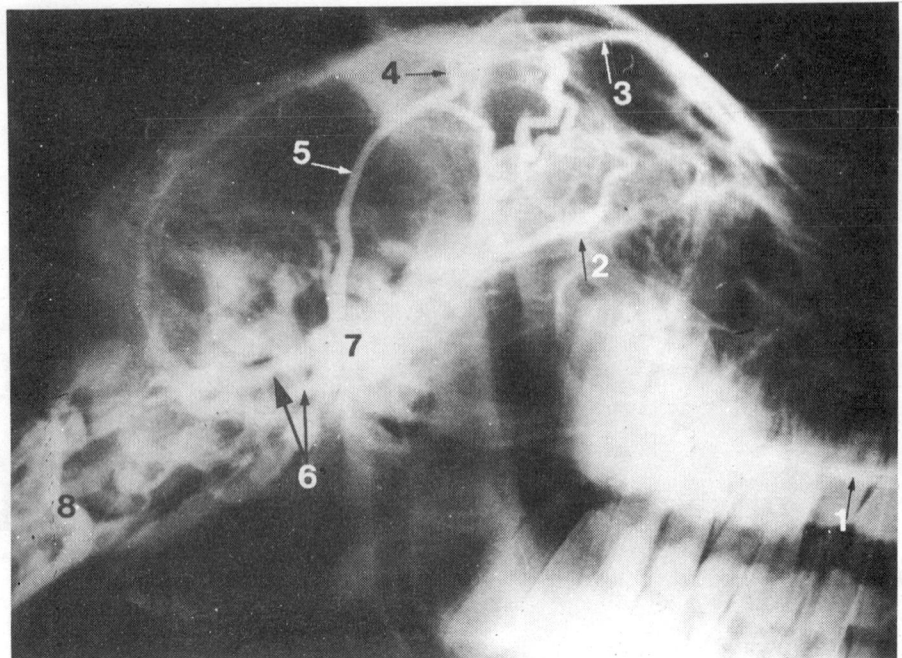

Fig. 12-23 : Lateral cavernous sinus venogram of a goat. 1 = Infraorbital vein, 2 = ophthalmic vein, 3 = Supraorbital vein. 4 = Cornual vein, 5 = Superficial temporal vein, 6 = Petrosal sinus, 7 = Cavernous sinus, 8 = Internal vertebral plexus.

Tip of the catheter can also be placed near the origin of renal artery, however, superimposition of other vessels does not allow good details. Fig. 12-25A shows normal selective renal angiogram of a goat. Fig. 12-25B is of an experimental dog.

INTRAOSSEOUS VERTEBRAL VENOGRAPHY

This technique permits visualisation of spinal and vertebral vencus drainage and can be used to identify mass lesions and vascular abnormalities of the area. It should not be used if osteomyelitis exists in the area.

Technique

Techniques for venography of the cervical and lumbar regions of calf are described here.

A) Cervical vertebral venography

   i) Anaesthetise the animal. Place pelvic region lower than the head.

  ii) Using aseptic procedures, introduce a 16 G bone marrow biopsy needle with stylet into the marrow cavity of either of the wings of first or second cervical vertebra with rotating movements. Introduce the needle till opposite cortex is felt and then slightly withdraw for proper placement into the marrow.

 iii) Test placement by aspirating bone marrow before proceeding further. If aspiration fails, flush marrow with saline and again aspirate. Spontaneous flow of marrow in the syringe indicates correct positioning. Improper placement of stylet allows plugging of the biopsy needle. The needle also gets blocked if its end is embedded into the opposite cortex.

iv) Test the correct placement of the needle by injecting 2-3 ml of water soluble iodine based contrast agent.

v) Rapidly inject 10-15 ml of the contrast agent into the marrow and obtain a radiograph during the last phase of the injection.

vi) For a ventrodorsal projection, shift the position of the animal, make another injection of the contrast agent and obtain a radiograph.

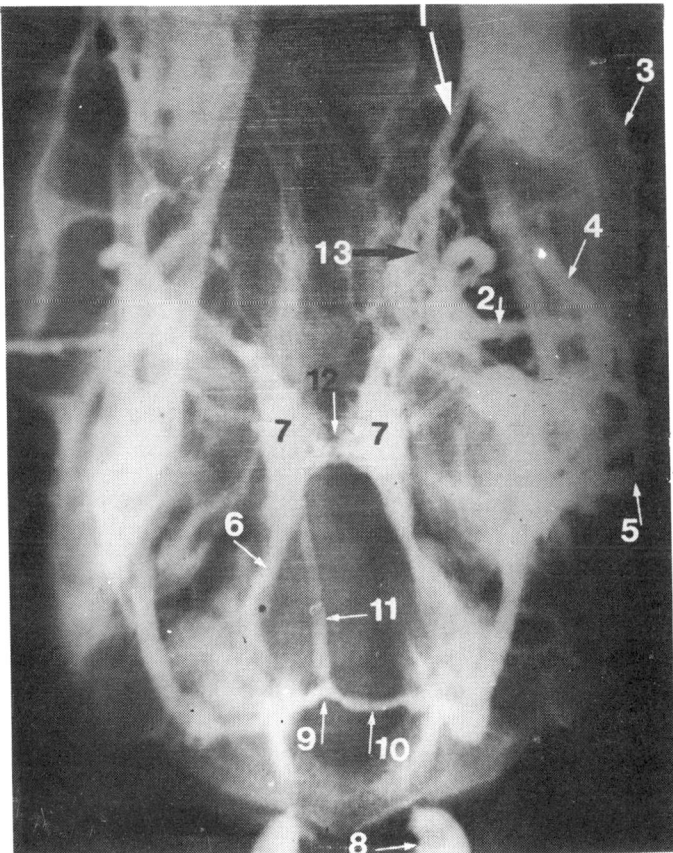

Fig. 12-24 : Ventrodorsal cavernous sinus venogram of a goat. 1 = Sphenopalatine vein, 2 = Ophthalmic vein, 3 = Supraorbital vein, 4 = Internal maxillary vein, 5 = Superficial temporal vein, 6 = Petrosal sinus, 7 = Cavernous sinus, 8 = Internal vertebral plexus, 9 = Sinum confluens, 10 = Transverse sinus, 11 = Dorsal longitudinal sinus, 12 = Intercavernous sinus, 13 = Ophthalmic plexus (Figs. 12-23 and 12-24 are from Chawla et al. Vet. Rad. **26,** 165, 1985. Used with permission).

B) Lumbar vertebral venography

i) Anaesthetise and control the animal with the head slightly lower than the pelvic region.

ii) Place the needle into the marrow of the last lumbar vertebra through a ventrolateral approach. Confirm placement of the marrow needle as discussed earlier.

iii) Inject 10-15 ml of contrast material rapidly and obtain lateral radiograph. Repeat the injection for ventrodorsal projection.

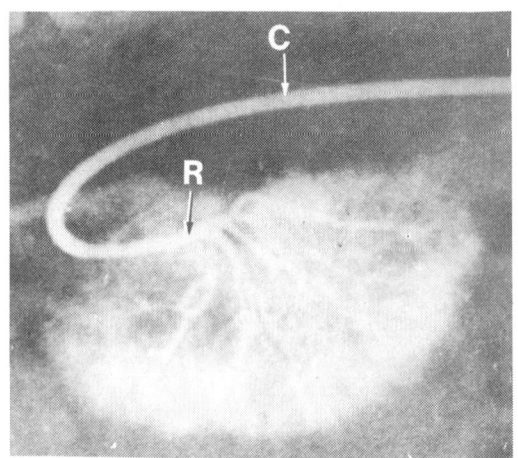

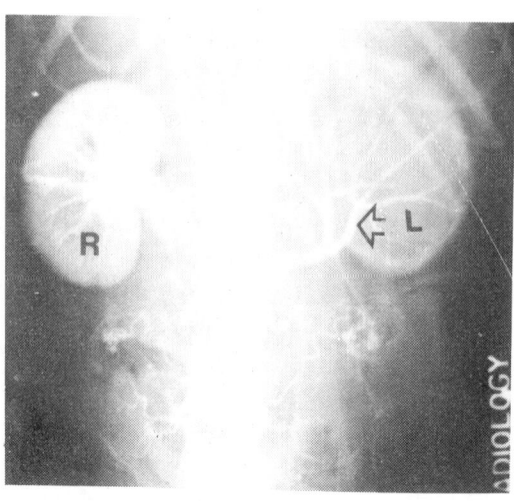

Fig. 12-25A : Normal selective renal angiogram of left kidney of a goat. C = Catheter, R = Main renal artery.

Fig. 12-25B : Renal angiogram of an experimental dog after ligation of left ureter. Note absence of clear cortex and splayed renal artery of left kidney (L). R = right kidney (Courtesy N.N. Balasubramanian).

Figs 12-26 and 12-27 show normal venous channels of a calf outlined using the described technique.

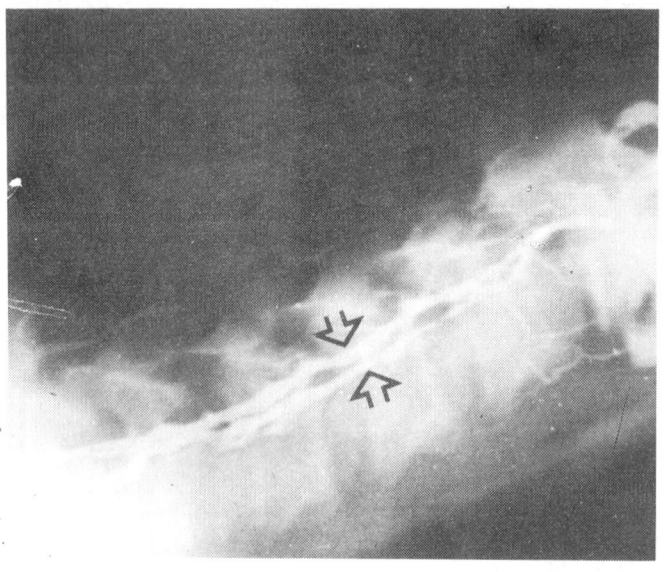

Fig. 12-26 : Cervical vertebral venogram of a calf, Arrows indicate cervical vertebral veins.

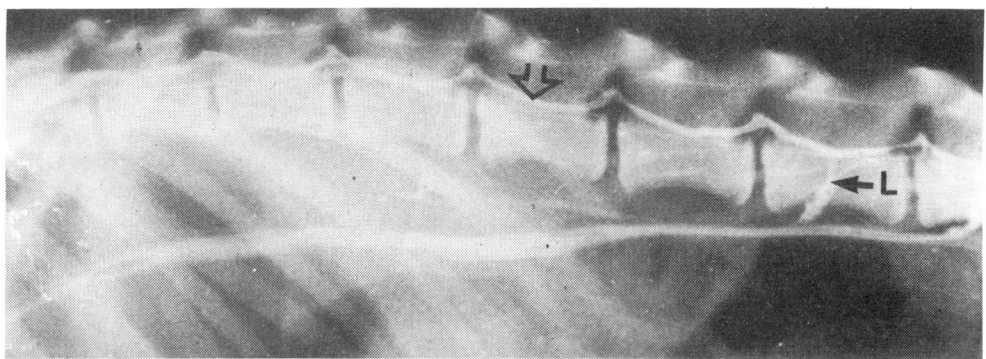

Fig. 12-27 : Lumbar vertebral venogram of a calf. Open arrow indicates lumbar vertebral vein L = Lumbar segmental vein.

MYELOGRAPHY

The technique refers to the contrast radiographic examination of the spinal cord and emerging spinal roots after injecting the contrast material into the subarchnoid space. It is indicated to diagnose intervertebral disc protrusion, intraspinal lesions, vertebral canal haemorrhage and spinal cord oedema. It should not be used in cases of meningitis, myelitis, myelomalacia and when myelography has been done in the recent past.

Technique

Any contrast material used for myelography should be non-irritating. Air can be used but does not provide adequate contrast. Metrizamide is an agent of choice being non-ionic, non-irritating, water soluble agent. The contrast column is also visualised for a longer time. Its dose is: dogs 0.2 to 0.4 ml/kg; neonatal calves 0.3 to 0.4 ml/kg.

For myelography, the animal should be anaesthetised and all aseptic precautions should be taken to inject the contrast medium. Survey radiographs should be obtained beforehand.

A. Cervical injection

   i) Control the animal in lateral or sternal recumbency with the head held in a hyperflexed position.
  ii) Introduce a 19 G spinal needle with a stylet into the subarachnoid space through the foramen magnum, confirmed by the free flow of cerebrospinal fluid (CSF)
 iii) Attach a syringe to the needle and aspirate CSF in equal quantity to that of the volume of the contrast material to be injected.
  iv) Slowly inject the contrast material (1 ml/minute) and withdraw the needle.
   v) Raise the head to an angle of 25° from the horizontal plane to prevent cranial flow of the contrast material.
  vi) In case of water soluble contrast agents, obtain lateral radiographs as soon as possible and then at 5 and 10 minutes. In case of oily agents, obtain radiographs at 10, 15 and 30 minutes.
 vii) If convulsions occur during or after injection of the contrast agent, anaesthetise the animal further.
viii) After obtaining radiographs, raise the head of the animal further to reduce risk of convulsions.

B. Lumbar injection

Only differences from the cervical injection are listed.

    i)   Control the animal in lateral recumbency and bring the hind limbs cranially to open the inter-arcuate space.

    ii)   Introduce the spinal needle into the spinal canal through the lumbosacral space.

    iii)   Remove CSF and inject contrast material.

    iv)   Obtain first radiograph without withdrawing the needle just after completion of the injection of the contrast material.

    v)   Subsequent radiographs be obtained at 5, 10 and 15 minutes. Figs 12-28 and 12-29A show myelograms of the calf. Fig. 12-29B shows myelogram of the dog.

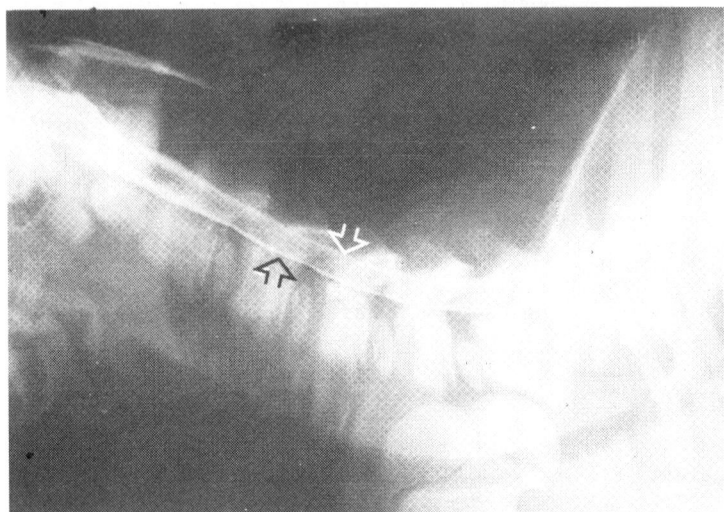

Fig. 12-28ˇ : Cervical myelogram of a normal 4 day old calf. White and black arrows indicate smooth dorsal and ventral contrast medium columns (Figs 12-28 and 12-29 are from Bargai, U., Vet. Rad. Ultrasound 34, 20, 1993. Used with permission).

ARTERIOGRAPHY

It refers to the contrast radiographic examination of the arterial system of an area. It is indicated to study the arterial pattern in normal subjects from anatomy point of view and to study variations in patterns due to a pathological process. It is also indicated to diagnose arterial occlusion. Water soluble contrast agents are used. Since these agents are irritating, the procedure should be used after anaesthetising the animal.

Technique

    i)   Anaesthetise the animal.

    ii)   Expose the desired vessel (Table 12-2) and catheterise.

    iii)   Inject contrast material using appropriate dose (The dose schedule for a calf is listed in Table 12-2).

    iv)   Obtain radiograph during the last phase of the injection.

Fig. 12-30 to 12-33A show arteriograms.

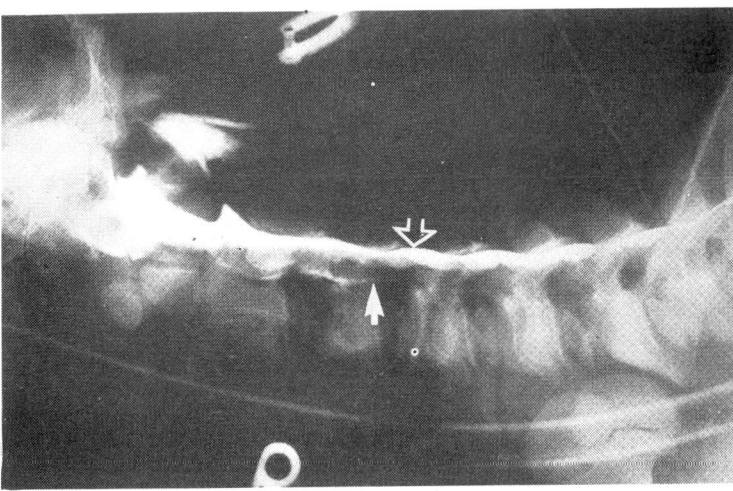

Fig. 12-29A : Cervical myelogram of a 5 days old calf. Open arrow indicates abnormally thick and convoluted dorsal column. Solid arrow indicates absence of ventral column after third cervical vertebra. The case had diffuse haemorrhage along the vertebral canal.

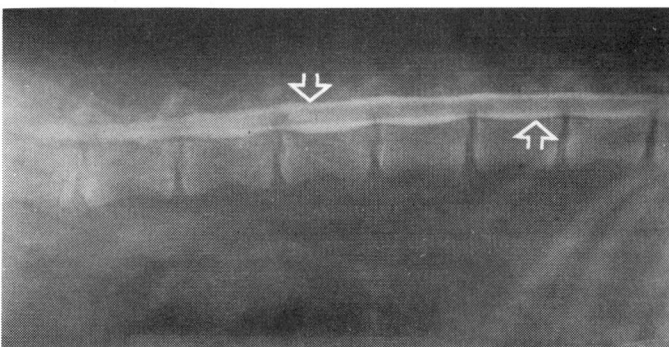

Fig. 12-29B : Normal lumbar myelogram of a dog (Courtesy N.N. Balasubramanian).

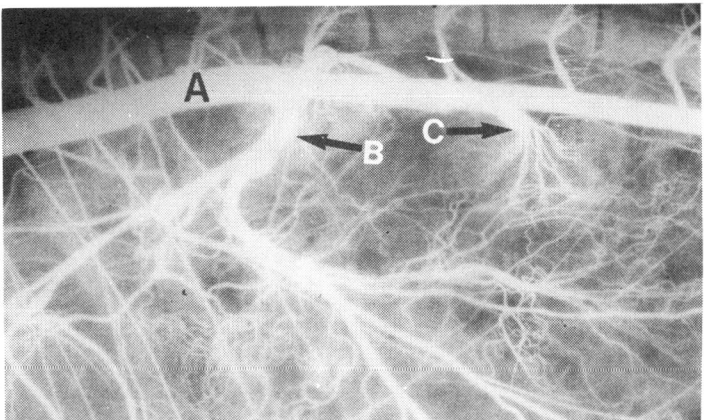

Fig. 12-30 : Abdominal aortogram of a goat. A = Aorta, B = Coeliomesenteric trunk, C = Renal artery.

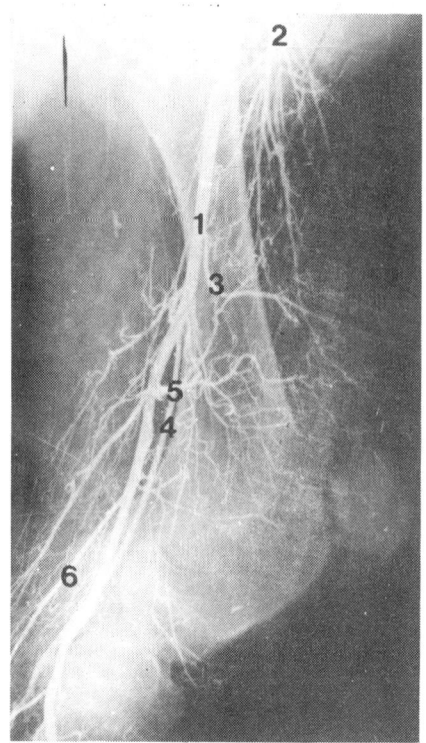

Fig. 12-31 : Normal arteriogram of femoral region of a calf. 1 = Femoral artery, 2 = Lateral circumflex iliac artery, 3 = Descending genicular artery, 4 = Saphenous artery, 5 = Nutrient artery, 6 = Caudal femoral artery (Courtesy Dr. Prem Singh).

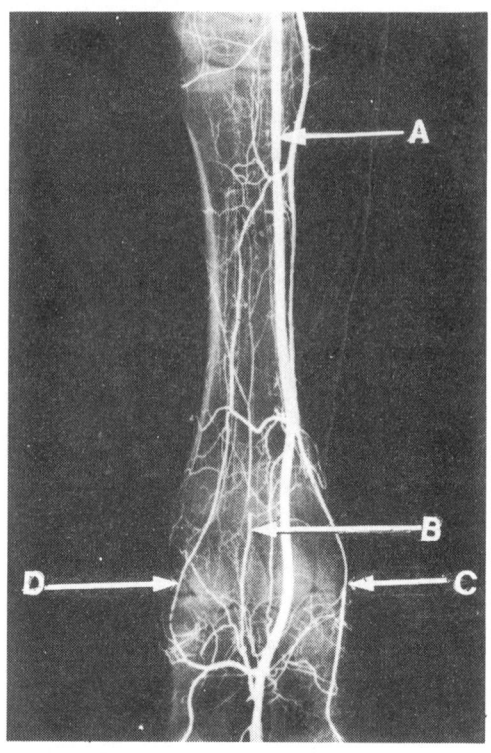

Fig. 12-32 : Normal arteriogram of metacarpal region in a buffalo calf. A = Great metacarpal artery, B = Dorsal metacarpal artery, C = Medial palmar metacarpal artery, D = Lateral palmar metacarpal artery (Courtesy Dr. R.L.N. Rao).

TABLE 12-2 : Artery to be catheterised and dose of contrast material for a calf

Area	Artery to be catheterised	Dose of sodium iothalamate (70% W/V)
Cerebral angiography	Common carotid	10-30 ml
Abdominal arteriography	Common carotid or femoral	60-80 ml
Femoral region	External iliac	30-40 ml
Metatarsal region	Tibial-cranial	15-20 ml
Hind digit	Dorsal metatarsal	10-15 ml
Fore digit	Dorsal metacarpal	10-15 ml
Metacarpal region	Median	15-20 ml

ARTHROGRAPHY

The technique is used to visualise structures of a diarthrodial joint after injection of a positive or negative contrast medium or a combination of both. The technique aids in diagnosis of various joint abnormalities or diseases. It should not be used if infection is suspected in the surrounding soft tissue. Pneumoarthrogram of a calf is shown in fig. 12-33B.

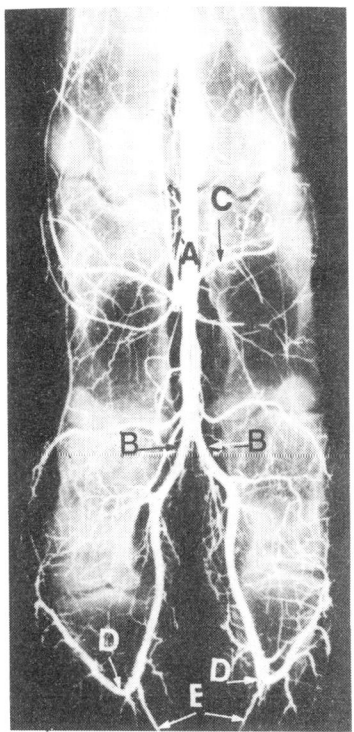

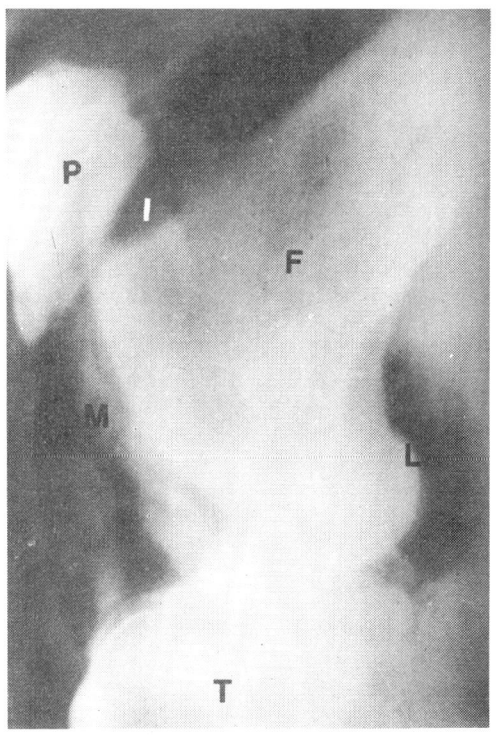

Fig. 12-33A : Dorsopalmar arteriogram of normal bovine digital region (foot). A = Common digital artery, B = Digital arteries, C = Distal palmar arch, D = Terminal arches, E = Laminar vessels (From Gogoi et al, Vet. Rad. 23, 171, 1982. Used with permission).

Fig. 12-33B : Normal pneumoarthrogram of stifle joint of calf. P = Patella, F = Femur, T = Tibia, M and L are medial and lateral femoro-tibial pouches, I = Infrapatellar pouch.

Technique

Double contrast study provides better results and is being discussed here. In such a study, positive contrast coats the joint margins while negative contrast demonstrates the whole joint. When used alone, positive contrast is rapidly resorbed and also has a tendency to mask intra-articular lesions. The negative contrast does not provide sufficient details.

 i)   Obtain a survey radiograph of the joint after controlling the animal in lateral recumbency.
 ii) Using aseptic technique, insert a 19 G hypodermic needle into the joint.
 iii) Remove as much synovial fluid as is possible.
 iv) Inject 2-3 ml of water soluble iodine based contrast agent and 10-15 ml of air into the joint.
 v)   Flex and extend the joint several times for thorough dispersion of the positive contrast agent.
 vi) Obtain radiographs of the joint.

Among contrast agents, meglumine and sodium diatrizoate, diluted as a 20 to 40% solution with saline or distilled water, provide better results.

In case of a positive contrast arthrogram, use 3-6 ml of the agent to visualise articular cartilage and 5-15 ml to outline tendon sheath and joint pouches.

FASCIAGRAPHY

It is a contrast radiographic study of tendons and associated structures. The technique can be used to diagnose adhesions, calcification and rupture of tendon and muscles.

Technique

i) Sedate the animal in lateral recumbency with the limb to be examined uppermost.
ii) Apply a tourniquet each proximal and distal to the site.
iii) Introduce a 16-18 G needle, using aseptic technique, between the muscle and tendon.
iv) Attach a three-way stop cock and a 50 ml glass syring to the needle.
v) Inject room air till area between the two tourniquets is moderately distended.
vi) Remove the needle.
vii) Obtain lateral projection.
viii) Remove air and then the tourniquets.

Fig. 12-34 shows normal fasciagram of the area of achillis tendon in a clalf.

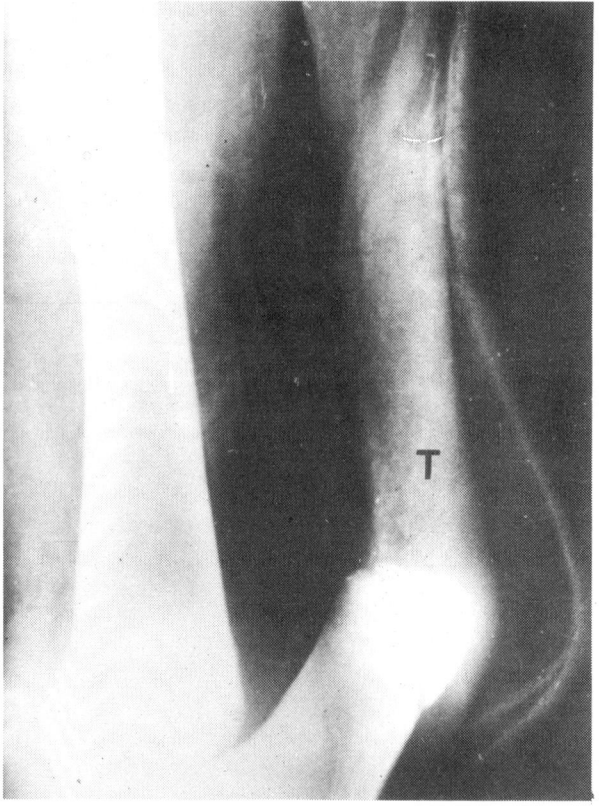

Fig. 12-34 : Normal fasciagram of the area of tendoachillis in a calf. T = Tendoachillis.

OSTEOMEDULLOGRAPHY (INTRAOSSEOUS PHLEBOGRAPHY)

The technique is used to visualise intraosseous and extraosseous venous channels of a long bone. It can be used to aid diagnosis of delayed and non-union of a fracture, degenerative arthritis and bone neoplasms. It has also been used to diagnose avascular necrosis.

Technique

    i)   Anaesthetise the animal with the limb to be examined lowermost.

    ii)   Expose the distal metaphyseal end of the test bone from the medial side using all aspetic precautions.

    iii)   Drill a hole with the help of an intramedullary pin in the metaphyseal region.

    iv)   Insert into the hole a 10 cm long spinal needle of which diameter is slightly greater than the pin used to drill the hole.

    v)   Ensure correct placement of the needle into the marrow cavity by following steps discussed under intraosseous vertebral venography.

    vi)   Place a tourniquet proximal to the test bone or use phlebocompression along the entire length of the bone using elastic bandage.

    vii)   Infuse rapidly 10-20 ml (dog, 5 ml) of water soluble iodine based contrast agent into the marrow.

    viii)   Obtain a lateal radiograph during the last phase of the injection and then at 2, 4, 6 minutes. Withdraw the needle. Figs 12-35 to 12-38 show the normal osteomedullograms of calves.

INTRAVENOUS PYELOGRAPHY (EXCRETORY UROGRAPHY)

Intravenous pyelography (IVP) refers to contrast radiographic examination of the kidneys and ureters after intravenous introduction of a positive contrast medium. Since water soluble iodine based contrast agents are excreted through urine, intravenous injection of these agents outlines the urinary tract. Apart from being an aid to diagnose abnormalities of urinary tract, the technique also serves as a rough index of kidney function. It should, however, not be used in severely dehydrated patients.

Technique

    i)   Prepare the animal the same way as described earlier for barium series.

    ii)   Create pneumoperitoneum in ruminants as described for peritoneography as it leads to ventral displacement of the rumen and other viscera to provide unobstructed view of the kidneys and ureters. It also provides a negative contrast, in the background of positive contrast in the urinary tract, to provide better visualisation of structures. In addition, exposure factors are reduced.

    iii)   Select either a bolus technique (low volume rapid infusion) or a drip technique (high volume drip infusion) to administer water soluble iodine based contrast agents. Sodium tri-iodinated organic compounds are generally recommended, however, sodium iothalamate (70% W/V) also gives satisfactory results. Usual dose is 2-3 ml/kg in small ruminants and 0.5 ml/kg in large ruminants. In dogs, total dose should not contain more than 35 g of iodine (refer to table 12-1 for iodine content of contrast materials). However, uraemic patients will require a higher dose.

    iv)   In **bolus technique,** inject intravenously total calculated dose as a bolus. In **drip technique,** dissolve the agent in normal saline to make a concentration of 23% (in case of sodium iothalamate 70% W/V). Inject entire volume intravenously over a period of about 10 minutes.

    v)   Obtain standing right lateral radiographs. In dogs, sheep and calves, ventrodorsal projection can also be made. Obtain one radiograph immediately after completion of injection of the contrast agent. Two films are required separately for left and right kidneys in adult ruminants. A 10 minutes radiograph shows both kidneys and ureters and contrast agent usually reaches urinary bladder in about 20 minutes.

Figs 12-39 to 12-41 show normal excretory urograms in sheep, calf, cow and cat.

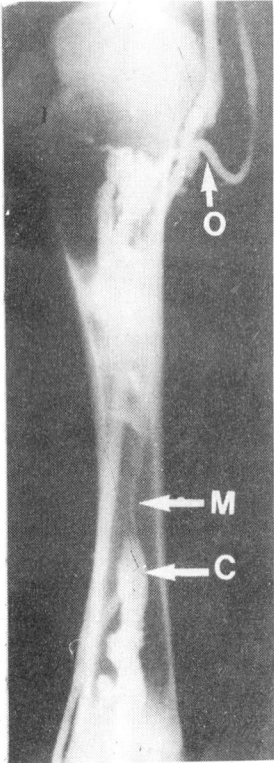

Fig. 12-35 : Normal osteomedullogram of the tibia of a buffalo calf. O = Extraosseous vessels, M = Central medullary vein, C = Deposited contrast material radiating upwards.

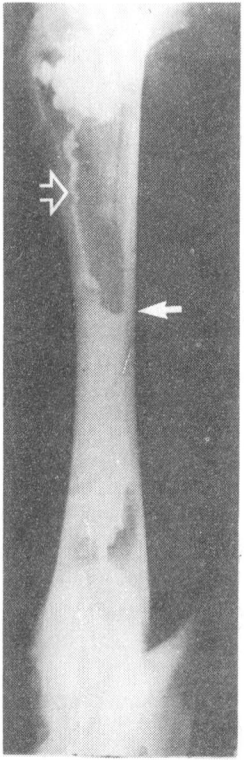

Fig. 12-36 : Osteomedullogram of a tibia of an experimental calf 10 days after introduction of infection in the bone. Note the fluid level indicated by solid arrow (purulent fluid mixed with contrast agent) 10 minutes after infusion of the contrast medium indicating delayed drainage. Open arrow indicates ascending medullary vein (Courtesy Dr. Yogender Sangwan).

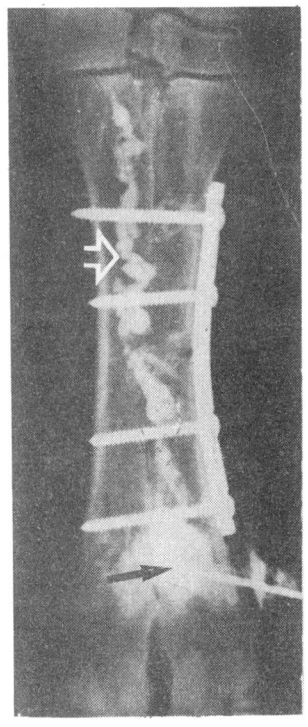

Fig. 12-37 : Osteomedullogram of the fractured metacarpal of a buffalo calf 6 weeks after its repair. Solid area indicates flow of contrast material into the proximal fragment (Courtesy Dr. R.L.N. Rao).

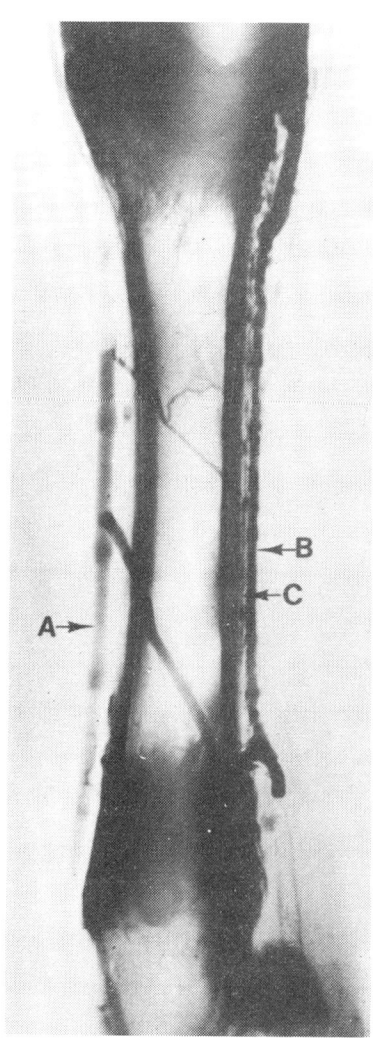

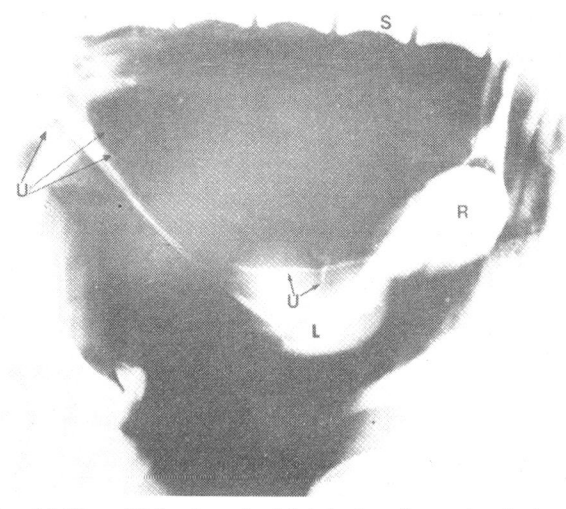

Fig. 12-39 : Right lateral abdominal radiograph of sheep 15 minutes after bolus intravenous injection of contrast medium and pneumoperitoneum. Right kidney is abnromally positioned in this animal. S = Vertebral column, R = Right kidney, L = Left kidney, U = Ureter (Figs 13-39 to 13-41 are from Singh et al. Vet. Rad. 24, 106, 1983. Used with permission).

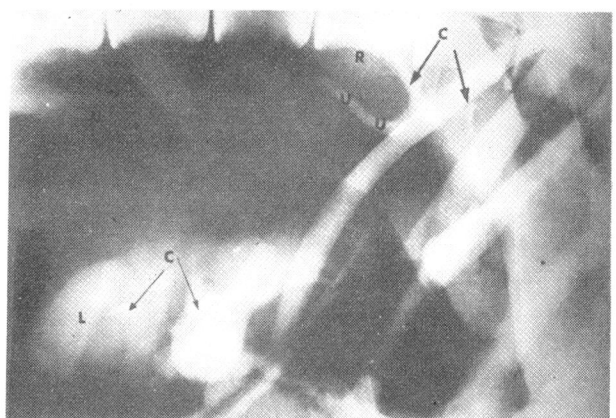

Fig. ,12-38 : Normal osteomedullogram of metatarsal of a buffalo calf. A = Dorsal common digital vein III, B and C = Plantar veins.

Fig. 12-40 : Right lateral abdominal radiograph of a calf 10 minutes after infusion method of intravenous pyelography. R = Right kidney, L = Left kideny, C = Calyces, U = Ureter.

URETHROGRAPHY

The technique is indicated to diagnose abnormalities of the urethra in the male. In the female, the urethra is short and wide and hence technique is inapplicable.

Technique

    i) Keep the animal off feed for 24 hours and off water for 12 hours. Anaesthetise or deeply sedate the animal.

    ii) Prepare the preputial area and in case of withers sinp off the urethral process (For rams and bucks use the procedure discussed at point VII)

    iii) Lubricate a catheter of appropriate diameter with lignocaine gel and insert into the urethra for a short distance through urethral opening.

    iv) Infuse 5-10 ml of 2% lignocaine into the urethra to desensitise urethral mucosa.

    v) Dilute water soluble iodine based contrast agent (For sodium iothalamate, dilute to a concentration of 15% by adding normal saline). Infuse 10-15 ml of the diluted agent in case of a calf or small ruminant.

    vi) Obtain a lateral radiograph during the last phase of infusion of the contrast agent.

    vii) In the case of buck and ram, do intravenous pyelography so that sufficient contrast material fills the urinary bladder. Press the abdomen externally till few drops of urine appear at the urethral opening. Take a radiograph.

Figs 12-42 and 12-43 show normal urethrograms in a buck and calf.

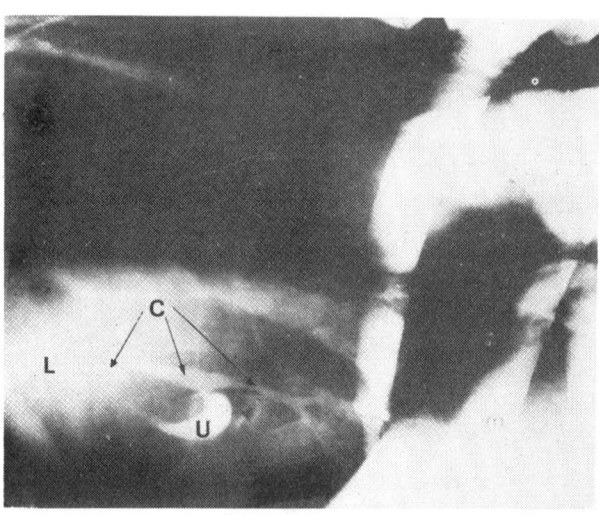

Fig. 12-41A : Right lateral abdominal radiograph of an adult cow 10 minutes after infusion method of intravenous pyelography. L = Left kidney. C = Calyces, U = Ureter.

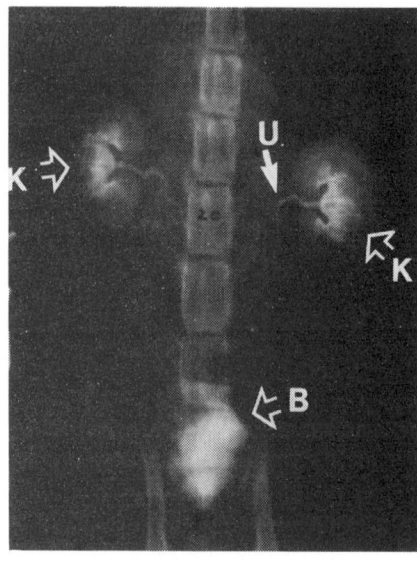

Fig. 12-41B : Normal intravenous pyelogram of a cat without using abdominal compression. K = Kidneys, U = Ureter, B = Urinary bladder.

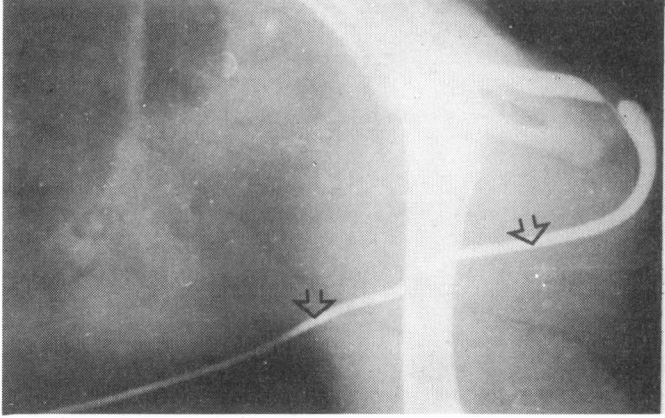

Fig. 12-42 : Urethrogram of a buck. Arrows indicate course of urethra.

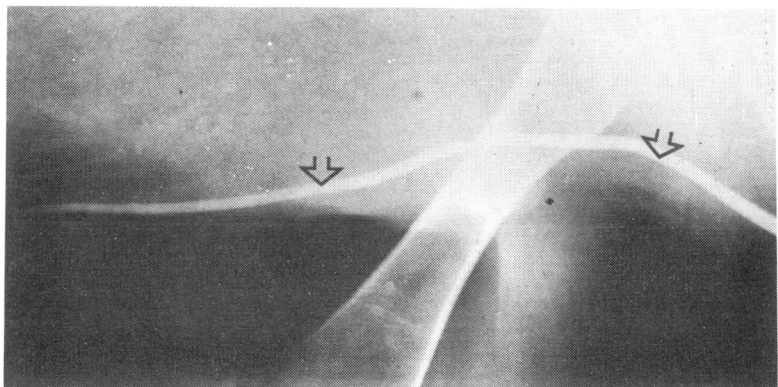

Fig. 12-43 : Urethrogram of a calf. Arrows indicate course of urethra.

CYSTOGRAPHY

Cystography refers to the contrast radiographic examination of the urinary bladder. The technique is indicated to diagnose structural abnormalities and diseases of the baldder. It should not be used if atony of the bladder is suspected. A positive or negative contrast agent can be used but double contrast or double contrast alongwith pneumoperitoneum provides better results.

Technique

i) Prepare the animal as described for barium series.

ii) Sedate the animal to avoid discomfort as a result of distension of the urinary bladder after introduction of the contrast agent. Restrain in lateral recumbency.

iii) Insert lubricated radiopaque catheter into the urethra. Advance the catheter into the bladder in ·the female. In the male, it is difficult to advance it into the bladder and so catheter is positioned at the neck of the bladder. Empty the bladder as much as possible by some abdominal compression.

iv) Infuse 2% liagnocaine solution (5-10 ml) into the bladder.

v) Remove air from the catheter by suction as air bubbles may produce artifacts on the radiograph.

vi) Introduce either a water soluble iodine based contrast agent ((e.g sodium iothalamate 70% W/V diluted to a concentration of 10-15%) or a negative contrast (room air, oxygen or carbon dioxide), amount required will depend upon the species and size of the animal. In small ruminants 80-120 ml of contrast material is required to fill the bladder. In dogs, the dose is 6-12 ml/kg.

vii) For a double contrast study, introduce 15-20 ml of the positive contrast, roll the animal and then introduce negative contrast to fill the bladder. For triple contrast, create pneumoperitoneum.

viii) Obtain recumbent right lateral and ventrodorsal projections. In case of triple contrast, obtain only standing right lateral projection. Remove the catheter.

Normal cystograms of buck are shown in Figs 12-44 to 12-47A. Positive contrast cystogram of a cat is shown in fig. 12-47B.

Double contrast cystography is considered superior as the mucosa and wall of the urinary bladder are clearly visualised. Triple contrast enhances the visualisation of the wall and mucosal and serosal surfaces.

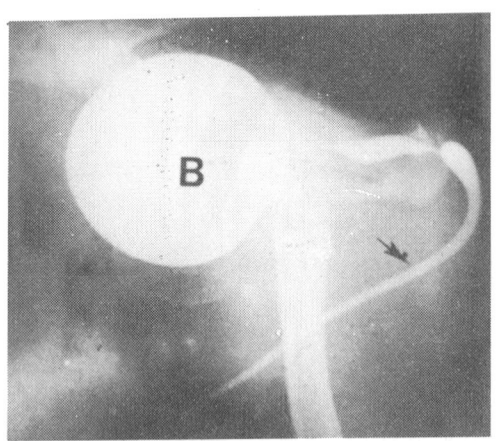

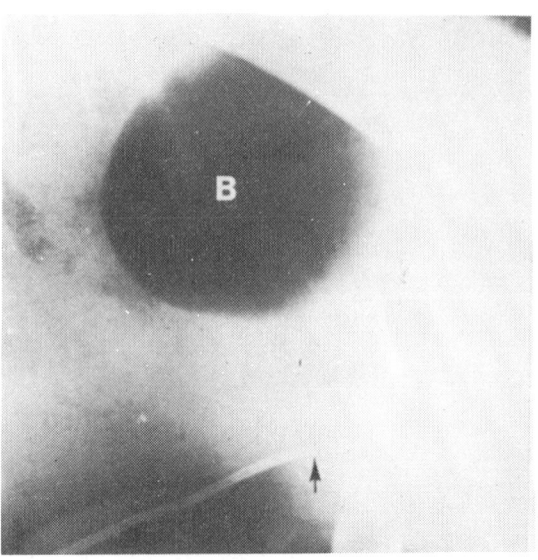

Fig. 12-44 : Normal positive contrast cystogram in a buck. B = Urinary bladder. Arrow indicates urethra (Figs 12-44 to 12-47A are from Tayal et al, Vet. Rad. 25, 260, 1984. Used with permission).

Fig. 12-45 : Normal negative contrast cystogram in a buck. B = Urinary bladder. Arrow indicates catheterised urethra.

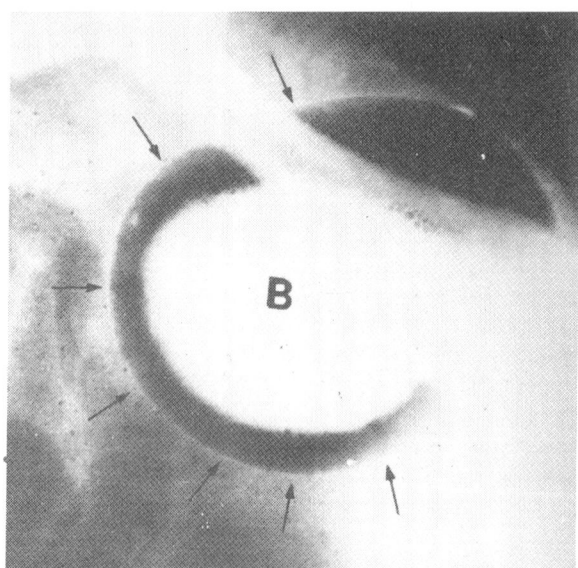

Fig. 12-46 : Normal double contrast cystogram in a buck. B = Urinary bladder. Arrows indicate wall of the urinary bladder.

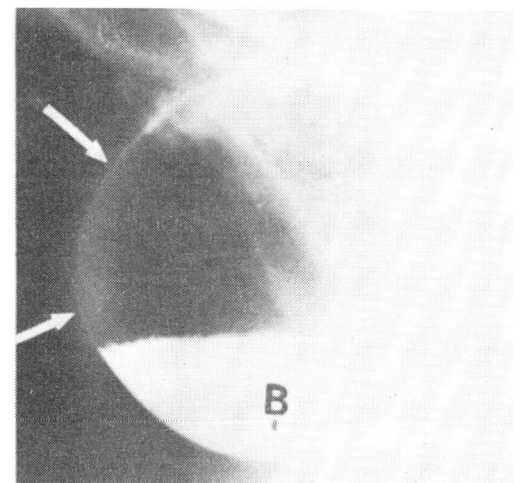

Fig. 12-47A : Double contrast cystogram alongwith pneumoperitoneum in a buck. B = Pooled positive contrast medium in the bladder. Arrows indicate wall of the urinary bladder.

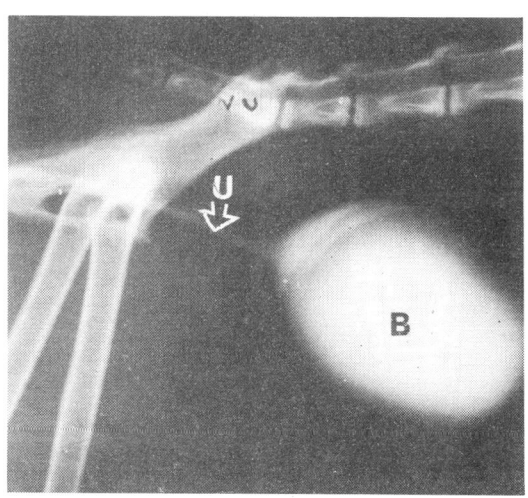

Fig. 12-47B : Positive contrast cystogram and voiding urethrogram of a cat using wooden spoons abdominal compression. B = Urinary bladder, U = Urethra (Courtesy N.N. Balasubramanian).

REFERENCES

Bargai, U. 1993. Myelography in neonatal calves. Vet. Rad. Ultrasound. **34**, 20.

Bhargava, A.K., Singh, A.P. and Singh, G.R. 1980. Positive contrast urethrography in goats. Indian J. Vet. Surg. **1**, 4.

Chawla, S.K., Singh, A.P., Chandna, I.S. and Nigam, J.M. 1982. Normal osteomedullography of tibia in sheep. Indian J. Anim. Sci. **52**, 1198.

Chawla, S.K., Singh, A.P., Sharma, D.N. and Chandna, I.S. 1985. Venography of the orbital venous system and cavernous sinuses in the goat. Vet. Rad. **26**, 156.

Dawson, P. and Howell, M. 1986. The non-ionic dimers: a new class of contrast agents. Brit. J. Radiol. **59**, 987.

Dennis. R., Herrtage, M.E. 1989. Low osmolar contrast media- a review. Vet. Rad. **30**, 2.

Dhar, S.K., Nigam, J.M. and Bhargava, A.K. 1974. Fasciagraphy: A technique for evaluation of adhesions around Achilles tendon of buffalo calves after tenorrhaphy. Am. J. Vet. Res. **35**, 135.

Douglas, S.W. and Williamson, H.D. 1980. Principles of Veterinary Radiography. 3rd edn., Williams and Wilkins Co., Baltimore.

Ducharme, N.G., Dill, S.G. and Rendano, V.T. 1983. Reticulography of the cow in dorsal recumbency: An aid in the diagnosis and treatment of traumatic reticuloperitonitis. J. Am. Vet. Med. Assoc. **182**, 585.

Gillette, E.L., Thrall, D.E. and Lebel, J..L. 1977. Carlson's Veterinary Radiology. 3rd edn., Lea and Febiger, Philadelphia.

Gogoi, S.N; Nigam, J.M. and Singh, A.P. 1982 Angiographic evaluation of bovine foot abnormalities. Vet. Rad. **23**, 171.

Heider, L., Wyman, M., Burt, J., Root, C. and Gardner, H. 1975. Nasolacrimal duct anomaly in calves. J. Am. Vet. Med. Assoc. **167**, 145.

Herrtage, M.E. and Dennis, R. 1987. Contrast media and their use in small animal radiology. J. Small Anim. Pract. **28**, 1105.

Kanwar, M.S., Chandna, I.S., Singh , A.P., Dahiya, Z.S. and Sharma, D.N. 1983. Contrast radiography of gastrointestinal tract of caprine. Indian J. Anim. Sci. **53**, 724.

Krishnamurthy, D., Peshin, P.K., Nigam, J.M., Sharma, D.N. and Kumar, R. 1981. Radiographic visualisation of the nasolacrimal duct in domestic animals. Indian J. Anim. Sci. **51**, 646.

Lesser, E.C. 1982. Contrast material reactors: pathogenecity and clinical aspects in 1981. In : Contrast Media in Radiology. Ed., Amiel, M., Springer-Verlang, New York.

Lee; R. 1978. Contrast media and technique-1. J. Small Anim. Pract. **19**, 589.

May, S.A., Wyn-Jones, G., Church, S., Brouwer, G. and Jones, R.S. 1986. Iopamidol myelogaphy in the horse. Equine Vet. J. **18**, 199.

Morgan, J.P. and Silverman, S. 1990. Techniques of Veterinary Radiography, 4th edn., Iowa State University Press, Ames.

Nayar, K.N.M., Venkataravanappa, N., Singh, A.P. and Chandna, I.S. 1983. Bronchography in sheep. Indian J. Anim. Sci. **53**, 1236.

Nigam, J.M., Krishnamurthy, D., Kumar, R., Sharma, D.N. and Peshin. P.K. 1981. Retrograde parotid sialography. H.A.U. J. Res. **11**, 143.

Pinet, A., Lyonnet, D., Maillet, P. and Grolean, J.M. 1982. Adverse reactions to intravenous contrast media in urography-a national survey. In: Contrast Media in Radiology. Ed., Amiel, M., Springer-Verlag, New York.

Rao, R.L.N., Singh, A.P., Nigam, J.M. and Chawla, S.K. 1985. Response of intraosseous venous circulation to metacarpal fracture healing in buffalo calves: An osteomedullographic study. Indian J. Anim. Sci. **55**, 632.

Sharma, S.K., Singh, A.P., Tayal, R., and Chandna, I.S. 1984. Contrast radiography of the ovine gastrointestinal tract. Vet. Rad. **25**, 17.

Singh, A.P., and Nigam, J.M. 1980. Intraosseous phlebography following allogenic bone transplantation in young buffalo-An experimental study. Cornell Vet. **70**, 103.

Singh, A.P. and Nigam, J.M. 1980. Radiography of bovine esophageal disorders. Mod. Vet. Pract. **61**, 867.

Singh, A.P., and Nigam, J.M. 1983. Vascular response to fracture healing in the bovine. Vet. Rad. **24**, 174.

Singh, A.P., Patil, D.B., Chawla, S.K., Peshin, P.K., Singh, J. and Sharifi, D. 1991. Contrast radiography of the caprine colon. Indian J. Anim. Sci, **61**, 969.

Singh, A.P., Peshin, P.K., Chawla, S.K., Chandna, I.S. and Singh J. 1985. Contrast radiography of alimentary tract in calves. Indian. J. Anim. Sci. **55**, 854.

Singh, A.P., Singh, G.R. Sharma, D.N., Nigam, J.M. and Bhargava, A.K. 1982. Arteriographic anatomy of the abdominal aorta in the goat, dog, pig and rabbit. Vet. Rad. **23**, 279.

Singh, A.P., Singh, J., Williamson, H.D., Peshin, P.K. and Nigam, J.M. 1983. Radiographic visualisation of the ruminant upper urinary tract by a double contrast technique. Vet. Rad. **24**, 106.

Singh, A.P., Vijay Kumar, D.S. and Nigam, J.M. 1982. Osteomodullography for the evaluation of fracture healing in calves. Bovine Pract. **3**, 8.

Singh, J., Peshin, P.K; Williamson, H.D. and Singh, K. 1981. Experimental studies on double contrast peritoneography in sheep and calves. Zbl. Vet. Med. A. **28**, 720.

Singh, P., Singh, J., Chawla, S.K; Peshin, P.K. and Singh, A.P. 1986. Vascular response to femoral fracture healing in cattle. Indian J. Anim. Sci. **56**, 633.

Tayal, R., Singh, A.P., Chandna, I.S. and Chawla, S.K. 1984. Contrast cystography in the goat. Vet. Rad. **25**, 260.

Tayal, R., Chandna, I.S., Singh, A.P. and Peshin, P.K. 1985. Selective renal angiography in goats. Indian J. Anim. Sci. **55**, 98.

Tayal, R., Singh, A.P., Chandna, I.S. and Nayar, K.N.M. 1986. Double contrast peritoneography in goats. Indian J. Anim. Sci. **56**, 11.

Ticer, J.W. 1984. Radiographic Technique in Veterinary Practice. 2nd edn., W.B. Saunders Co., Philadelphia.

SELECTED QUESTIONS

1. Define the following terms:
    i)   Positive contrast medium
    ii) Negative contrast medium
    iii) Dacrocystorhinography
    iv) Sialography
    v)   Myelography.
2. Discuss the following:
    i)   Systemic reactions of contrast media
    ii) Low osmolarity contrast media
    iii) Cholycystapaques
3. Discuss the importance of pneumoperitoneography in veterinary diagnostic radiology.
4. Classify positive contrast media and discuss briefly water soluble iodine preparations.
5. Discuss: i) Reticulography, ii) Intravenous pyelography.

CONTRAST MATERIALS
FILL COLOUR IN X-RAYS

IMPROPER
EXPOSURE FACTORS
OVERESTIMATE OR UNDERESTIMATE
THE JOB

CHAPTER

13 RADIOGRAPHIC POSITIONING

A.P. SINGH
P.K. PESHIN

The basic objective of maintaining a correct position of the animal during radiographic procedure is to obtain accurate radiographic details of the part being examined. The position should be such that it provides proper restraint of the patient and immobilisation of the area to be radiographed without causing discomfort to the patient.

The positioning procedures described in this chapter aim to help the reader to obtain quality radiographs but enough scope may still exist to improve these techniques. There are certain principles applicable to all positioning procedures, irrespective of the species of the animal, and should be followed as far as possible.

PREPARATION OF THE ANIMAL

The main objective of the preparation of the animal is to avoid formation of undesirable shadows on the radiograph which may otherwise create confusion in radiographic interpretation. The density of sand, mud, faeces, urine, any iodinated medicament or any other material is sufficient to produce undesirable shadows on a radiograph. Every effort should, therefore, be made to clean and dry the hair coat of the part to be radiographically examined, especially if the area is located distally towards the digital region. The foot should be critically examined and mud, grit, stones, faecal material etc should be removed. In the sheep, dirty wool may produce radiographic densities and, therefore, if possible, wool should be clipped before radiography is done.

Any kind of supportive material used to immobilise the injured part may be sufficient to produce radiographic denisities and to affect radiographic interpretation. Therefore, splints and plaster casts should be removed before the part is radiographed unless there is otherwise a definite medical reason not to do so. After removal of a plaster cast, the hair coat should be thoroughly cleaned because plaster fragments may remain attached to the hair to cast shadows on a radiograph.

RESTRAINT

Proper restraint of the animal during radiographic procedures is essential in veterinary practice due to many reasons. Radiation safety precautions require that, as far as possible, persons should not be present in the room where exposure is made. If the animal is not restrained, motion of the part being examined during an exposure will not allow production of a diagnostic radiograph. This will result in repeated exposures and wastage of film and other material. In routine practice, ruminants do not pose much problem and adequate physical restraint is sucfficient. However, nervous or vicious animals may require some form of chemical restraint. Light sedation is usually sufficient. Deep sedation or even general anaesthesia may become necessary under some circumstances e.g. to obtain projections with the animal controlled in ventrodorsal position or to obtain projections of the head region. Ruminants pose inherent problems in general anaesthesia and so it should be avoided as far as possible, despite several advantages. When minimal chemical restraint is used and it becomes necessary for the assistants to be present in the room during the exposure, all personal radiation safety precautions must be followed. Dogs may be deeply sedated or general anaesthesia be used. Light sedation is usually sufficient in case of horses. However, in some situations general anaesthesia may have to be used. Positioning aids have already been discussed in chapter 7. The use of such aids facilitate radiographic positioning.

THE BOVINE (CATTLE AND BUFFALO)

THE SHOULDER JOINT

Radiographic examination of the shoulder joint is not a routine requirement in cattle and buffaloes. However, whenever required, lateral recumbency should be preferred to ensure visualisation of the entire joint. A grid must be used to reduce scatter radiation.

Mediolateral View (ML)

Position: The animal is allowed to stand with the limb to be examined pulled as far cranial as possible. The cassette in holder is positioned lateral to the shoulder joint (Fig. 13-1). If the animal is in lateral recumbency, the joint is positioned lowermost and the limb is pulled as far cranial as possible. The upper limb is pulled in caudal direction. The cassette is placed beneath the shoulder joint.

Beam centre: In the standing animal, the X-ray beam is directed from the medial side parrallel to the ground and centered on the shoulder joint. If the animal is in lateral recumbency, the beam is directed from the medial side vertically at a right angle to the cassette.

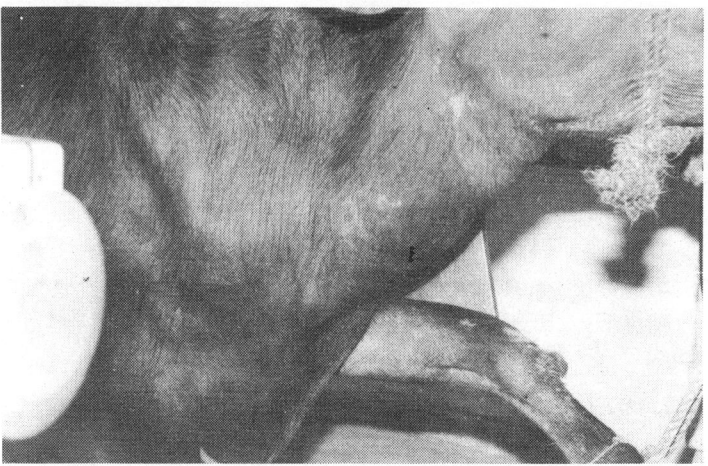

Fig. 13-1 : Buffalo. Position for mediolateral view of shoulder joint

THE HUMERUS

Mediolateral View (ML)

Position: The animal is placed in lateral recumbency with the limb to be radiographed lowermost and drawn cranially. The cassette is placed beneath the humeral region. the upper limb is drawn caudally.

Beam centre: The X-ray beam is directed vertically and centered at the mid diphyseal region of the humerus. The beam should be collimated and a large sized film used to include both proximal and distal joints. A grid should be used.

THE ELBOW JOINT

Radiographic examination of the elbow joint is difficult in standing adult cattle and buffaloes because of the size involved and the problem of positioning of the cassette to obtain either a true mediolateral or a true craniocaudal projection. The craniocaudal view is usually slightly oblique. A grid should be used.

Craniocaudal View (CrCd)

Position: The animal should stand normally with the limb to be radiographed pulled as far cranial as possible. The cassette placed in a holder is positioned at the caudal aspect of the olecranon process of the ulna and held in such a way that it is perpendicular to the X-ray beam (Fig. 13-2). Craniocaudal view can also be obtained with the animal in a lateral recumbent position and with the limb to be examined uppermost. The limb is fully extended and pulled cranially to a small degree.

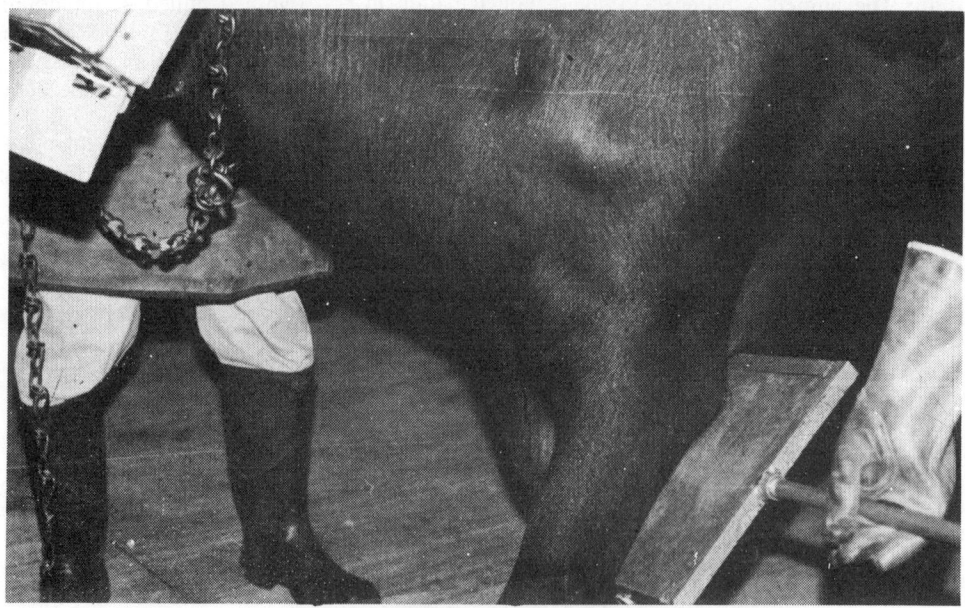

Fig. 13-2 : Buffalo. Position for craniocaudal view of elbow joint.

Beam centre: The X-ray beam is directed from the cranial aspect of the joint perpendicualr to the face of the cassette, as far as possible. In the lateral recumbent animal, the beam is directed horizontally at right angle to the cassette and centered on the cranial aspect of the elbow joint.

Mediolateral View (ML)

Position: This view can be obtained both in standing and lateral recumbent positions. A grid should be used. In a standing animal, the limb to be radiographed is pulled cranially. The cassette in a holder is positioned against the lateral aspect of the elbow Joint (Fig. 13-3). In recumbent position, the limb is placed lowermost, with the cassette held under the elbow joint, and pulled cranially. The upper limb is pulled in a caudal direction.

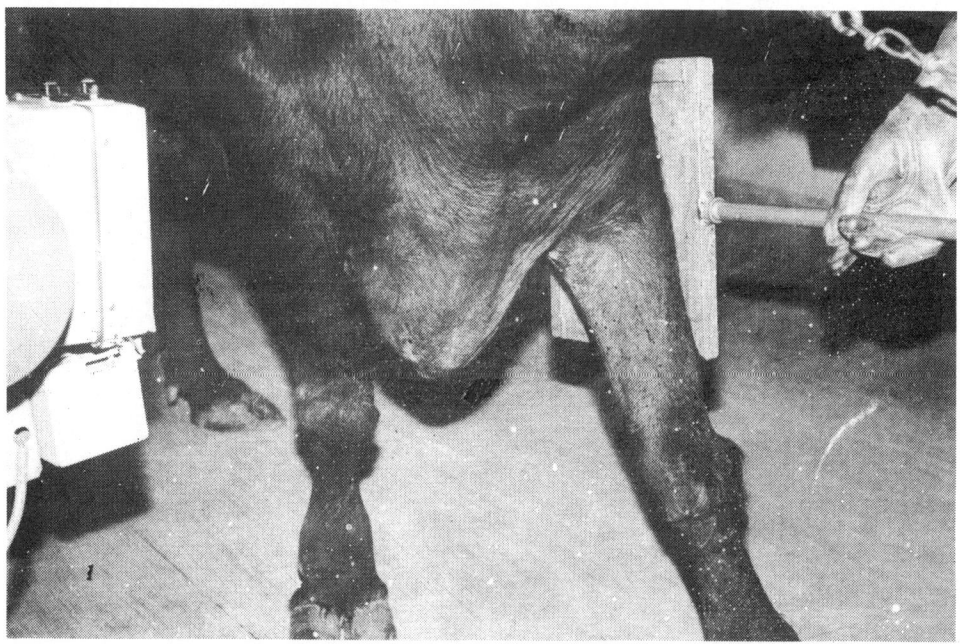

Fig. 13-3 : Buffalo. Position for mediolateral view of elbow joint.

Beam centre: The X-ray beam is directed horizontally from the medial to the lateral side parallel to the ground and centered on the elbow joint. The beam is directed more caudally if injury to the olecranon is suspected. If the animal is recumbent, a vertically directed beam, at right angle to the cassette, is used to obtain the radiograph.

THE RADIUS

Craniocaudal View (CrCd)

Position: Radiography of the radius can be done with the animal standing (Fig. 13-4). However, to ensure inclusion of the elbow joint and olecranon process, it is advisable to restraint the animal in lateral recumbency with the limb to be radiographed uppermost. The cassette in a holder is held against the caudal aspect of the radius. A grid is not required.

Beam centre: The X-ray beam is directed horizontally from the cranial aspect and centered at the mid region of the radius. The beam should be collimated and a large sized film used to include both the elbow and carpal joints.

Lateromedial View (LM)

Position: The animal stands normal with the limb to be radiographed drawn slightly forward (Fig. 13-5). The cassette in a holder is placed against the medial aspect of the radius.

Beam centre: The X-ray beam is directed horizontally and is centered at the mid region of the radius. The beam should be collimated and a large size cassette be used to include both the elbow and carpal joints.

THE CARPAL JOINT

Radiographic examination of the carpal joint is routinely required in cattle and buffaloes. In routine practice, dorsopalmar and lateromedial or mediolateral projections are made. Flexed lateromedial or oblique view may also be obtained if required.

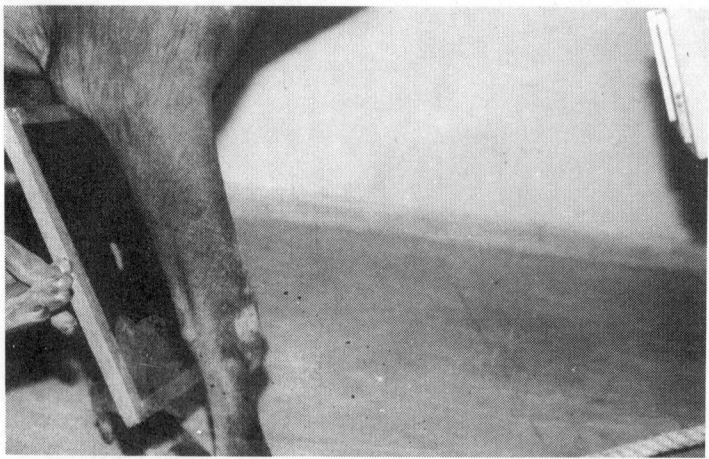

Fig. 13-4 : Buffalo. Position for craniocaudal view of radius.

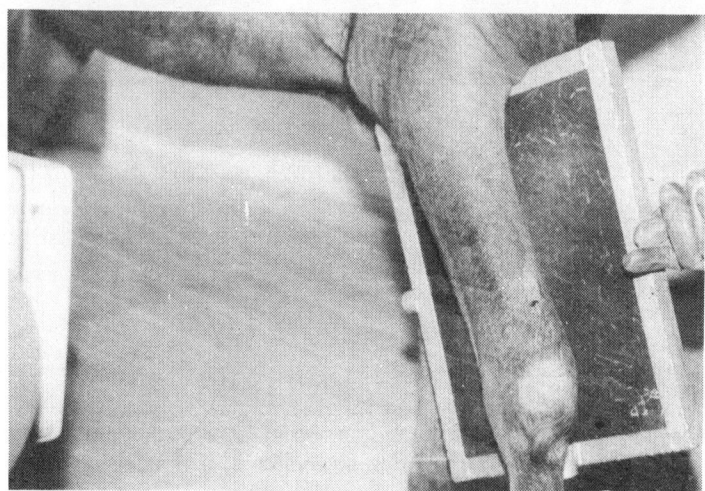

Fig. 13-5 : Buffalo. Position for lateromedial view of radius.

Dorsopalmar View (DPa)

Position: The animal is allowed to stand 'normal' and the cassette in a holder is held against the palmar surface of the carpal joint (Fig. 13-6).

Beam centre: The X-ray beam is directed parallel to the ground and centered on readily palpable intercarpal joint space along the midsagittal plane of the dorsal surface. However, in those animals which stand with the limb slightly rotated out, the DPa view should be made with the X-ray tube positioned slightly lateral to the longitudinal axis of the animal. A grid is usually not used but the requirement may change in case of swollen carpal joint of heavy adult animals.

Lateromedial View (LM)

Position: The animal is allowed to stand 'normal'. The cassette in a holder is placed firmly against the medial surface of the joint (Fig. 13-7).

Fig. 13-6 : Buffalo. Position for dorsopalmar view of carpus.

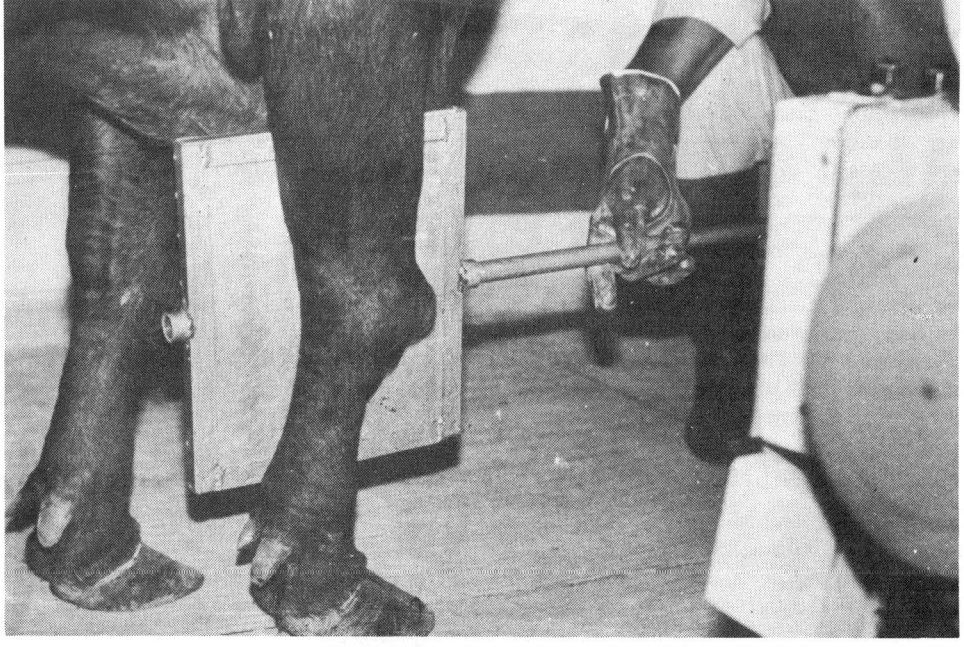

Fig. 13-7 : Buffalo. Position for lateromedial view of carpus.

Beam centre: The X-ray beam is directed parallel to the ground and centered at the midportion of the lateral aspect of the joint.

Flexed Lateromedial View (LM flexed)

Position: The animal is allowed to stand 'normal' and the limb to be radiographed is lifted and the carpal joint is flexed (Fig. 13-8). The cassette in a holder is positioned against the medial aspect of the flexed joint.

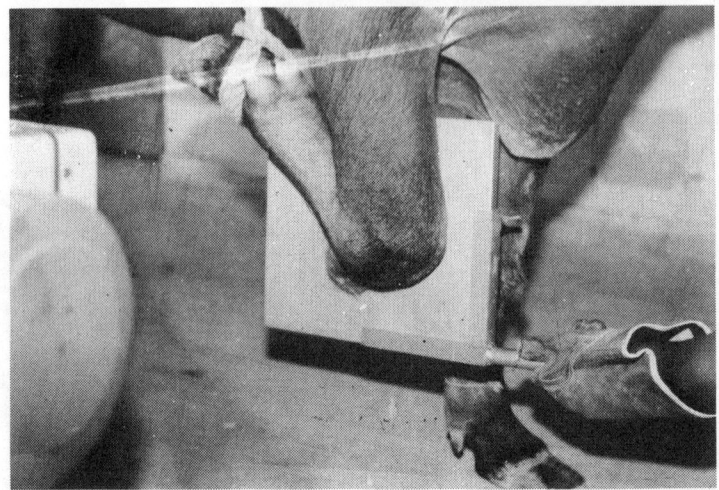

Fig. 13-8 : Buffalo. Position for lateromedial view of flexed carpus.

Beam centre: The X-ray beam is directed parallel to the ground and centered at the midportion of the lateral aspect of the flexed joint.

THE METACARPUS

The radiographic study of the metacarpus is routinely done in bovines. A large cassette is required to allow the study of entire metacarpal bone including the proximal and distal joints. Usually a cassette of 12" × 15" or 14" × 17" size is used. The radiographs are obtained with the animal in a weight bearing position. Grid is not required.

Dorsopalmar View (DPa)

Position: The animal should stand in a comfortable position and the cassette in a holder is held against the palmar surface of the metacarpus (Fig. 13-9).
Beam centre: The X-ray beam is directed horizontally at right angle to the cassette and centered at the mid metacarpal region or at the point of interest. The beam should be collimated and a large sized film used to include both proximal and distal joints.

Lateromedial View (LM)

Position: The animal should stand 'normal' and the cassette in a holder is placed against the medial aspect of the metacarpal region (Fig. 13-10).
Beam centre: The X-ray beam is directed horizontally parallel to the ground and centered at the midpoint of the lateral surface of the metacarpus. The beam should be collimated and a large sized film used to include both the carpal and fetlock joints.

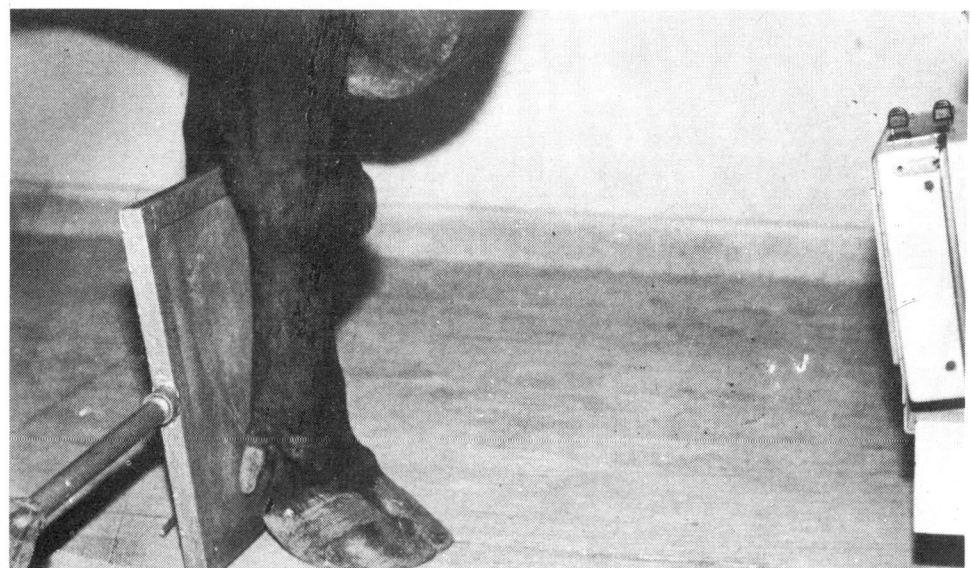

Fig. 13-9 : Buffalo. Position for dorsopalmar view of metacarpus.

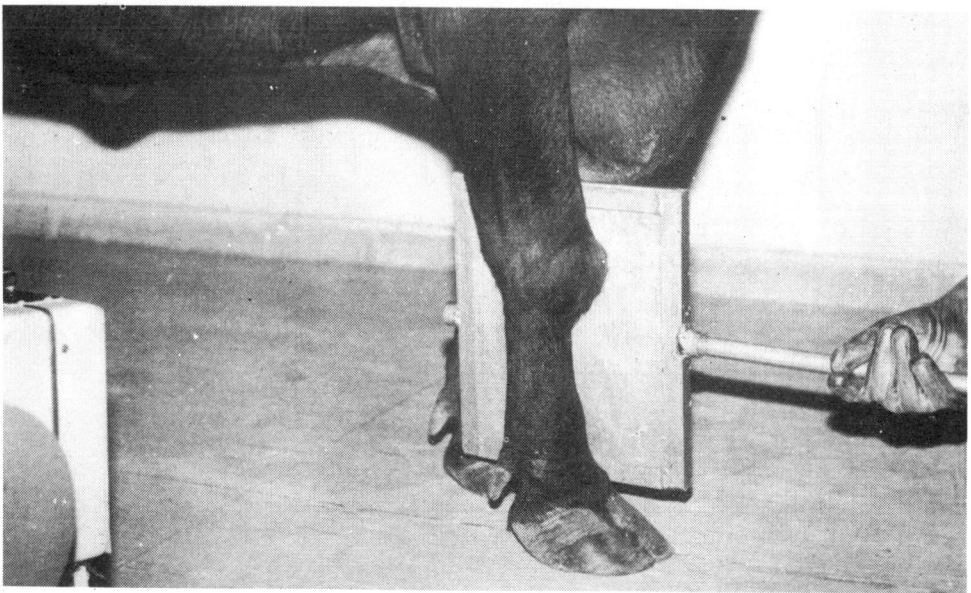

Fig. 13-10 : Buffalo. Position for lateromedial view of metacarpus.

THE FETLOCK JOINT

The fetlock joint includes the metacarpal/metatarsal-phalangeal area and the proximal sesamoids. In routine, dorsopalmar/dorsoplantar and lateromedial or mediolateral views are made. A flexed lateromedial view can also be obtained for better visualisation of the articular surface. Grid is not required.

Dorsopalmar/Dorsoplantar (DPa/DPl) View

Position: The animal should stand 'normal' with the foot rested on a wooden block. The cassette in a holder is placed against palmar/plantar aspect of the fetlock (Fig. 13-11).

Fig. 13-11 : Buffalo. Position for dorsopalmar view of fetlock joint.

Beam centre: The X-ray beam is directed horizontally at right angle to the cassette and centered on the dorsal surface of the fetlock joint.

Lateromedial View (LM)

Position: The animal should stand 'normal' with the foot rested on a wooden block. A cassette in a holder, which rests on the ground, is held against the medial aspect of the fetlock joint (Fig. 13-12).

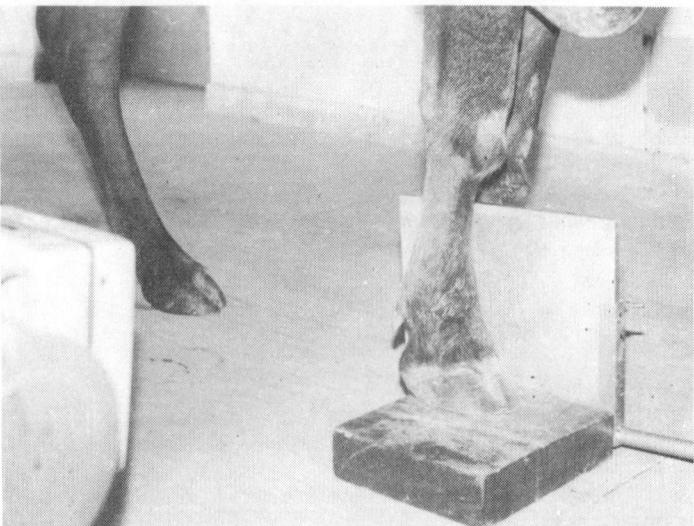

Fig. 13-12 : Buffalo. Position for lateromedial view of fetlock joint.

Beam centre: The X-ray beam is directed parallel to the ground and at right angle to the cassette. The beam is centered on the lateral surface of the fetlock joint.

THE DIGITS

Radiographic study of the digital region is the most common radiographic procedure in bovines. Radiography of digits is relatively easy to do with minimal errors in positioning. Dorsopalmar/ dorsoplantar and lateromedial views are usually obtained, however, sometimes oblique view is also made. Cleanliness of the foot is very important to ensure a good quality radiograph. Wooden blocks are useful for positioning of the digits. A grid is not required.

i) FIRST AND SECOND PHALANGES

Dorsopalmar View (DPa)

Position: In a 'normal' standing animal, the foot to be radiographed is positioned on a wooden block with a cassette slot. The block elevates the foot sufficiently to direct the X-ray beam parallel to the ground. It also allows extension of the cassette distal to the third phalanx. The positioning is almost similar to that for the fetlock joint. If the block is not available, the foot can be rested on the ground provided that the X-ray tube can be postioned that low. It is important to ensure that the foot is placed as straight as possible and is not deviated. The cassette is placed against the palmar or plantar surface of the digit.

Beam centre: The X-ray beam is directed in a mid sagittal plane at right angle to the cassette from dorsal aspect of the foot. The central beam is parallel to the ground and is centered at the proximal interphalangeal joint or at the area of interest.

Lateromedial View (LM)

Position: The animal stands 'normal' with the foot to be radiographed positioned on a wooden block preferably with a cassette slot. The foot can also be rested on the ground, if the X-ray tube can be dropped that low. For a true LM view, the foot should be as straight as possible. The cassette is held vertically on the medial aspect of the digit.

Beam centre: The X-ray beam is directed from the lateral side at right angle to the face of the cassette. The beam should be parallel to the ground and centered at the proximal interphalangeal joint or at the area of interest.

ii) THE THIRD PHALANX

Dorsopalmar View (DPa)

Position: The animal stands in a 'normal' position. The foot is then placed on a weight bearing cassette holder with a cassette inside (Fig. 13-13).

Beam centre: The X-ray beam is directed at an angle of approximately 45° to the ground from the dorsal aspect of the foot. The beam is centered at or just below the coronary band.

THE HIP JOINT

Radiographic examination of the pelvis of adult cattle and buffaloes is rarely done. The limiting factors include the size of the pelvis of an adult animal, difficulties in positioning of the animal in dorsal recumbency, positioning of the cassette and grid, centering of the X-ray beam on the grid correctly and possibility of motion of the animal during a long exposure. Above all, a powerful X-ray machine is necessary to penetrate the massive tissue thickness of the area. The udder in dairy cows and buffaloes further increases the tissue thickness. Even if all these limitations could be overcome, the extent of scatter radiation may not permit a diagnostically useful radiograph. However, a radiograph of the pelvis of a young animal can be obtained. A grid should always be used.

Fig. 13-13 : Buffalo. Position for dorsopalmar view of third phalanx using a weight bearing cassette.

Ventrodorsal (VD) View

Position: A deeply sedated animal is placed in dorsal recumbency with both the hind limbs extended laterally as far as possible (Fig. 13-14). The cassette placed inside a weight bearing cassette holder is positioned beneath the pelvic region.

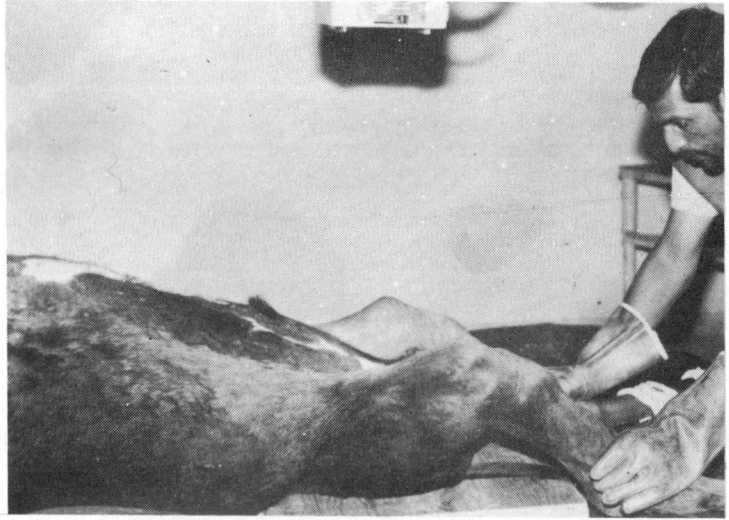

Fig. 13-14 : Calf. Position for ventrodorsal view of hip joints.

Beam centre: The X-ray beam is directed vertically at the centre of the cassette and at right angle to it. The beam should be collimated and a large sized film used to include the entire pelvic region.

Lateral View

Position: The animal is deeply sedated and restrained in left or right lateral recumbency. The cassette is placed beneath the pelvic area with the lower limb pulled cranially (Fig. 13-15).

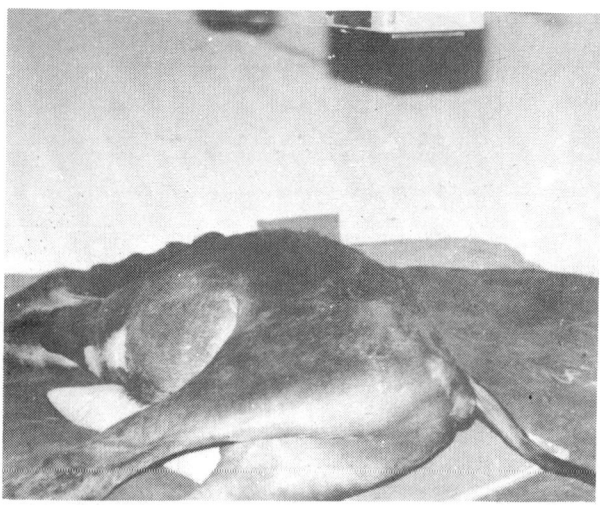

Fig. 13-15 : Calf. Position for lateral view of hip joints.

Beam centre: The X-ray beam is directed vertically at right angle to the cassette and centered at the centre of the cassette. The beam should be collimated and a lage sized film used to include the entire pelvic region.

THE FEMUR

Mediolateral View (ML)

Position: The animal is placed in lateral recumbency with the limb to be radiographed lowermost. The upper limb is drawn away from the area of the primary beam. The cassette is placed beneath the femoral region.

Beam centre: The X-ray beam is directed vertically and centered at the mid femoral region or at the area of interest. The primary beam should be slightly angled from distal to proximal direction. The beam should be collimated and a large sized film used to include proximal and distal joints. A grid should be used.

THE STIFLE JOINT

Radiographic study of the stifle joint is relatively not a routine procedure in adult cattle and buffaloes. Diagnostic lateromedial and caudocranial projections can be obtained even in standing animal, if correct positioning is done. However, in caudocranial view thickness of the tissues of the region does not allow clear visualisation of the area proximal to the joint space. Therefore, in such cases it is advisable to reduce the focal film distance for better penetration of the tissue thickness. A grid should also be used.

Caudocranial View (CdCr)

Position: The animal stands 'normal' with the X-ray tube positioned caudal to the stifle joint. The cassette in a holder is placed on the cranial aspect of the joint (Fig. 13-16). The cassette should be pushed firmly in a proximal and medial direction to ensure inclusion of the entire stifle joint.

Beam centre: The X-ray beam is directed in a mid-sagittal plane of the limb in a caudocranial direction parallel to the ground and centered at the stifle joint or at the readily palpable patella.

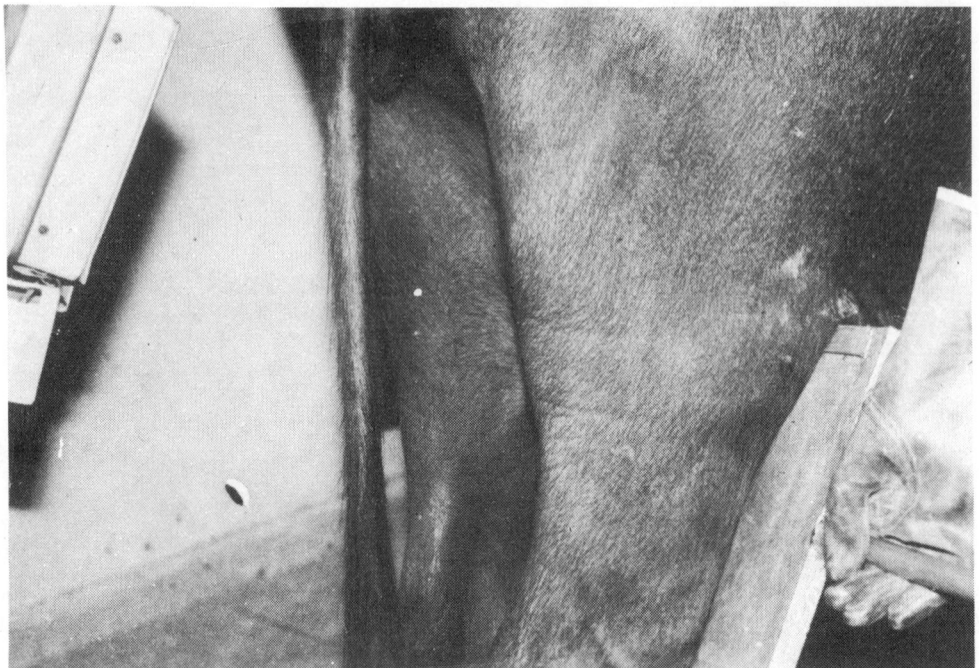

Fig. 13-16 : Buffalo. Position for caudocranial view of stifle joint. The cassette should be pushed as proximally as possible.

Lateromedial View (LM)

Position: The animal stands in a 'normal' position. The cassette in a holder is placed medial to the stifle joint (Fig. 13-17) and pushed as far proximally as possible between the limb and abdominal wall. In the cow and she buffalo, large udder poses difficulties in positioning the cassette high enough. In such cases, often the femoral condyles and patella are not visualised fully. The area being sensitive, the animal should be sedated to facilitate positioning of the cassette.

In noncooperative animals, the mediolateral view may be made with the animal in a lateral recumbent position and with the limb to be examined in the lowermost position. The upper limb is flexed and kept away from the area of the exposure.

Beam centre: In the standing animal, the X-ray beam is directed towards the cassette at right angle, the central beam being parallel to the ground. The beam should be centered on the readily palpable joint space. In the recumbent animal, the X-ray beam is directed vertically and centered on the medial surface of the stifle joint.

THE TIBIA

Caudocranial View (CdCr)

Position: The animal stands 'normal' and the cassette in a holder is placed against the cranial aspect of the tibia (Fig. 13-18). The cassette should be large enough to include the entire tibia alongwith distal and proximal joints.

Beam centre: The X-ray beam is directed horizontally from the caudal aspect of the tibia at right angle to the cassette and centered at the mid tibial region. The beam should be collimated and a large sized film used to include both the stifle and tarsal joints. A grid should be used.

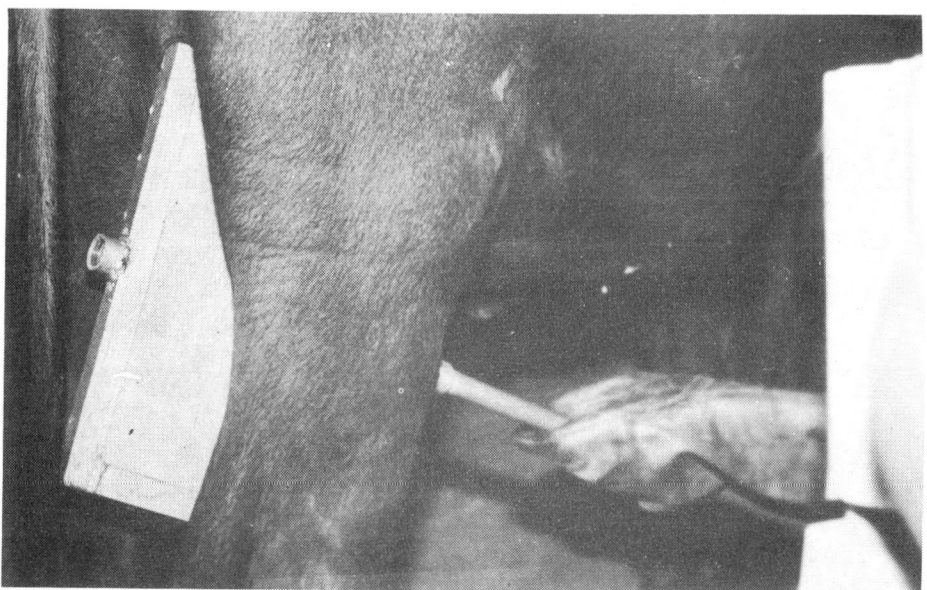

Fig. 13-17 : Buffalo. Position for lateromedial view of stifle joint. The cassette should be pushed as proximally as possible.

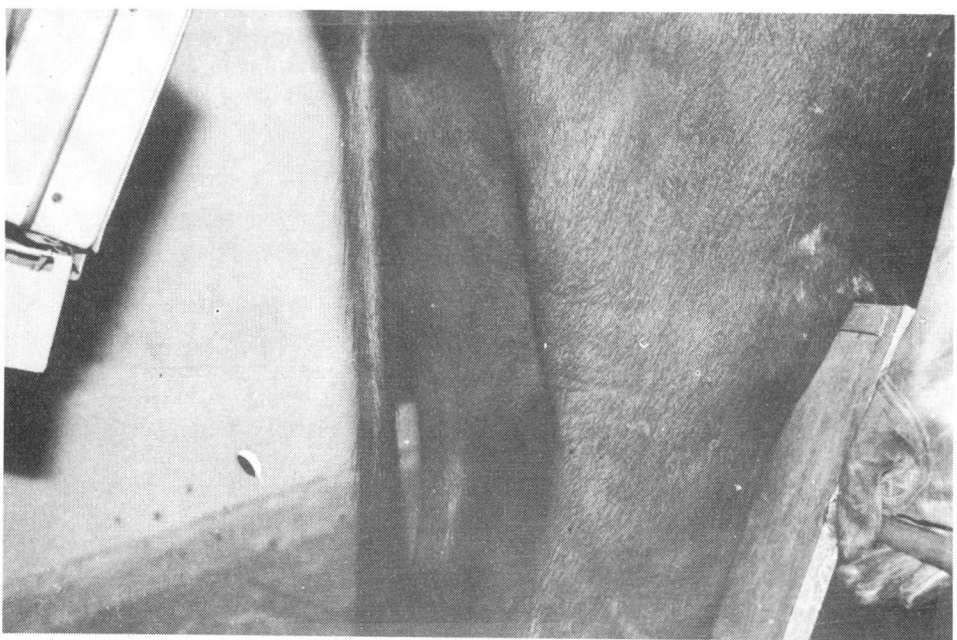

Fig. 13-18 : Buffalo. Position for caudocranial view of tibia.

Lateromedial View (LM)

Position: The animal stands 'normal'. The cassette in a holder with adjustable leg is held against the medial aspect of the tibia (Fig. 13-19).

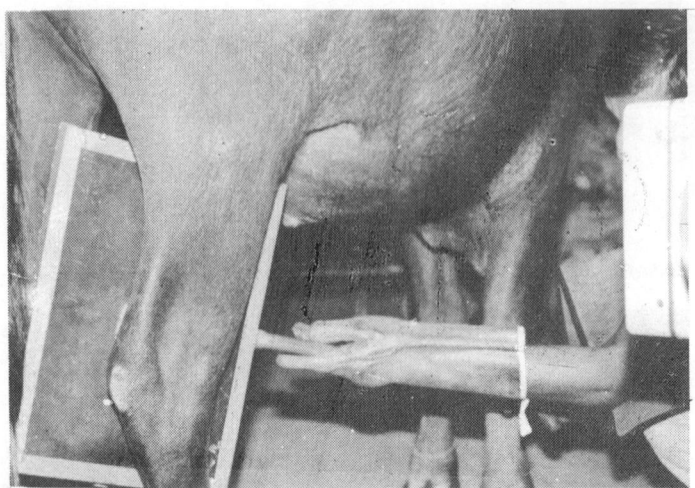

Fig. 13-19 : Buffalo. Position for lateromedial view of tibia.

Beam centre: The X-ray beam is directed horizontally and centered at the lateral aspect of the mid tibial region. The beam should be collimated and a large sized film used to include both the stifle and tarsal joints. A grid is not used.

THE TARSAL JOINT

To visualise the tarsal joint, dorsoplantar and lateromedial views are usually made in routine clinical practice. A cassette holder with a adjustable leg is of great value in positioning the cassette and in eliminating risk of motion of the cassette. A grid may be used in heavy adult animals especially if the joint is grossly swollen.

Dorsoplantar View (DPl)

Position: The animal stands 'normal' and the cassette in a holder with adjustable leg is positioned against the plantar aspect of the tarsal joint (Fig. 13-20). The cassette should be parallel to the calcaneous.

Beam centre: The beam is directed in a sagittal plane parallel to the ground and centered at the mid point of the dorsal surface of the tarsal joint.

Lateromedial View (LM)

Position: In a standing animal, the cassette in a holder is placed against the medial aspect of the tarsal joint (Fig. 13-21).

Beam centre: The X-ray beam is directed parallel to the ground at the cassette and centered on the lateral sufrace of the tarsal joint.

THE METATARSUS

Radiography of the metatarsus is done essentially in the same manner as of the metacarpus. It is advisable to include both the distal and proximal joints in the radiograph and if not feasible then atleast one joint must be included for the purpose of orientation.

THE VERTEBRAL COLUMN

Most radiographic projections of the vertebral column are made in a standing animal without any special preparation or devices. The radiographic examination of the vertebral column is limited to only lateral views except in the cervical region, where ventrodorsal view can be made on young calves. However, ventrodorsal views are of limited value because of the low tissue desnsity of the vertebrae and the large amount of overlying tissues (mediastinum and sternum in thoracic region, gas and ingesta in lumbar region). Even on lateral view, abdominal viscera may superimpose the vertebrae and results in poor quality radiograph. However, in immature young calves, such problems are not observed. Difficulties are also likely to arise in directing the central beam on the cassette because the X-ray tube is positioned on one side of the animal while the cassette is on the other side. A grid is usually not required for the cervical region but is essential for the thoracolumbar vertebrae. The use of pneumoperitoneum as a negative contrast enhances the visualisation of the lumbar vertebral column especially in young animals.

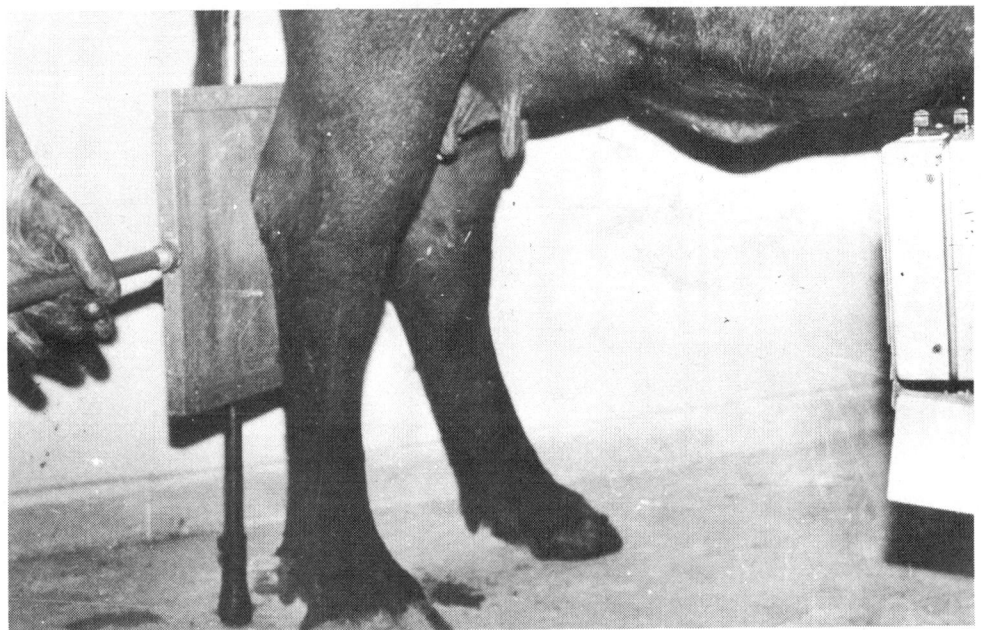

Fig. 13-20 : Buffalo. Position for dorsoplantar view of tarsal joint.

i) The cervical vertebrae (Lateral view)

Position: The animal should be in 'normal' standing position. The cassette preferably in a holder is placed on the lateral surface of the cervical region to be radiographed. For routine survey examination, atleast three overlapping radiographs are required. Additional radiographic studies may be made with the neck in extended or flexed position. In the latter case, the animal should be deeply sedated and controlled in a lateral recumbent position. Position for cervical vertebral radiography in a recumbent animal is shown in Fig. 13-22.

Beam centre: In a standing animal, the X-ray beam is directed horizontally at right angle to the cassette and centered at the cervical region of interest. For survey examination, the beam is centered at C_2, C_4 and C_6 to produce three overlapping radiographs. If the animal is in lateral recumbency the beam is directed vertically.

ii) The thoracic vertebrae (Lateral view)

Position: The animal stands 'normal' and the cassette in a holder is placed horizontally on the lateral surface of the thoracic vertebral region (Fig. 13-23). It is not however, always possible to place the cassette in close contact with the vertebral column and, therefore, some distortion of the radiographic image may occur. Alternatively an assistant has to hold the cassette in position and in such a case assistant's hands must be kept outside the primary X-ray beam.

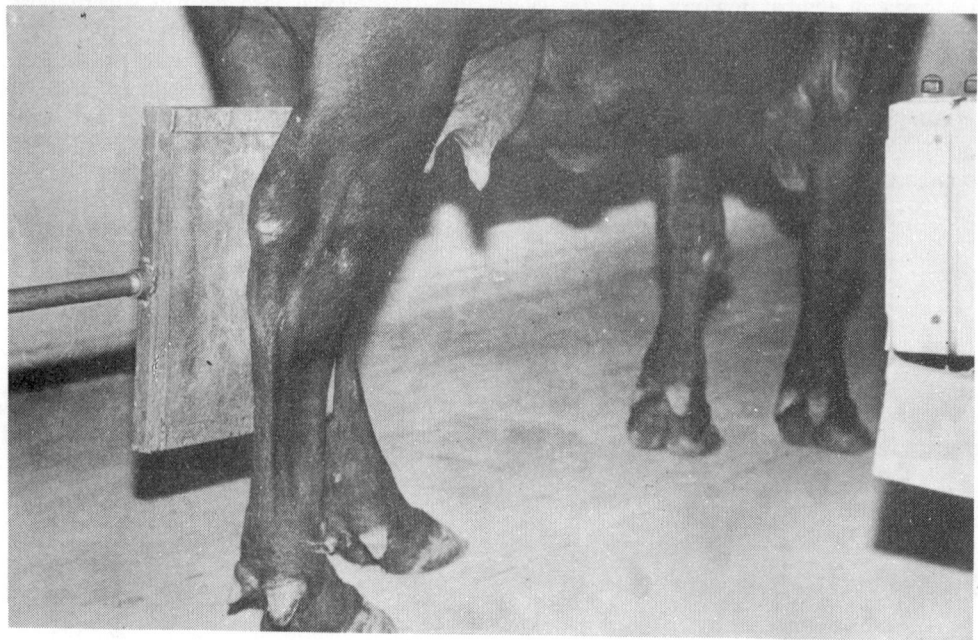

Fig. 13-21 : Buffalo. Position for lateromedial view of tarsal joint.

Beam centre: The X-ray beam is directed horizontally at right angle to the cassette and centered at the area of interest on the lateral surface opposite to the cassette. A grid should be used.

iii) The lumbar vertebrae (Lateral view)

Position: The animal stands 'normal' and the cassette in a holder is placed horizontally on the lateral surface of the lumbar region (Fig. 13-24).

Beam centre: The X-ray beam is directed horizontally at right angle to the cassette and centered at the midpoint of the lumbar vertebral region on the lateral surface opposite to the cassette. A grid should be used.

iv) The coccygeal vertebrae (Dorsoventral view)

Position: The animal stands 'normal' and the cassette preferably in a holder is placed beneath the extended tail in the area of interest (Fig. 13-25).

Beam centre: The X-ray beam is directed vertically from the dorsal aspect of the tail in a mid sagittal plane and is centered at the area of interest or at the centre of the cassette. A grid is not required.

THE SKULL

Radiography of the head of cattle and buffaloes presents greater difficulties than that of the horse and small ruminants. Because of the size of the structure, it may not be possible to reproduce whole area of the bovine skull on a single film. Thickness of the skull and density of the dental structures interfere with clearcut visualisation of a considerable area. Moreover, because of variations

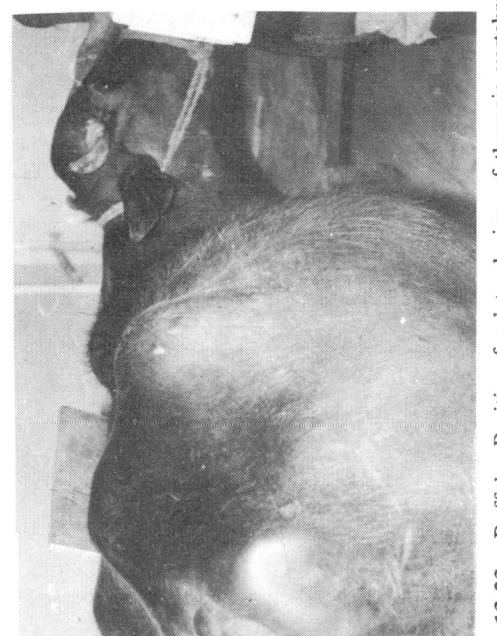

Fig. 13-23 : Buffalo. Position for lateral view of thoracic vertebrae.

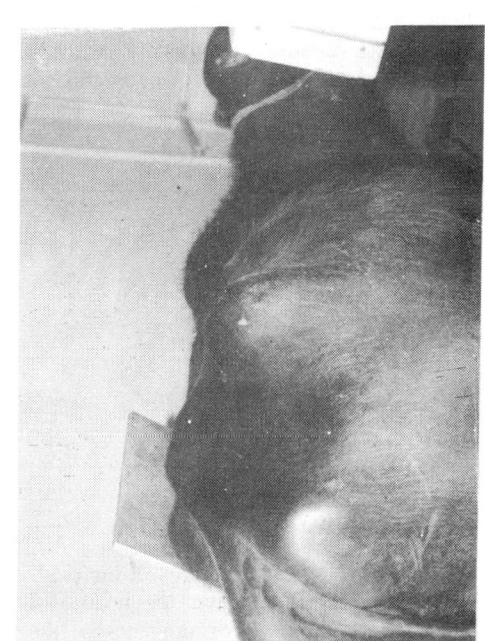

Fig. 13-24 : Buffalo. Position for lateral view of lumbar vertebrae.

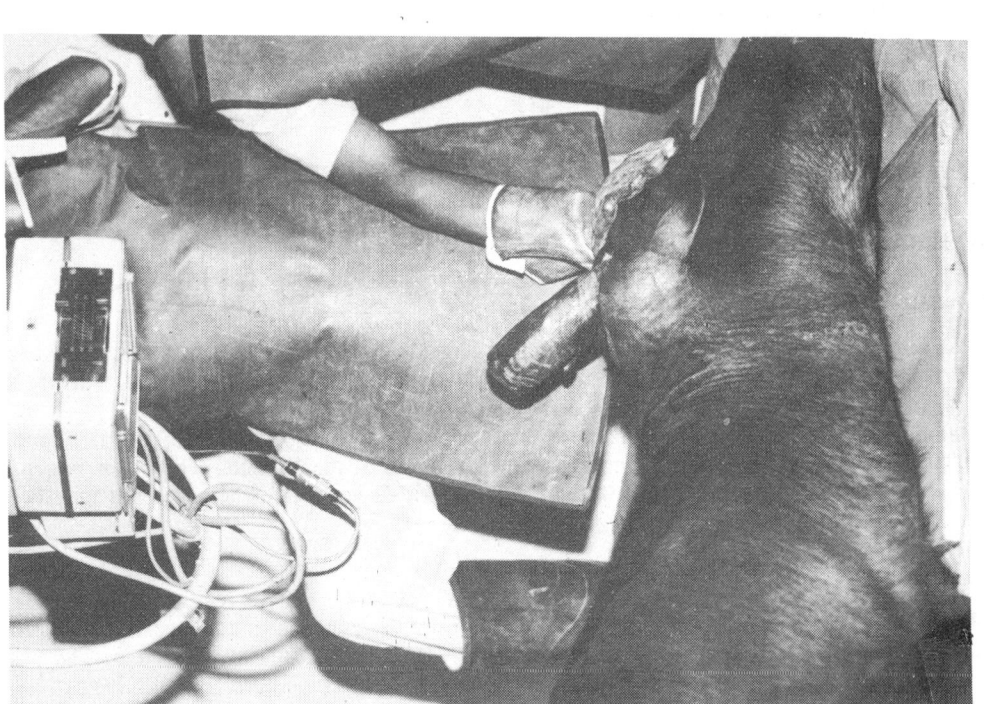

Fig. 13-22 : Buffalo. Position for lateral view of the cervical region.

in the densities of various structures of the head region, variable exposure factors are required for different structures. Above all, the horns do not allow perfect positioning of the cassette for lateral projections of the skull. Dorsoventral or ventrodorsal view may be of greater diagnostic value but is difficult to make because of thickness of the tissue of the area and difficulties of positioning of the adult animal and the cassette. However, this view can be made on young or medium sized animals.

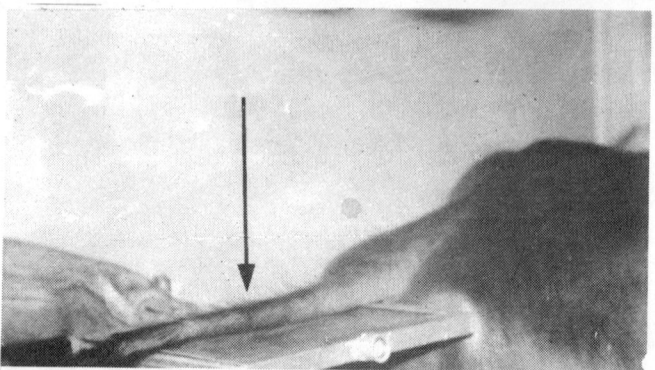

Fig. 13-25 : Buffalo. Position for dorsoventral view of coccygeal vertebrae. Arrow indicates beam direction.

Most radiographic examinations of the skull can be made on standing animal. However, difficulties may be experienced in keeping the head still and of positioning the cassette, apart from risk of radiation to the assistants. For this reason, the animal should suitably be controlled within a stock and tranquilised if necessary. It is also advisable to cover the eyes of the animal if the cassette is to be placed in the near vicinity. The cassette holder should be used as far as possible.

For practical reasons, the radiography of the bovine skull can be divided as follows:

    A. The mandible.
    B. The turbinates.
    C. The frontal bone/sinuses.
    D. The cranium.

A. MANDIBLE

i) ROSTRAL PART

Dorsoventral View (DV)

Position: The animal stands 'normal' with the head lowered as far as possible. The cassette is placed beneath the rostral part of the mandible in close contact with the ventral surface of both rami (Fig. 13-26).

Beam centre: The X-ray tube is positioned on the dorsal aspect of the rostral region of the head. The central beam is directed vertically at right angle to the cassette. The beam is centered on the dorsal midline of the head rostral to the orbits. A grid is not usually required but may be used to reduce scatter radiation.

Lateral view

Position: The animal should stand 'normal' with the head in near natural position. Alternatively, the animal may be controlled in lateral recumbency with the neck slightly extended. The cassette is placed adjacent to the lateral aspect of the rostral part of of the mandible (Fig. 13-27)

Beam centre: The X-ray beam is directed horizontally in standing animal and vertically in lateral recumbent animal at right angle to the face of the cassette from the opposite side and centered on the area of interest or on the lateral surface of the rostral part of the mandible. A grid is not required.

Intra-oral projection

Position: The animal is sedated and secured in a lateral recumbent position. The mouth is opened with a nose lead and the cassette in a holder is pushed into the animal's mouth as far as possible. The front of the cassette should face ventrally.

Beam centre: The X-ray beam is directed horizontally at the cassette from the ventral aspect of the head with the central beam at right angle to the cassette. A grid is not used. A ventrodorsal view is thus obtained.

ii) THE HORIZONTAL PART

Lateral View

Position: The animal stands 'normal' and the cassette is placed against the lateral surface of the mandible (Fig. 13-28). The cassette should be positioned according to the area of the mandible to be radiographed and should be parallel to the mid sagittal plane of the head.

Beam centre: The X-ray tube is positioned on the opposite side of the cassette and the beam is directed in horizontal plane at right angle to the cassette. The primary beam is centered at the midpoint of the horizontal or vertical ramus of the mandible. A grid should be used. Alternatively, the animal may be restrained in lateral recumbency with the side to be radiographed lowermost. The cassette is placed beneath the mandible and the X-ray beam is directed vertically at right angle to the cassette.

B. THE TURBINATES (LATERAL VIEW)

Position: The animal stands 'normal' and the cassette is held against the lateral surface of the turbinates region parallel to the mid-sagittal plane of the head (Fig. 13-29).

Beam centre: The X-ray tube is positioned on the opposite side. The X-ray beam is directed perpendicular to the face of the cassette and centered on the turbinate area. A grid is not used.

C. THE FRONTAL BONE/SINUSES (LATERAL VIEW)

Position: The animal stands 'normal' and the cassette is placed adjacent to the frontal region on lateral side of the head (Fig. 13-30).

Beam centre: The X-ray beam is directed horizontally towards the cassette from the opposite side in a plane parallel to the ground and centered on the area of frontal region (sinuses).

D. THE CRANIUM (LATERAL VIEW)

Position: The animal stands 'normal' and the cassette is placed adjacent to the lateral surface of the cranial region and held in a near vertical manner

Beam centre: The X-ray beam is directed in a plane parallel to the ground and as perpendicular to the face of the cassette as possible and centered on the cranial region. Grid should be used.

LARYNGOPHARYNGEAL REGION (LATERAL VIEW)

Position: The animal stands 'normal' and the cassette is placed against the lateral aspect of the laryngopharyngeal region (Fig. 13-31). Non-cooperative animals may be restrained in lateral recumbency and the cassette is placed beneath the laryngopharyngeal region.

Beam centre: The X-ray beam is directed at right angle to the cassette and centered on the laryngopharyngeal region. A grid is not required.

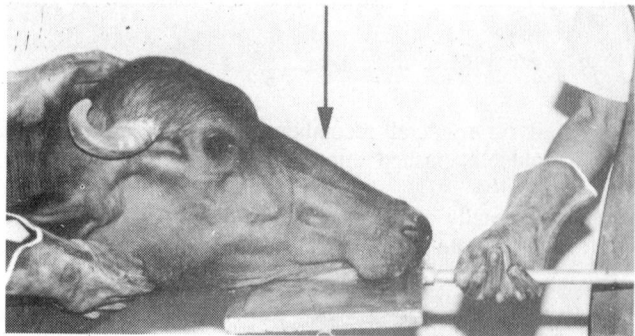

Fig. 13-26 : Buffalo. Position for dorsoventral view of rostral part of mandible.

Fig. 13-27 : Buffalo. Position for lateral view of rostral part of mandible.

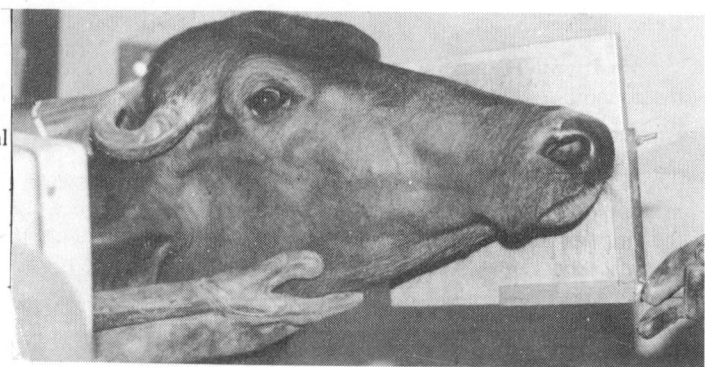

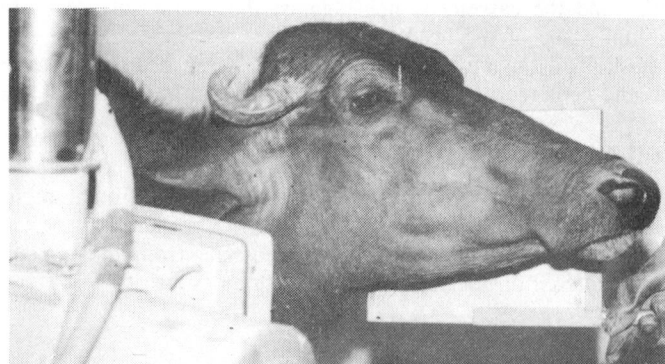

Fig. 13-28 : Buffalo. Position for lateral view of horizontal part of mandible.

Fig. 13-29 : Buffalo. Position for lateral view of turbinates.

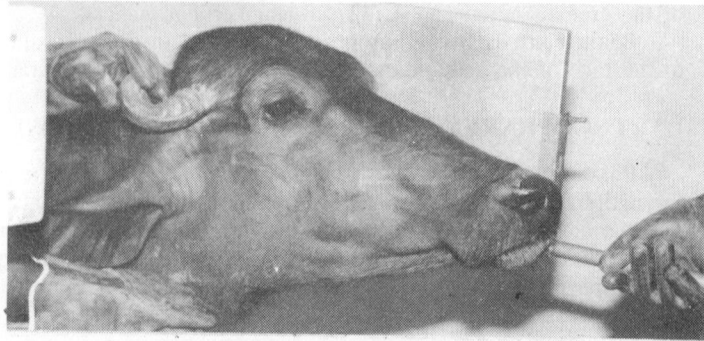

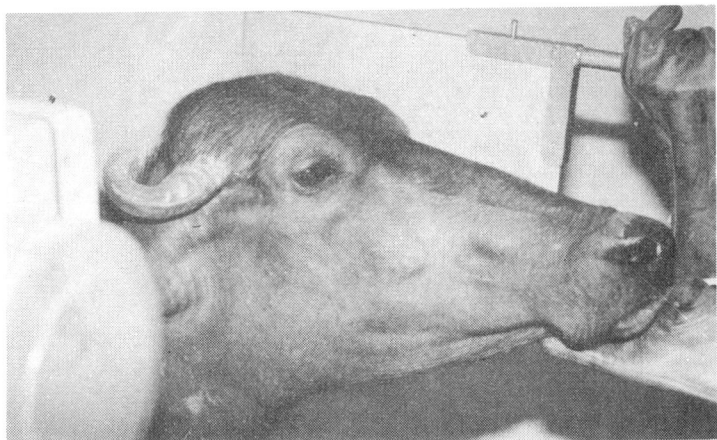

Fig. 13-30 : Buffalo. Position for lateral view of frontal bone/sinus.

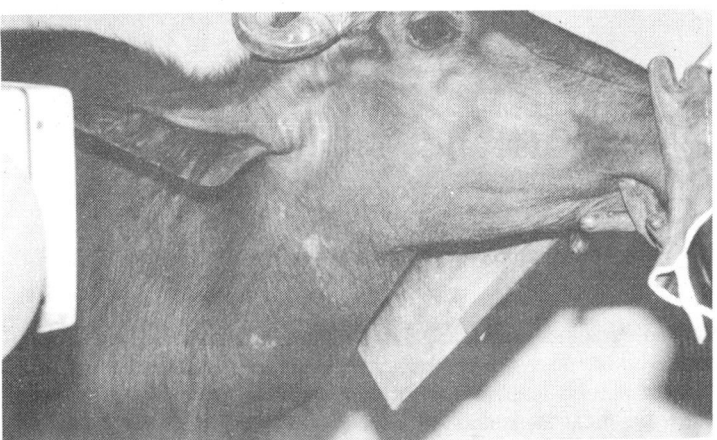

Fig. 13-31 : Buffalo. Position for lateral view of laryngopharyngeal region.

THE THORAX

Radiographic study of the thorax of cattle and buffaloes is a routine practice these days. This is partly because of the availability of more powerful X-ray machines and partly because of greater awareness about the diseases of the thorax. In young calf, it may be possible to obtain the entire lateral view of the throrax on a single large sized film but in adult bovines three to four overlapping radiographs would be necessary (Fig. 13-32). The areas to be radiographed include dorsocranial, dorsocaudal, ventrocranial and ventrocaudal lung fields. However, since little information can be obtained from the area cranial to the heart due to superimposition of cranioventral lung field by triceps musculature, only three radiographs are considered sufficient. The dorsocranial and dorsocaudal fields are examined with the cassette oriented horizontally (35 cm height) and a third radiograph is made of ventrocaudal area including central lung field with the cassette oriented vertically (42 cm height). Exposure factors should be reduced for radiography of the dorsocaudal field. In routine practice, a single lateral radiograph of the dorsocaudal or central lung field is made on a large sized cassette in the first instance because this is considered often sufficient in animals with extensive lung disease. Dorsoventral or ventrodorsal view is not feasible because of the thickness of the thorax of adult bovine.

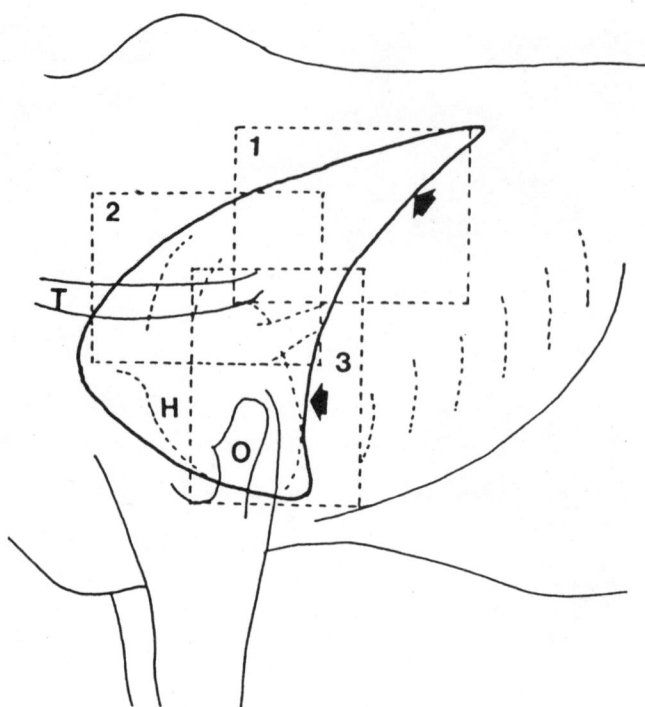

Fig. 13-32 : Placement of cassettes for lateral view of lungs 1, 2, 3 = Cassettes, T = Trachea, H = Heart, O = Olecranon. Arrows indicate line of diaphragm (Adapted from Bargai, et al., 1989).

In routine, the radiographic examination of the lung is done with the animal standing. However, some type of travis should be used to restraint the animal. The side bars of the travis should be adjustable so that these can be removed or lowered to avoid any hinderence in the field of exsposure. For the heart area, the animal is usually positioned in lateral recumbency.

The cassette should be placed on the side of the lung with suspected lesions. The thickness of the thorax is such that the lesions/area away from the film are greately magnified due to large object-film distance and may be blurred beyond recognition.

A. Lung (Lateral view)

Position: The animal should stand 'normal'. The cassette in a holder is positioned against the area of interest. The position is shown in Fig. 13-33.

Beam centre: The X-ray beam is directed horizontally from the opposite side of the animal and centered at the area of interest. The beam should be perpendicular to the face of the cassette. The beam should be centered on the tenth rib for the dorsocaudal, on the fifth for the central and on the sixth for the ventrocaudal lung fields.

B. Heart area (Lateral view)

Position: The animal is placed in the left lateral recumbency with the thoracic limb extended forward. The cassette is oriented vertically and is positioned beneath the animal in the cardiac region (Fig. 13-34). The cassette should be placed in such a way that its two-third part should be cranial to the sixth sternebra.

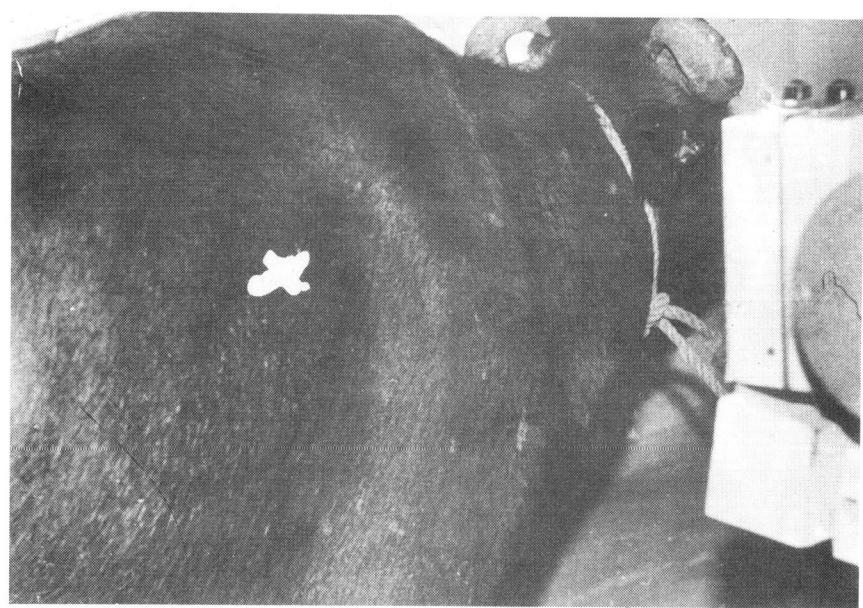

Fig. 13-33 : Buffalo. Position for lateral view of lungs. White mark indicates X-ray beam centre.

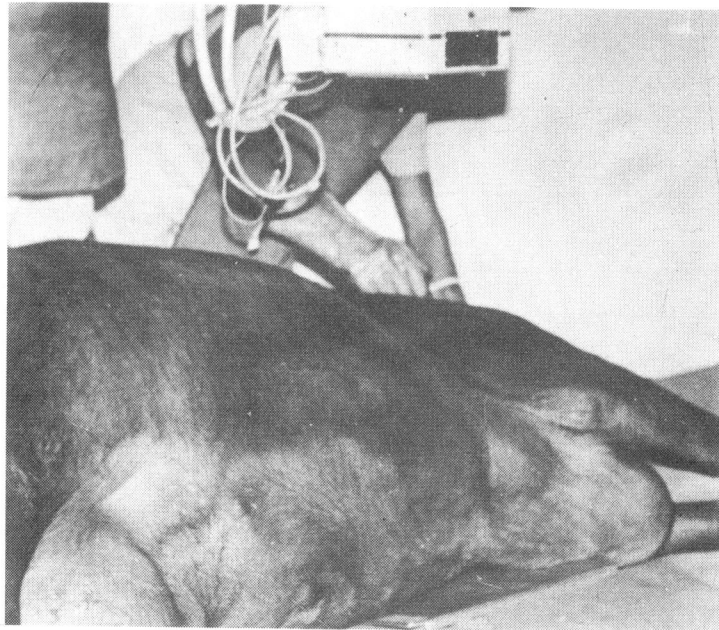

Fig. 13-34 : Buffalo. Position for lateral view of cardiac region.

Beam centre: The X-ray beam is directed vertically at right angle to the cassette and is centered at the fifth rib about 15 cm above the sternum. A grid should be used.

ABDOMEN

Radiographic examination of the bovine abdomen is mainly done for diagnosis of the disorders of the reticulum. Therefore, only the cranioventral area of the abdomen is radiographed. This area can be visualised by making a single lateral view on a large sized cassette (35 x 42 cm).

Position: Radiography of the carnioventral abdomen may be done with the animal either standing or in lateral recumbency (Fig. 13-35). Lateral view may also be made with the animal in the dorsal recumbent position (Fig. 13-36). The thoracic limbs should be pulled forward moderately. The cassette is oriented vertically and positioned on the left side of the animal. The cassette should be placed in such a way that its two-third part should be caudal to the sixth sternebra.

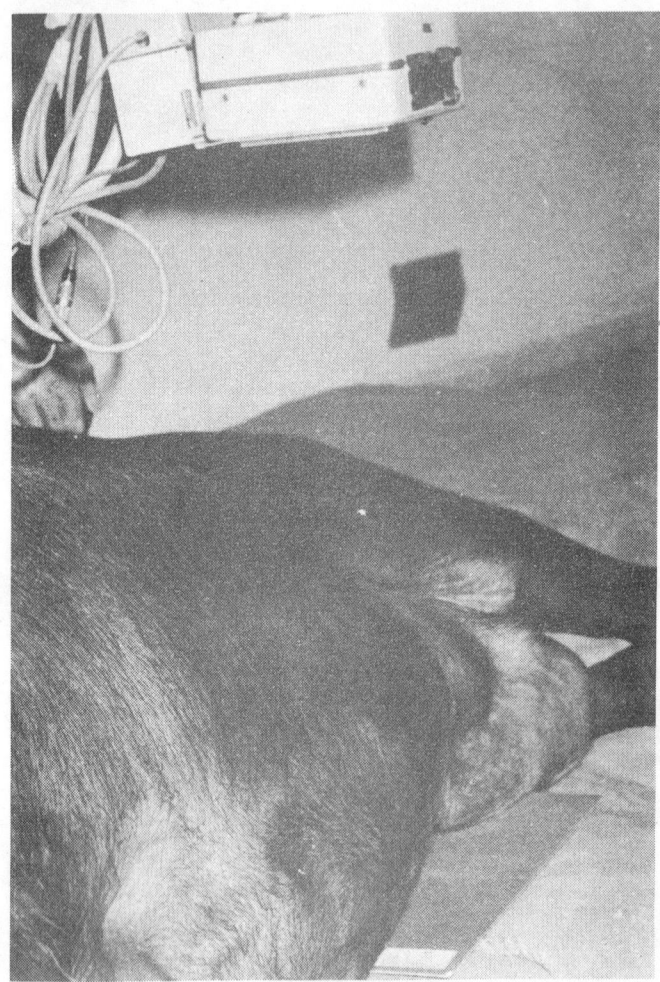

Fig. 13-35 : Buffalo. Position for lateral recumbent view of cranioventral abdomen.

Beam centre: The X-ray beam is directed horizontally at right angle to the face of the cassette in case the animal is standing or is in a dorsal recumbent position. If the animal is positioned in the lateral recumbency, the beam is directed vertically at the cassette. The beam is centered about 15 cm dorsal to the seventh sternebra. A grid should be used.

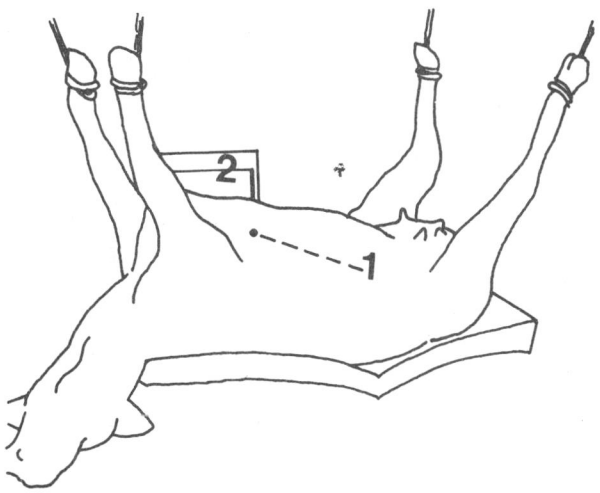

Fig. 13-36 : Dorsal recumbent position for lateral view of cranioventral abdomen and caudal pericardial region. 1 = Beam centre, 2 = Cassette (From Ducharme et al, J. Am. Vet. Med. Assoc. 182, 585, 1983. Used with permission).

THE CAMEL

Radiography of the camel follows the same protocol as that described for bovine with slight variations in some instances. In most studies, two projections are made at right angle to each other.

Radiography of the thoracic and pelvic limbs may be done with animal standing while sternal recumbent position is required for the examination of the head, vertebral column and thoracic region. Radiograph of the cranial abdomen may be obtained either with the animal standing or in lateral recumbency. Various radiographic positions for various parts are shown in figs 13-37 to 13-61.

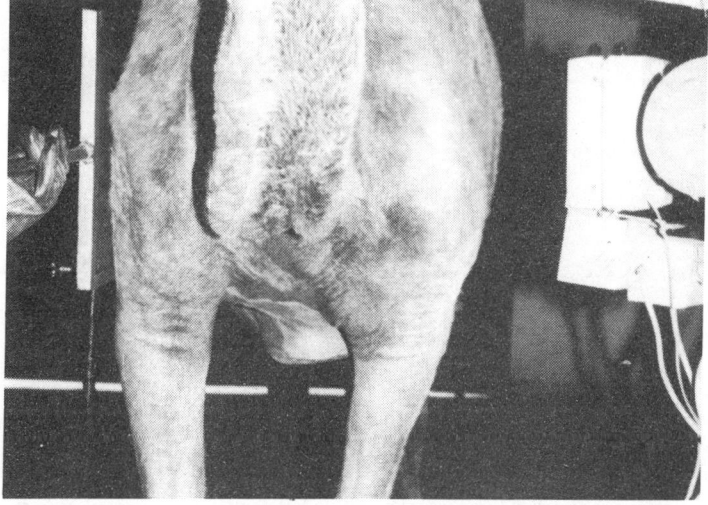

Fig. 13-37 : Camel. Position for mediolateral view of shoulder joint.

Unlike bovines, difficulty is not encountered to obtain a caudocranial view of the stifle joint since enough space is available between the joint and flank to push the cassette proximally. It allows clear visualisation of the parts proximal to the joint including the patella. However, metacrapus and metatarsus of the camel are much longer and hence it is difficult to include the entire bone and proximal and distal joints on the available large sized films. If the cassette is held obliquely against the bone, it may become possible to include the entire bone on a single film.

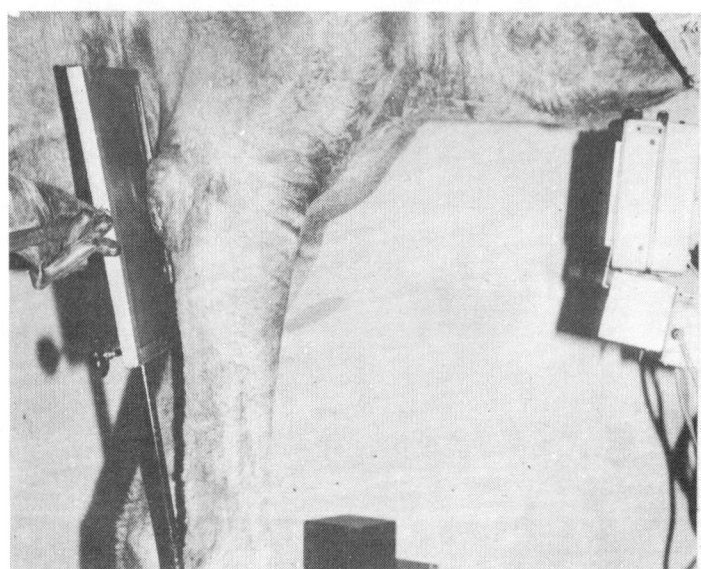

Fig. 13-38 : Camel. Position for craniocaudal view of elbow joint.

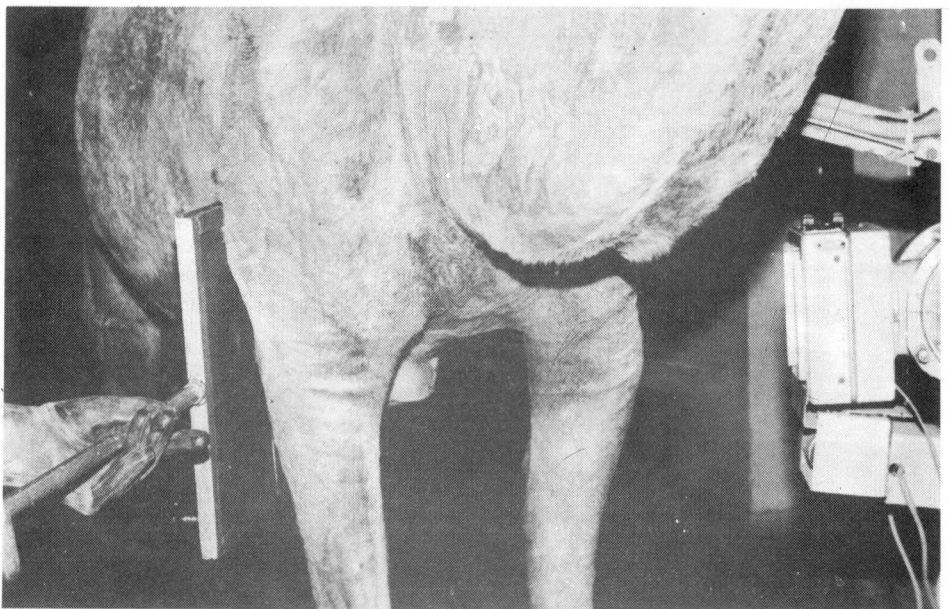

Fig. 13-39 : Camel. Position for mediolateral view of elbow joint.

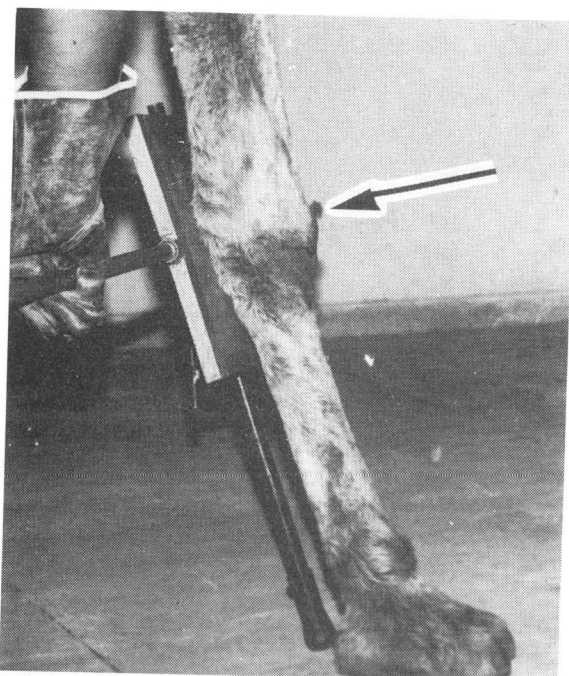

Fig. 13-40 : Camel. Position for dorsopalmar view of carpus. Note the slightly forward placement of the limb in comparison to bovine. Arrow indicates X-ray beam direction.

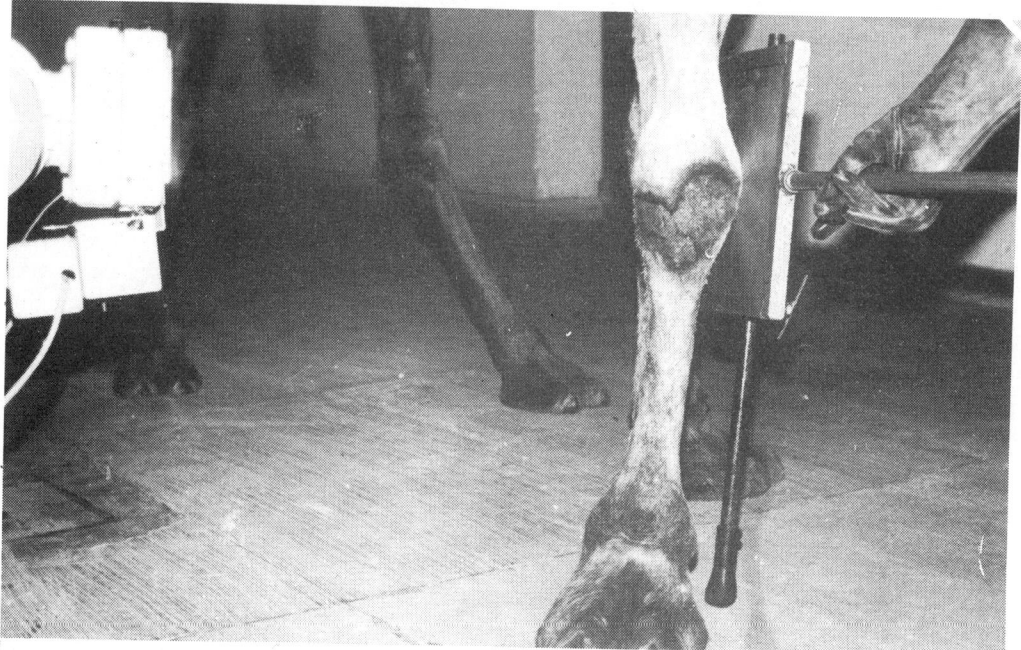

Fig. 13-41 : Camel. Position for lateromedial view of carpus.

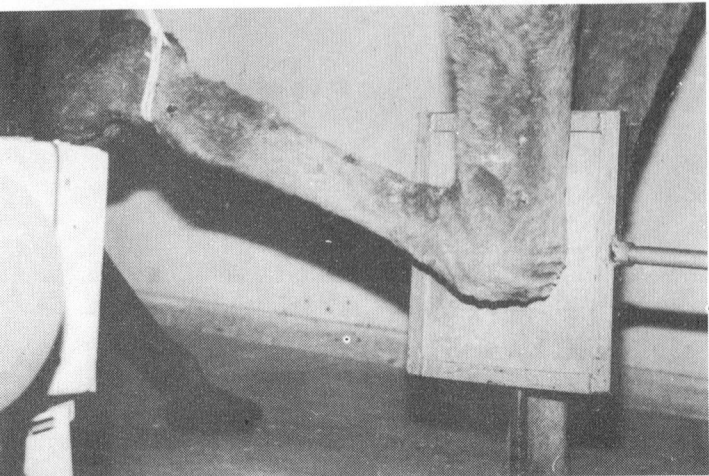

Fig. 13-42 : Camel. Position for lateromedial view of flexed carpus.

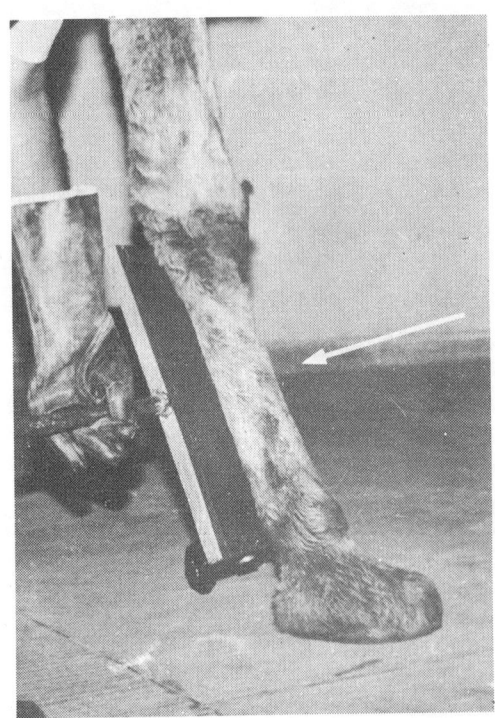

Fig. 13-43 : Camel. Position for dorsopalmar view of metacarpus. Note slightly forward placement of limb in comparison to bovine. Arrow indicates X-ray beam direction.

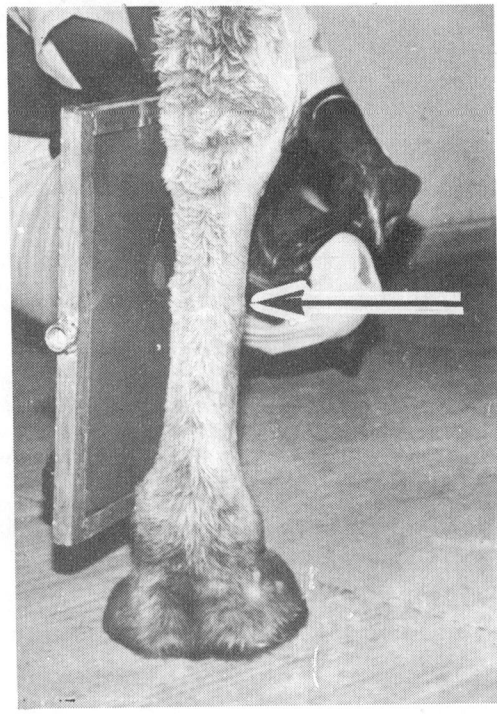

Fig. 13-44 : Camel. Position for mediolateral view of metacarpus. Arrow indicates X-ray beam direction.

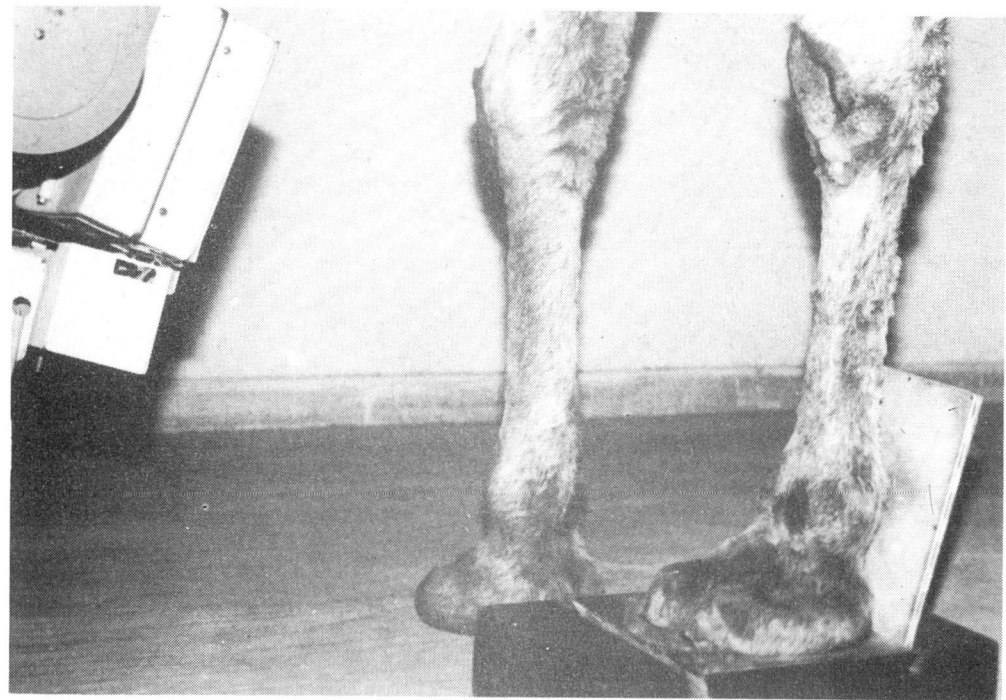

Fig. 13-45 : Camel. Position for dorsopalmar view of fetlock using wooden block with a slot for cassette.

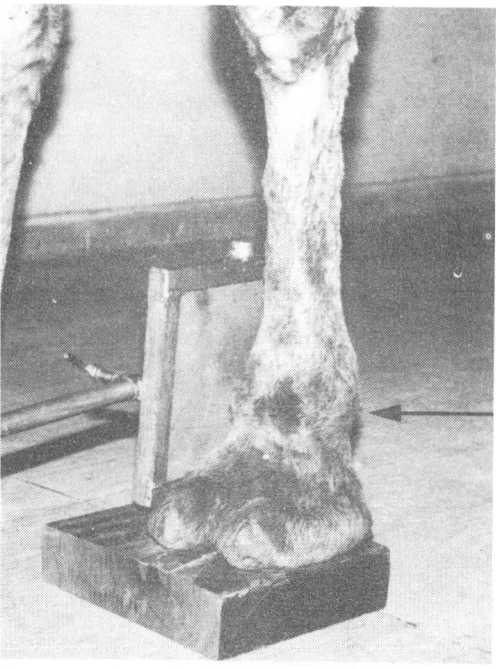

Fig. 13-46 : Camel. Position for lateromedial view of fetlock. Arrow indicates X-ray beam direction.

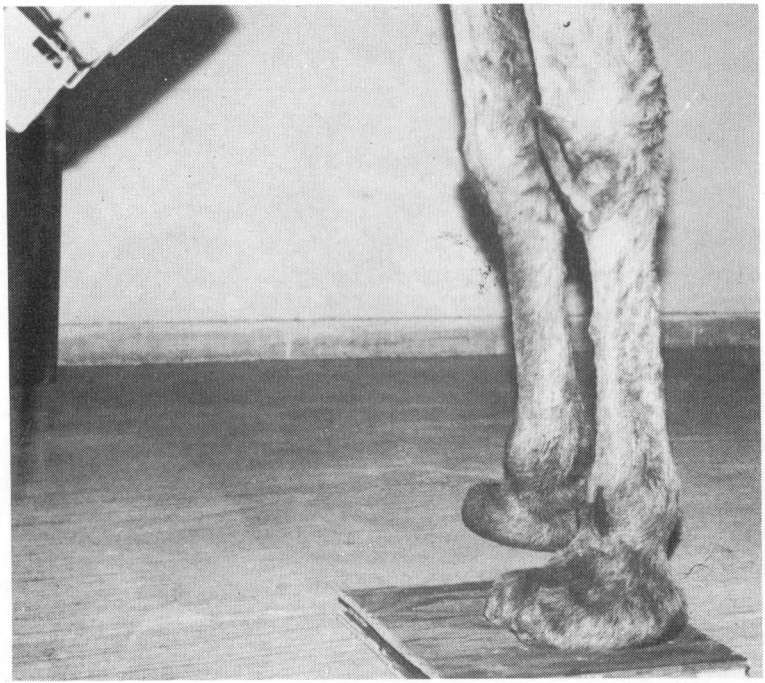

Fig. 13-47 : Camel. Position for dorsopalmar view of third phalanx using weight bearing cassette holder.

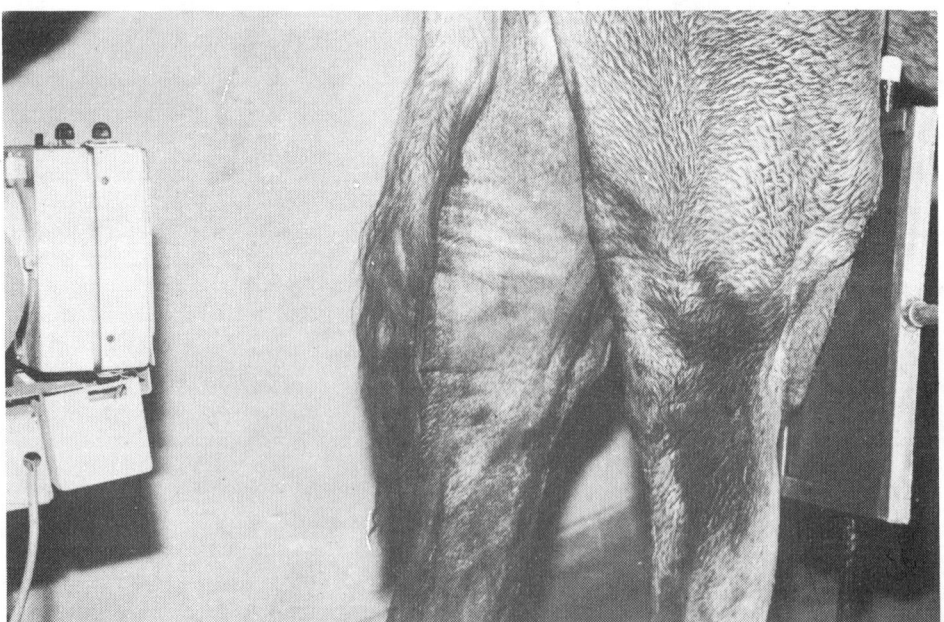

Fig. 13-48 : Camel. Position for caudocranial view of stifle joint.

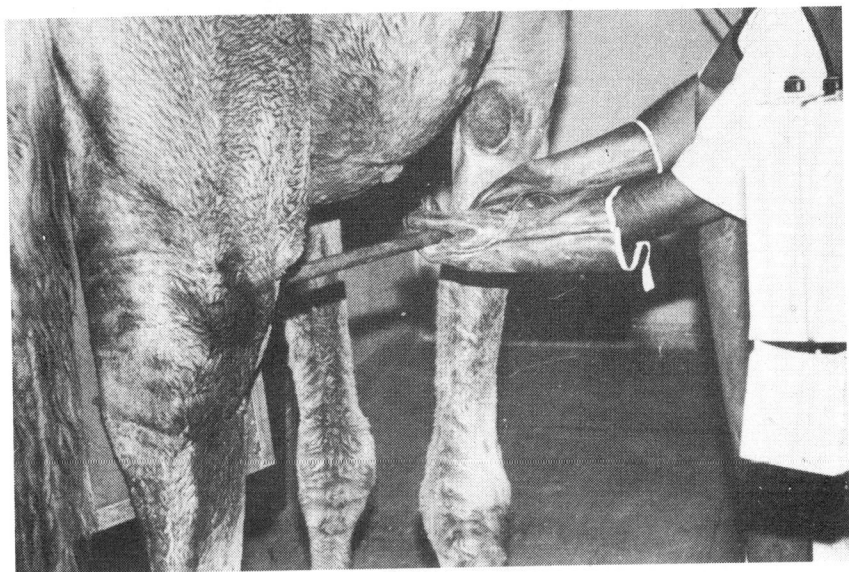

Fig. 13-49 : Camel. Position for lateromedial view of stifle joint. Note that in comparison to bovine (Fig. 13-17) cassette can be pushed easily proximally to cover entire area of interest.

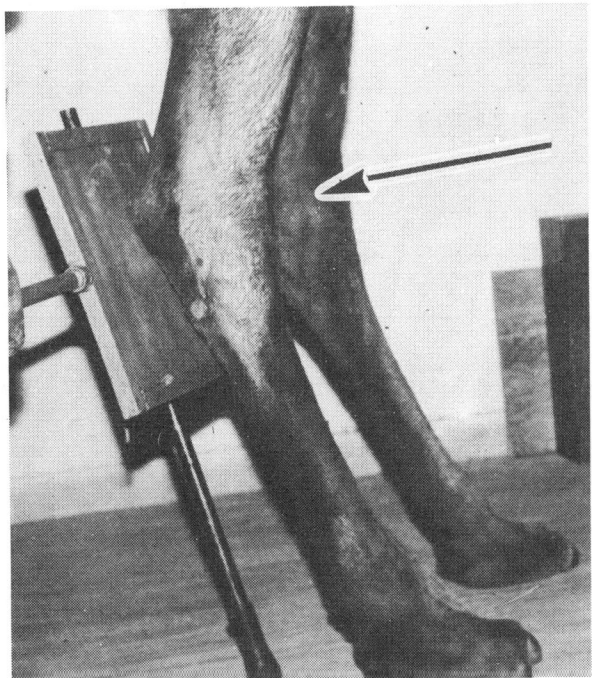

Fig. 13-50 : Camel. Position for dorsoplantar view of tarsus. Arrow indicates X-ray beam direction.

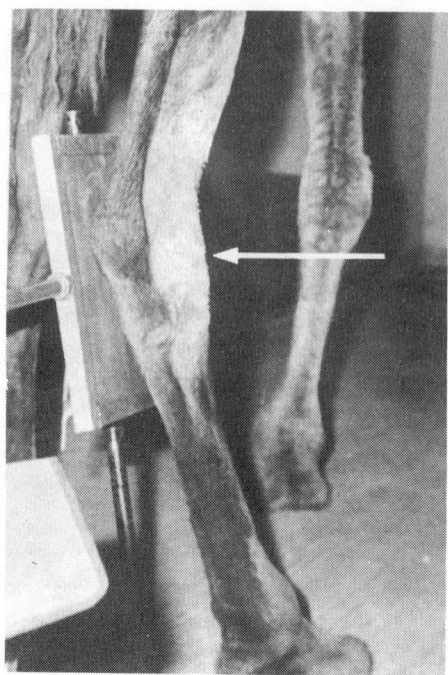

Fig. 13-51 : Camel. Position for lateromedial view of tarsus. Arrow indicates X-ray beam direction.

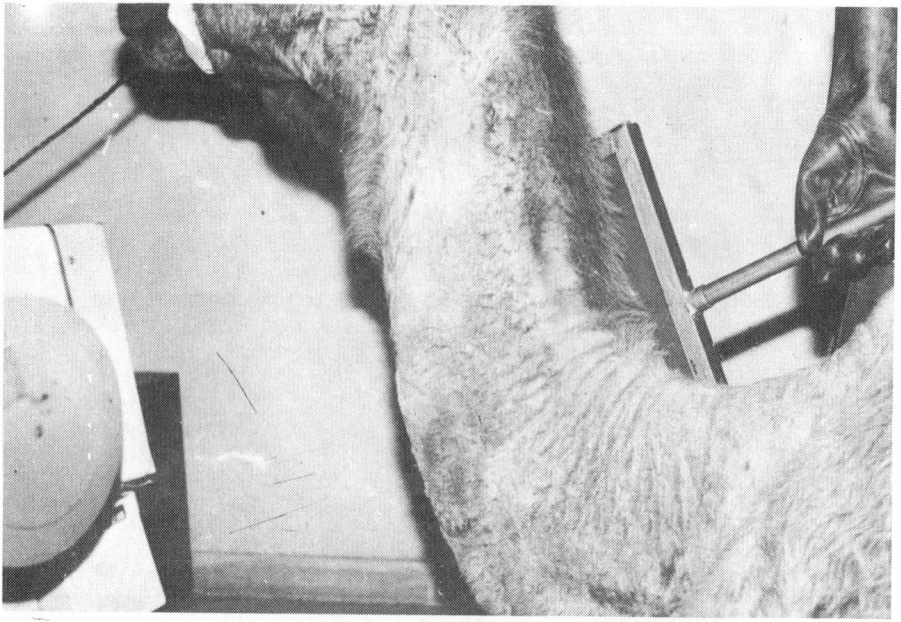

Fig. 13-52 : Camel. Position for lateral view of cervical vertebrae with animal in sternal recumbency.

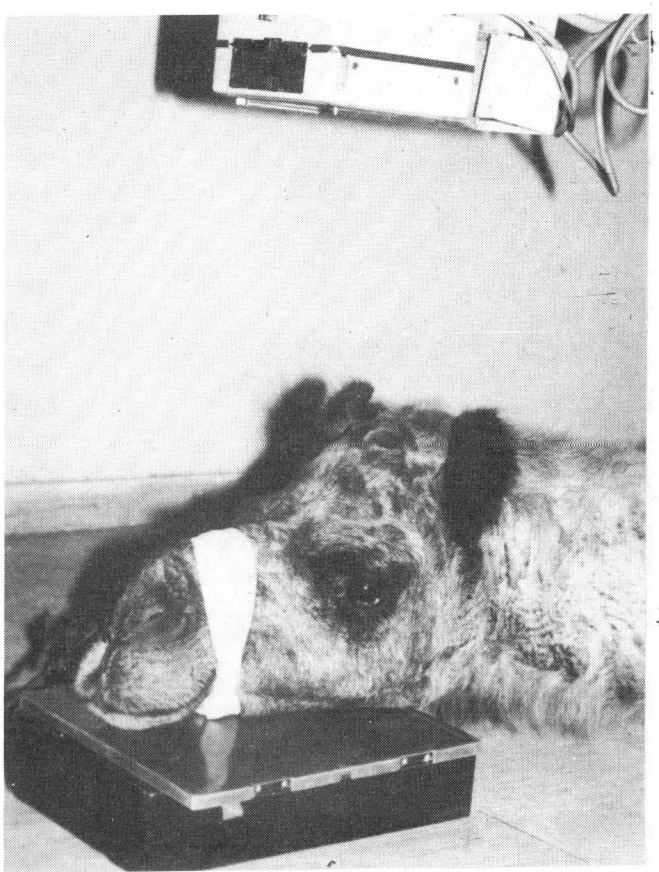

Fig. 13-53 : Camel. Position for dorsoventral view of rostral part of mandible with the cassette rested on a wooden block.

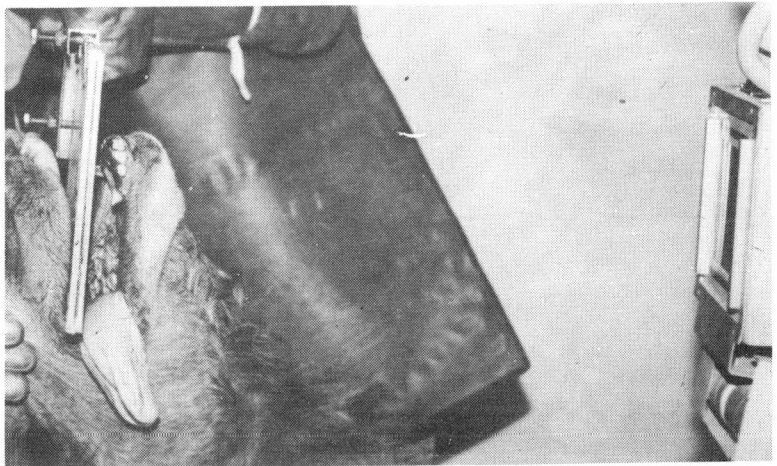

Fig. 13-54 : Camel. Position for ventrodorsal view of rostral part of mandible with intraoral position of cassette. The neck is held upright.

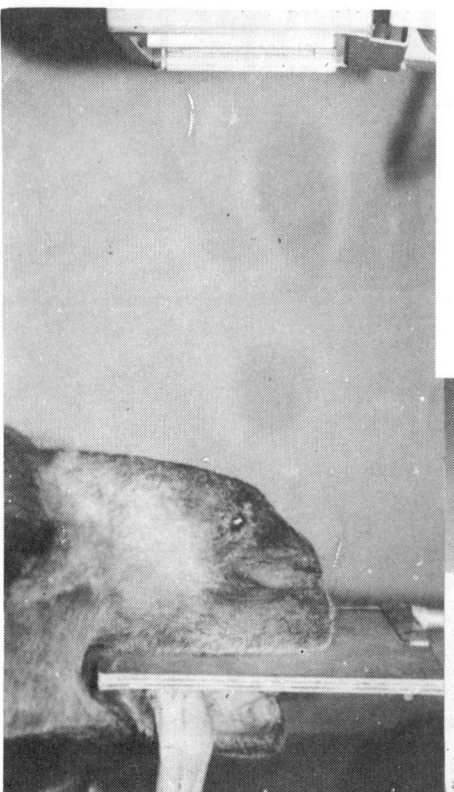

Fig. 13-55 : Camel. Position for dorsoventral view of maxilla with intraoral position of cassette. The neck is held in a natural position.

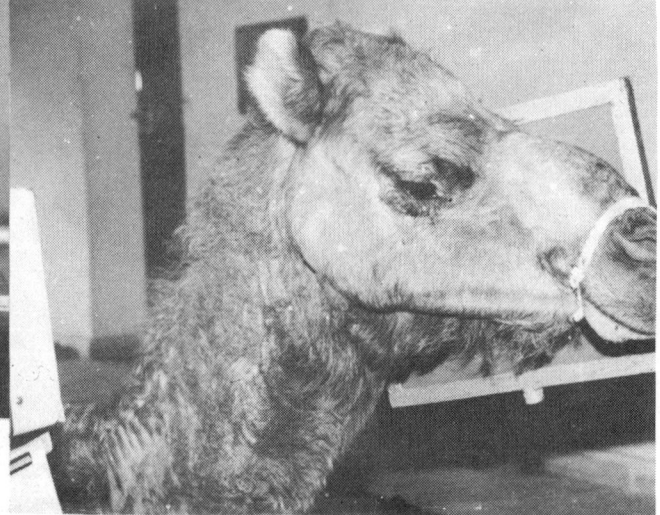

Fig. 13-56 : Camel. Position for lateral view of horizontal part of mandible with animal in sternal recumbency.

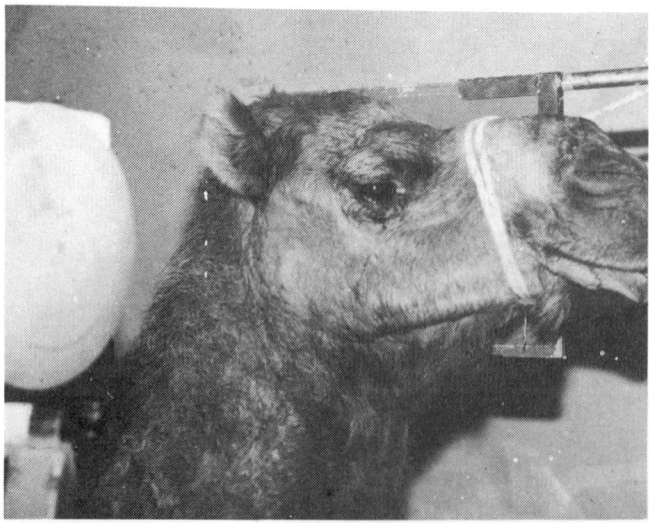

Fig. 13-57 : Camel. Position for lateral view of turbinates with animal in sternal recumbency.

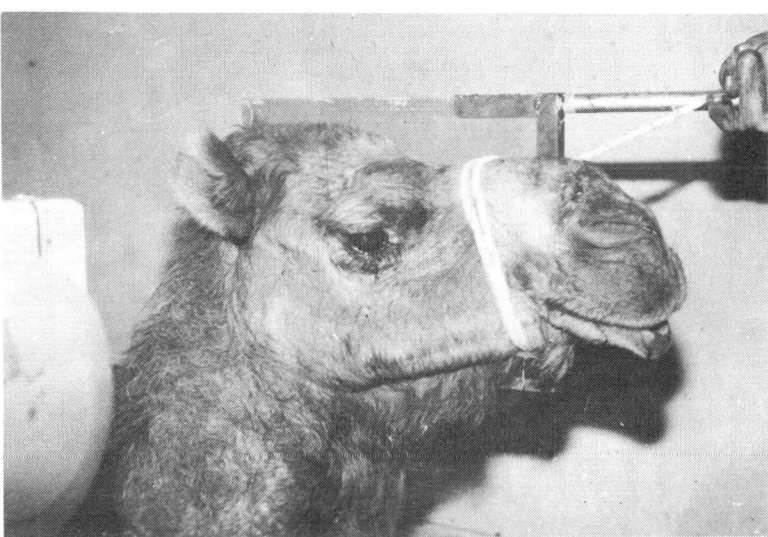

Fig. 13-58 : Camel. Position for lateral view of frontal bone/sinus with animal in sternal recumbency.

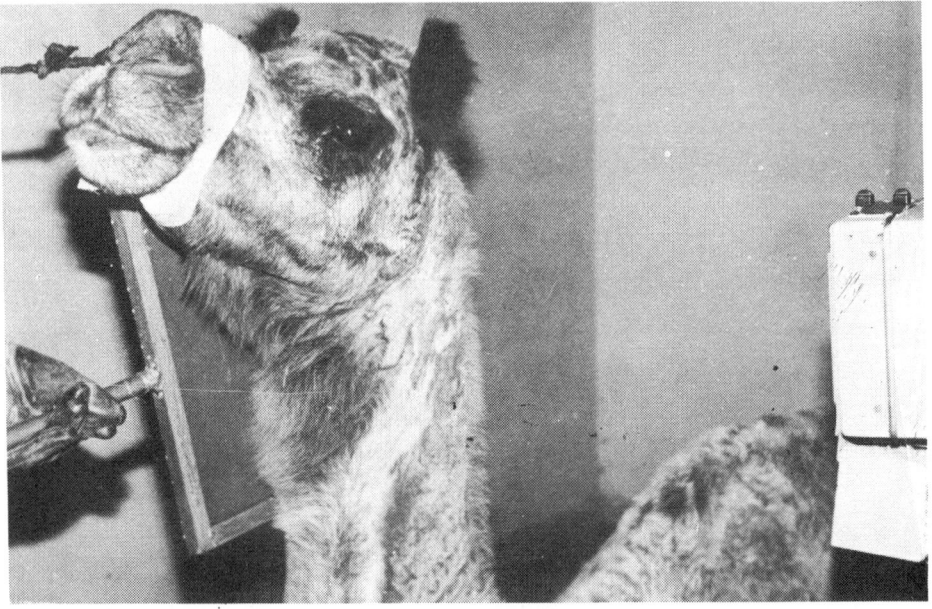

Fig. 13-59 : Camel. Position for lateral view of laryngopharyngeal region.

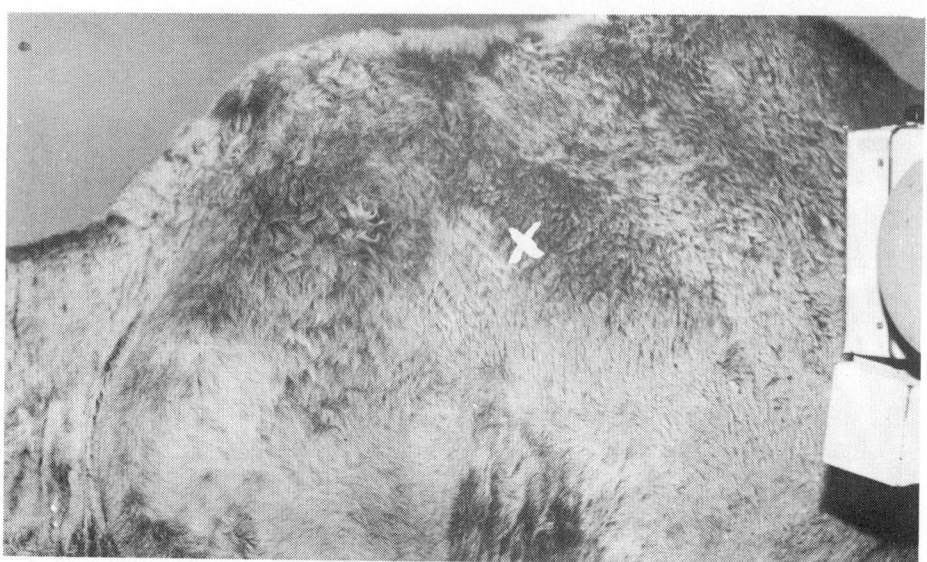

Fig. 13-60 : Camel. Position for lateral view of thorax. White mark indicates site for beam centring. Animal in sternal recumbency.

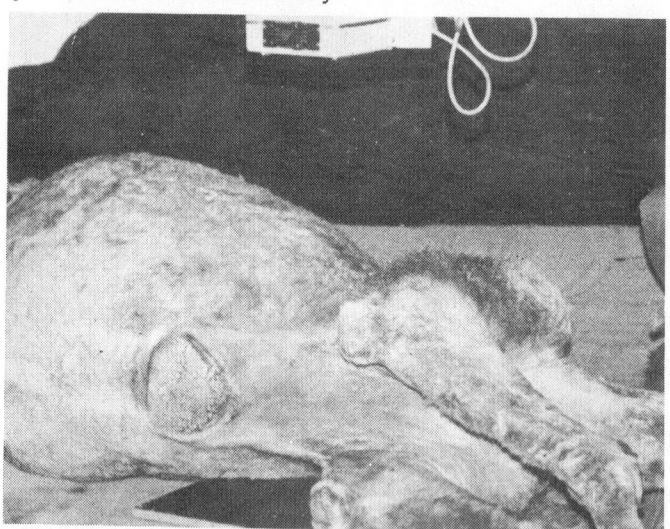

Fig. 13-61 : Camel. Position for lateral view of cardiac region with animal in lateral recumbent position.

THE SHEEP

THE SHOULDER JOINT

Caudocranial View (CdCr)

Position: The animal is secured in dorsal recumbency with the thoracic limbs extended cranially (Fig. 13-62). The head should be held firmly between the extended limbs to avoid rotation. This position is best achieved when the animal is deeply sedated. The shoulder joint of interest should be centered over the cassette.

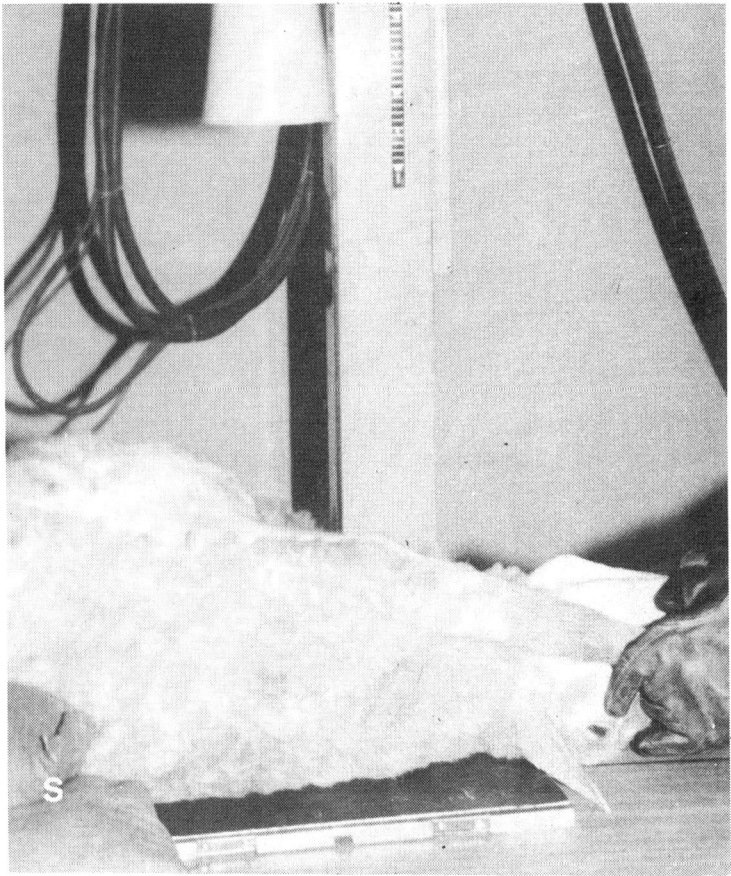

Fig. 13-62 : Sheep. Position for caudocranial view of shoulder joint.
S = Sand bags.

Beam centre: The X-ray beam is directed vertically and centred at the joint. The collimation of the beam and the size of the cassette should be large enough to cover the scapula and shoulder joint.

Mediolateral View (ML)

Position: The animal is placed in lateral recumbency with the limb to be radiographed lowermost. The thoracic limb of interest should be extended slightly cranially and a cassette is placed under the shoulder region (Fig. 13-63). The upper limb is drawn back over the thorax and the neck is extended.

Beam centre: The X-ray beam is directed vertically and centered on the shoulder joint.

THE HUMERUS

Caudocranial View (CdCr)

Position: The animal is placed in dorsal recumbency with the limb to be radiographed extended cranially (Fig. 13-64). The sternum is rotated slightly away from the area of interest.

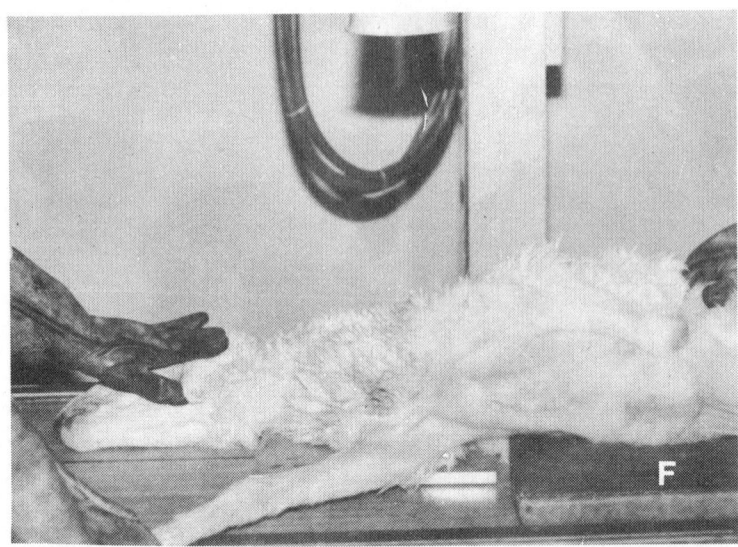

Fig. 13-63 : Sheep. Position for mediolateral view of shoulder joint. F = Foam block.

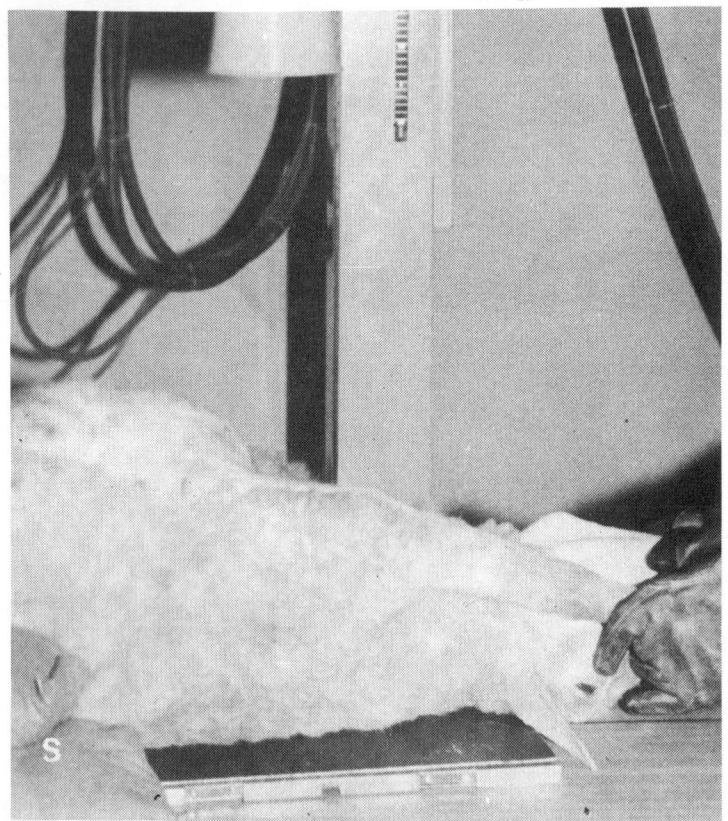

Fig. 13-64 : Sheep. Position for caudocranial view of humerus. S = Sand bags.

Beam centre: The X-ray beam is directed vertically and centered at the mid humeral region. The beam should be collimated to include both the shoulder and elbow joints.

Mediolateral View (ML)

Position: The sheep is placed in lateral recumbency with the limb to be exposed lowermost and extended moderately forward (Fig. 13-65). The upper limb is drawn back and held over the thorax. The sternum is elevated with a suitable foam block. Cassette is placed under the humeral region to be radiographed.

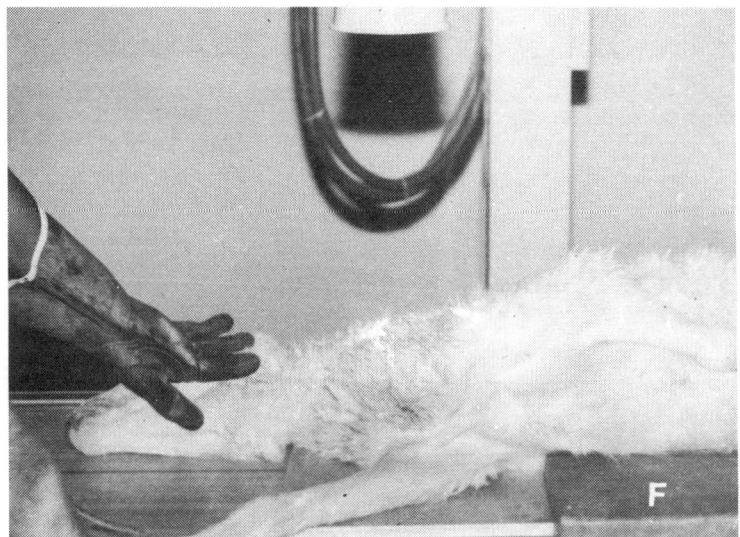

Fig. 13-65 : Sheep. Position for mediolateral view of humerus.
F = Foam block.

Beam centre: The X-ray beam is directed vertically and centered at the mid humeral region. The beam should be collimated and a large sized film used to include both the shoulder and elbow joints.

THE ELBOW JOINT

Caudocranial View (CdCr)

Position: The animal is placed in dorsal recumbency with the thoracic limbs extended cranially (Fig. 13-66). Sand bags and/or foam blocks should be used to maintain a true ventrodorsal position. Such a procedure also eliminates rotation of the area of interest. In this region, there is some magnification due to part film distance with which one has to compromise. The cassette is positioned below the elbow joint.

Beam centre: The X-ray beam is directed vertically and centered at the caudal aspect of the elbow joint.

Mediolateral view (ML)

Position: The sheep is placed in lateral recumbency with the limb to be radiographed lowermost and extended slightly forward. The upper limb is flexed and held back away from the X-ray beam. The head is extended and rested on a foam block. The cassette is positioned beneath the lower limb with the elbow in the centre (Fig. 13-67).

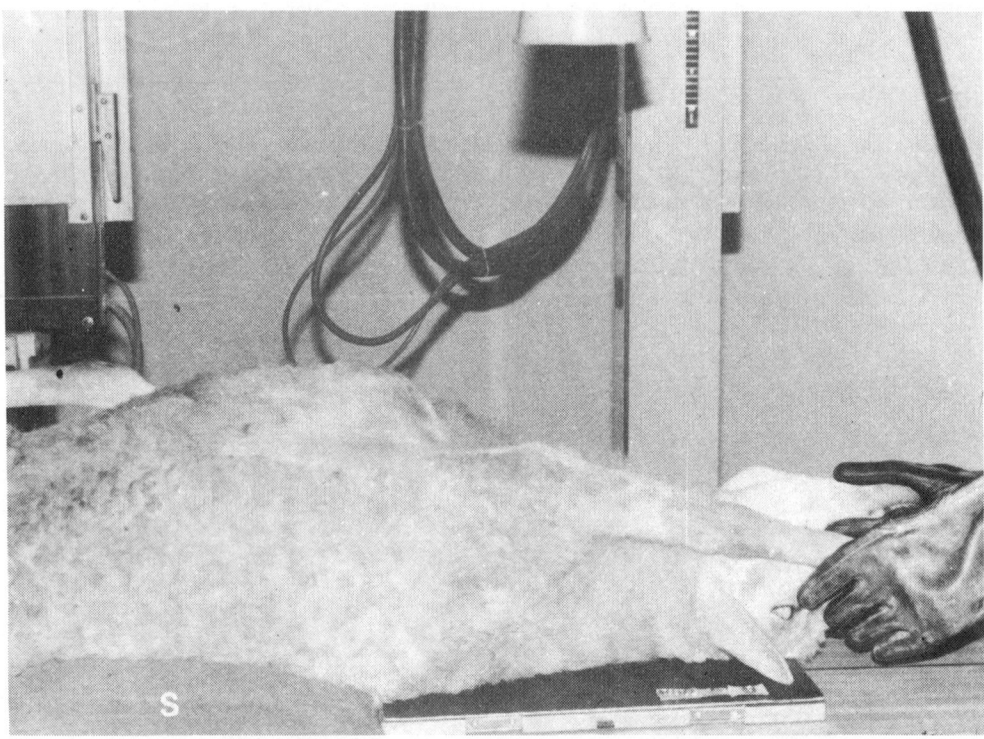

Fig. 13-66 : Sheep. Position for caudocranial view of elbow joint. S = Sand bags.

Beam centre: The X-ray beam is directed vertically and is centered at the medial surface of the elbow joint.

THE RADIUS

Craniocaudal View (CrCd)

Position: The animal is controlled in sternal recumbency with the limb to be radiographed extended cranially and held firmly with a gloved (lead) hand. The head is rotated towards the contralateral side or away from the limb to be examined. The cassette is placed beneath the radius (Fig. 13-68). Alternatively, the animal is placed in lateral recumbency with the limb to be examined uppermost and fully extended. The cassette in a holder is held against the caudal surface of the radius.

Beam centre: In sternal recumbency, the X-ray beam is directed vertically and centered on the mid diaphyseal region of the radius. The beam should be collimated and a large sized film used to include both the upper and lower joints. In lateral recumbency, the beam is directed from the cranial aspect, parallel to the table and centered at the mid diaphyseal region of the radius.

Mediolateral View (ML)

Position: The animal is placed in lateral recumbency with the limb to be examined lowermost. If necessary, foam pad should be placed under the head to maintain proper alignment (Fig. 13-69). The upper limb is flexed and held away from the field of X-ray beam. The cassette is placed beneath the radius.

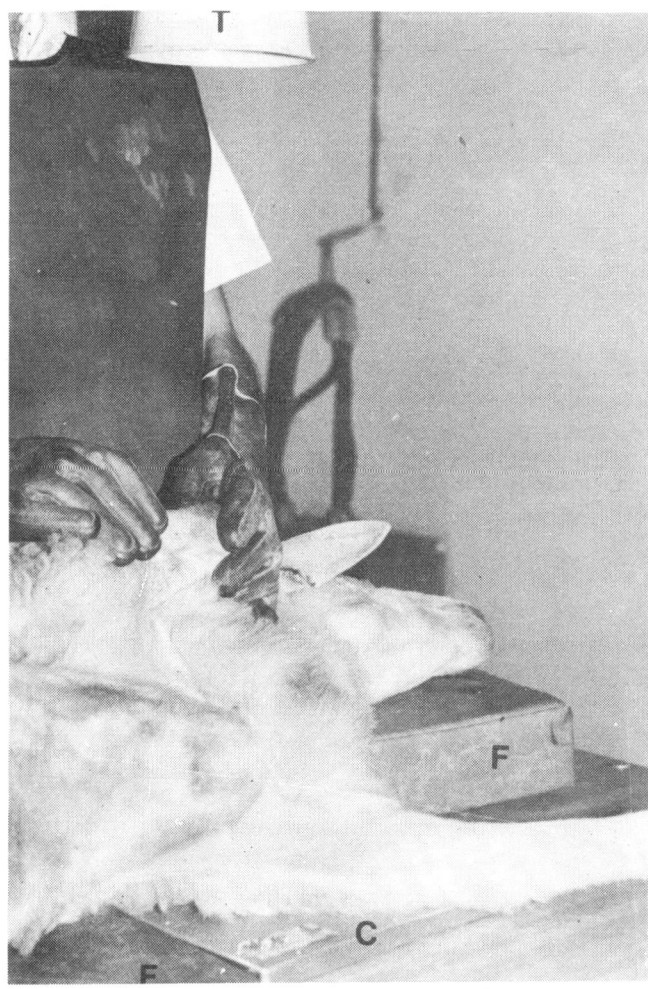

Fig. 13-67 : Sheep. Position for medio-
lateral view of elbow joint.
F = Foam blocks.   C = **Cassette.**
T = X-ray tube.

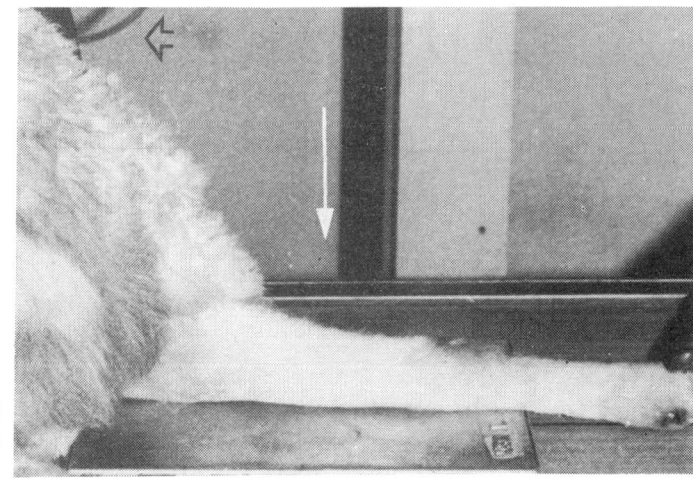

Fig. 13-68 : Sheep. Position for cranio
caudal view of radius. Arrow indicates
side of rotation of head.

Beam centre: The X-ray beam is directed vertically and centered on the medial mid diaphyseal region of the radius. The beam should be collimated to include the elbow and carpal joints.

THE CARPUS

Dorsopalmar View (DPa)

Position: The sheep is controlled in sternal recumbency with the limb to be radiographed extended and placed on a cassette (Fig. 13-70). The other limb is flexed and held away from the area of interest. The head is rotated away from the path of the X-ray beam.

Beam centre: The X-ray beam is directed vertically at right angle to the cassette and centered at the dorsal surface of the carpus.

Mediolateral View (ML)

Position: The animal is placed in lateral recumbency with the limb to be radiographed lowermost. The limb is extended and placed over the cassette. The upper limb is flexed and held away from the area of interest. A foam block is placed under the head to maintain proper alignment (Fig. 13-71).

Beam centre: The X-ray beam is directed vertically at right angle to the cassette and centered at the medial aspect of the carpus.

Flexed Mediolateral View (ML Flexed)

Position: The sheep is placed in lateral recumbency with the limb to be examined lowermost and pulled forward. The carpus is flexed and placed over the cassette (Fig. 13-72).

Beam centre: The X-ray beam is directed vertically at right angle to the cassette and is centered at the medial aspect of the flexed carpal joint.

THE METACARPUS

Dorsopalmar View (DPa)

Position: The animal is controlled in sternal recumbency with the limb to be radiographed extended forward and placed over the cassette (Fig. 13-73). The contralateral limb is flexed and the head is rotated away from the X-ray beam.

Beam centre: The X-ray beam is directed vertically at right angle to the cassette and centered at the dorsal mid diaphyseal region of the metacarpus. The beam should be collimated to include both the carpal and fetlock joints.

Mediolateral View (ML)

Position: The sheep is placed in lateral recumbency with the limb to be radiographed extended and placed over the cassette (Fig. 13-74). The upper limb is flexed and held away from the area of interest.

Beam centre: The X-ray beam is directed vertically at right angle to the cassette and centered at the medial aspect of the mid metacarpal region. The beam should be collimated to include both the carpal and fetlock joints.

THE FETLOCK JOINT

Dorsopalmar View (DPa)

Position: The sheep is maintained in sternal recumbency. The limb to be examined is extended and the distal part of the limb is placed flat on the cassette (Fig. 13-75).

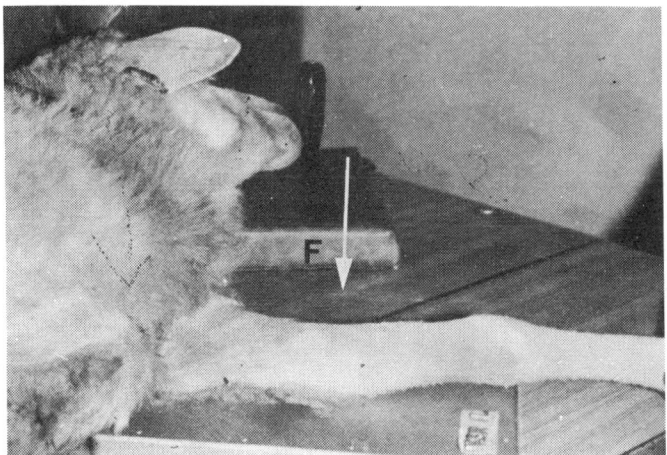

Fig. 13-69 : Sheep. Position for mediolateral view of radius.
F = Foam block.

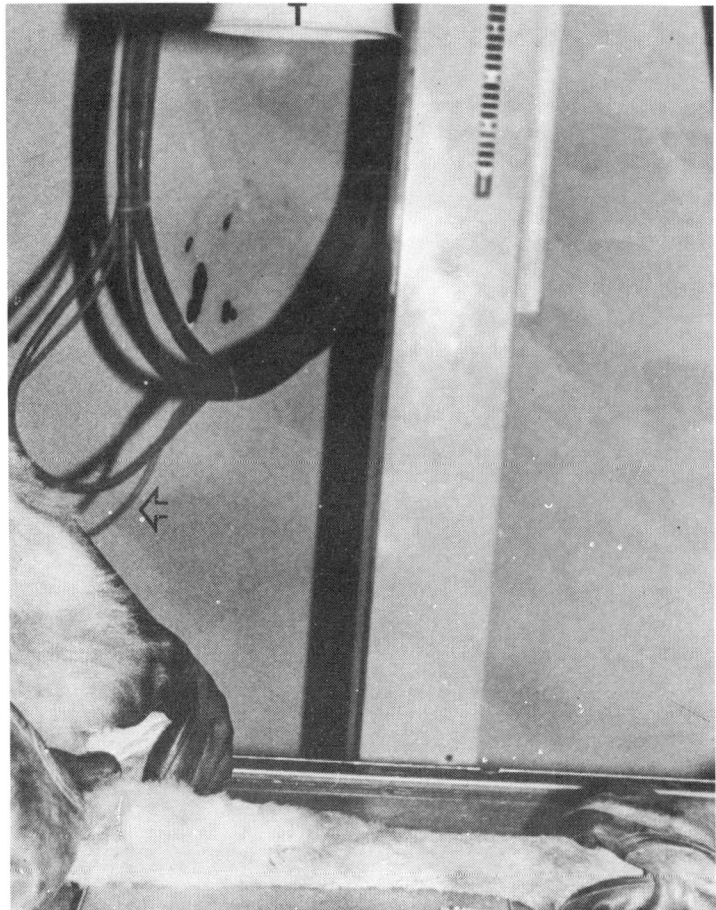

Fig. 13-70 : Sheep. Position for dorsopalmar view of carpus,
T = X-ray tube. Arrow indicates side of rotation of head.

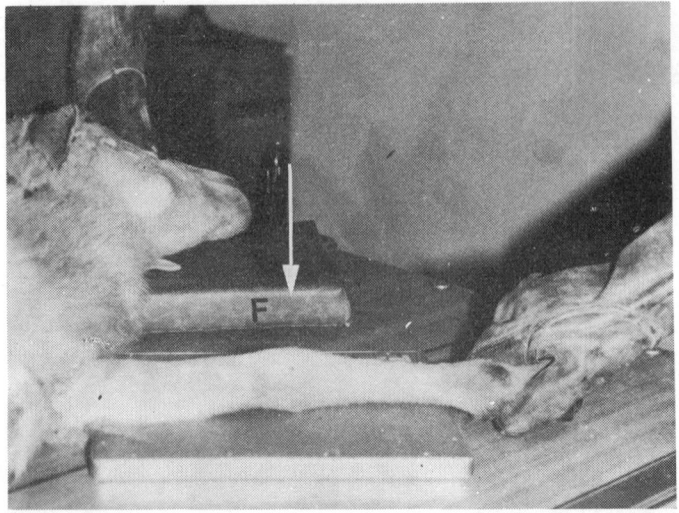

Fig. 13-71 : Sheep. Position for mediolateral view of carpus. F = Foam block.

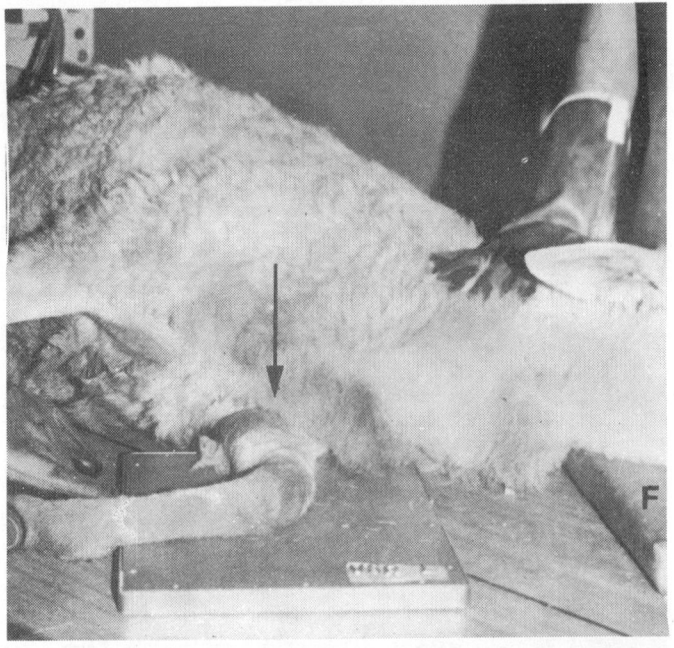

Fig. 13-72 : Sheep. Position for mediolateral view of flexed carpus. F = Foam block for resting head.

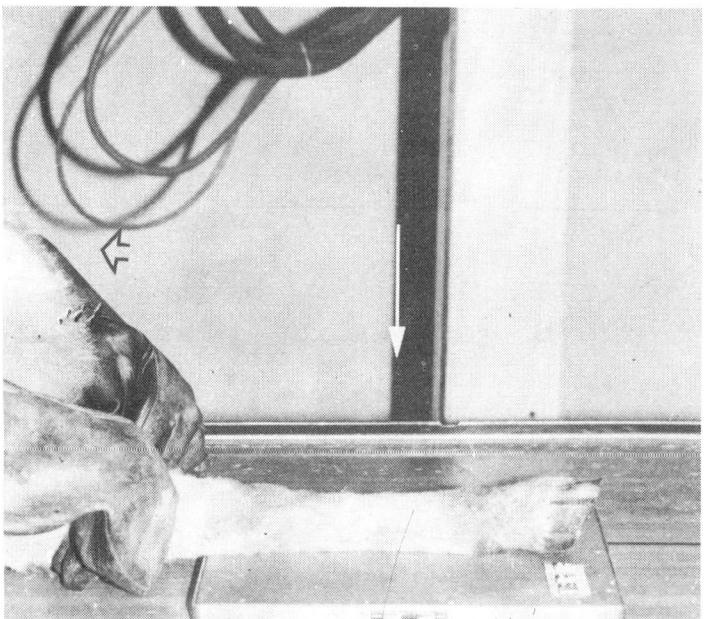

Fig. 13-73 : Sheep. Position for dorsopalmar view of metacarpus. Arrow indicates side of rotation of head.

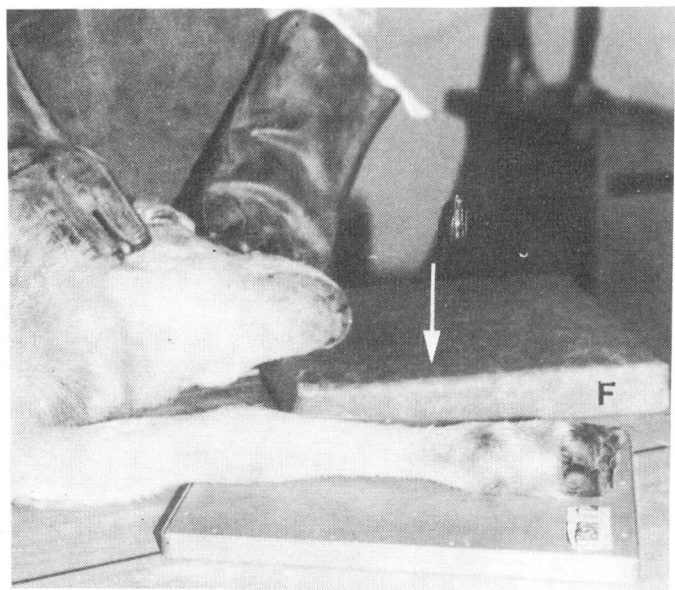

Fig. 13-74 : Sheep. Position for mediolateral view of metacarpus. F = Foam block.

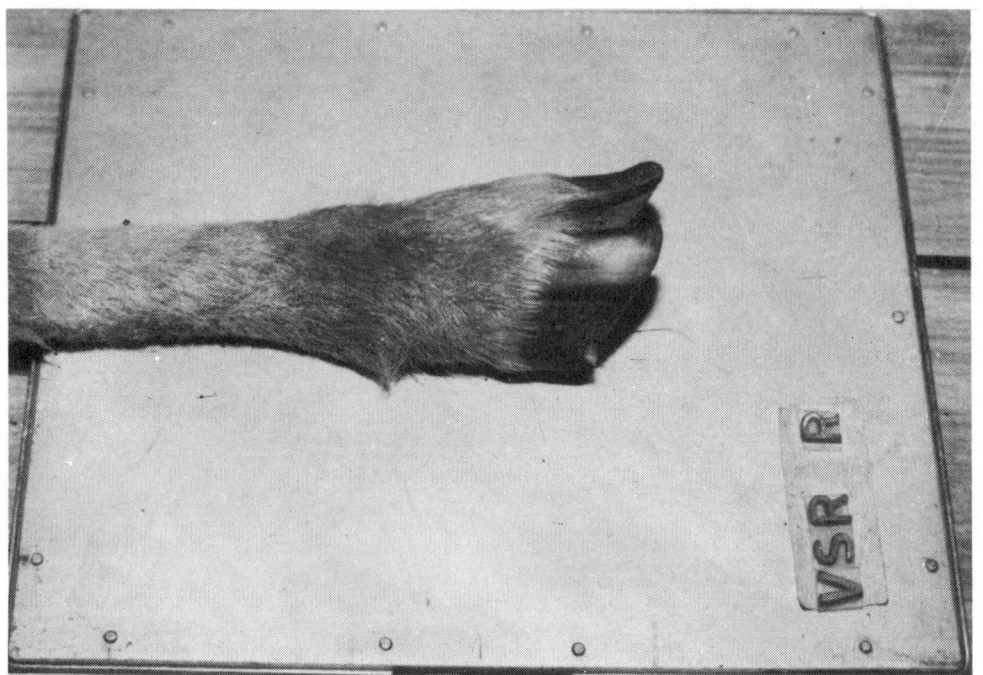

Fig. 13-75 : Sheep. Position for dorsopalmar view of fetlock. Animal in sternal recumbency.

Beam centre: The X-ray beam is directed vertically at right angle to the cassette and centered at the dorsal aspect of the fetlock joint.

Mediolateral View (ML)

Position: The animal is placed in lateral recumbency with the limb to be radiographed lowermost and over the cassette with the fetlock joint in its centre (Fig. 13-76). The upper limb is pulled back away from the area of interest.

Beam centre: The X-ray beam is directed vertically at right angle to the cassette and centered at the medial aspect of the fetlock joint.

THE DIGITS

The protocol for radiographic examination of the digits is same as that described for DPa and ML views of the fetlock joint except that the X-ray beam is centered on the digits.

THE HIP JOINT

Ventrodorsal View (VD)

Position: After controlling the animal in dorsal recumbency, sand bags are placed on both sides of the abdominal and thoracic walls to maintain a true ventrodorsal postition. This also eliminates rotation of the area of interest. Each pelvic limb is held firmly at the level of distal tibia and extended so that both limbs are parallel to each other and also in relation to vertebral column (Fig. 13-77). The cassette is placed in such a way that its centre lies at the level of acetabulum. A grid should be used. Deep sedation or general anaesthesia facilitates correct positioning of the animal for this view.

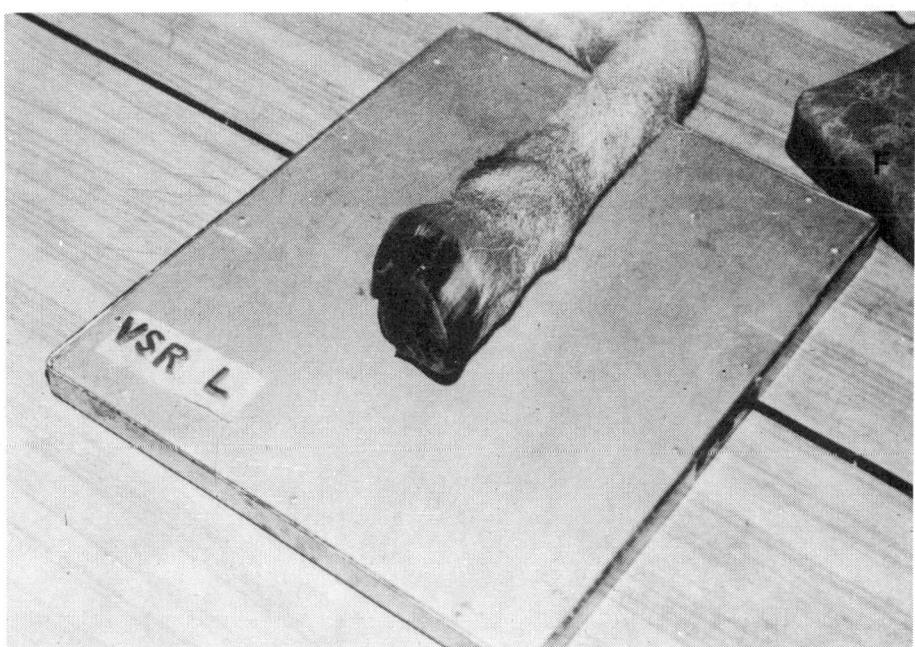

Fig. 13-76 : Sheep. Position for mediolateral view of fetlock. Animal in lateral recumbency with head rested on a foam block (F).

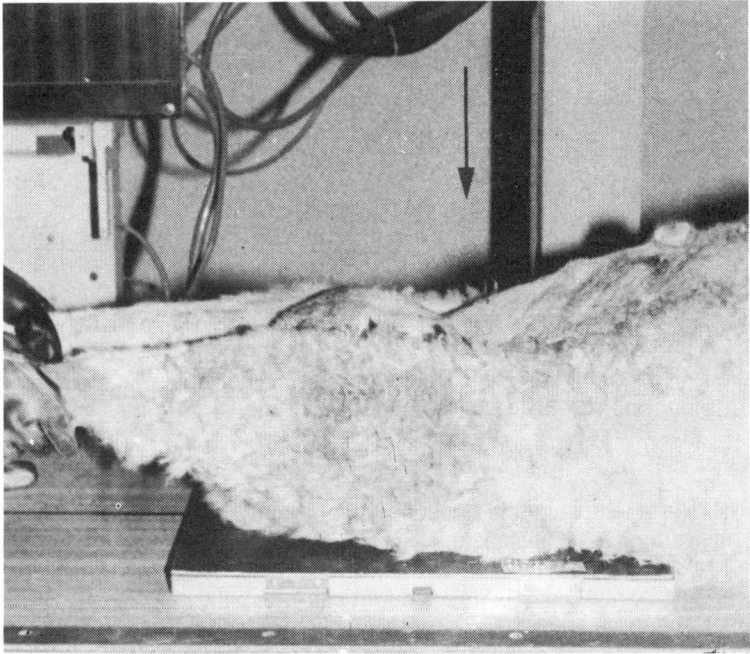

Fig. 13-77 : Sheep. Position for ventrodorsal view of hip joints.

Beam centre: X-ray beam is directed vertically at right angle to the cassette with primary beam at the level of the acetabulum. The beam is collimated to include the hip joints as well as the pelvis.

Lateral View

Position: The sheep is placed in lateral recumbency with the side of interest near to the cassette. A foam block of appropriate thickness is placed between the stifles to help eliminate rotation of the pelvis (Fig. 13-78). The pelvic limbs are placed on top of each other and the pelvis is centered over the cassette.

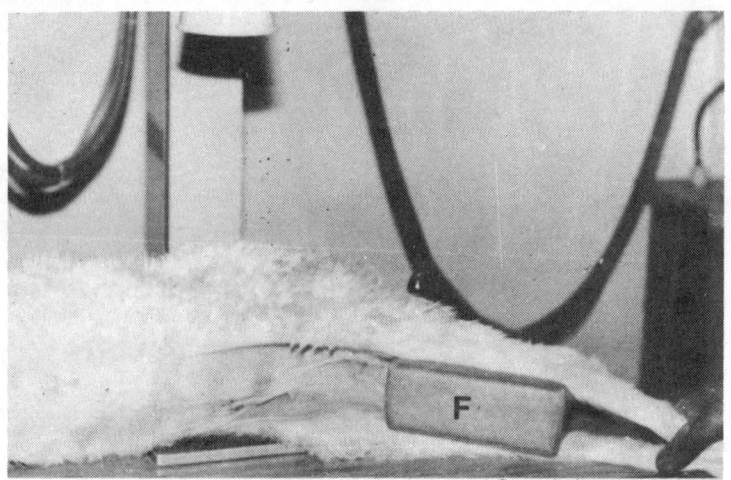

Fig. 13-78 : Sheep. Position for lateral view of hip joints. F = Foam block.

Beam centre: The X-ray beam is directed vertically at right angle to the cassette and centered at the level of acetabulum.

THE FEMUR

Carniocaudal View (CrCd)

Position: The animal is controlled in dorsal recumbency and maintained in this position with sand bags placed on either side of the thoracoabdominal region. The pelvic limbs are grasped from the distal end of the tibia, extended and slightly abducted for a true CrCd view. A cassette, large enough to include the hip and stifle joints, is placed beneath the femoral region to be radiographed (Fig. 13-79).

Beam centre: A vertically directed X-ray beam is centered at the cranial aspect of the mid femur. The beam should be collimated to include the hip and stifle joints.

Mediolateral View (ML)

Position: The sheep is placed in lateral recumbency with the pelvic limb to be examined lowermost. The upper limb is abducted, rotated and held approximately upright away from the area of interest (Fig. 13-80). The cassette is placed under the femur to be radiographed.

Beam centre: The X-ray beam is directed vertically at right angle to the cassette and centered at the medial aspect of the mid femur. The size of the film and collimation of the beam should be such that the hip as well as stifle joints are included in the radiograph.

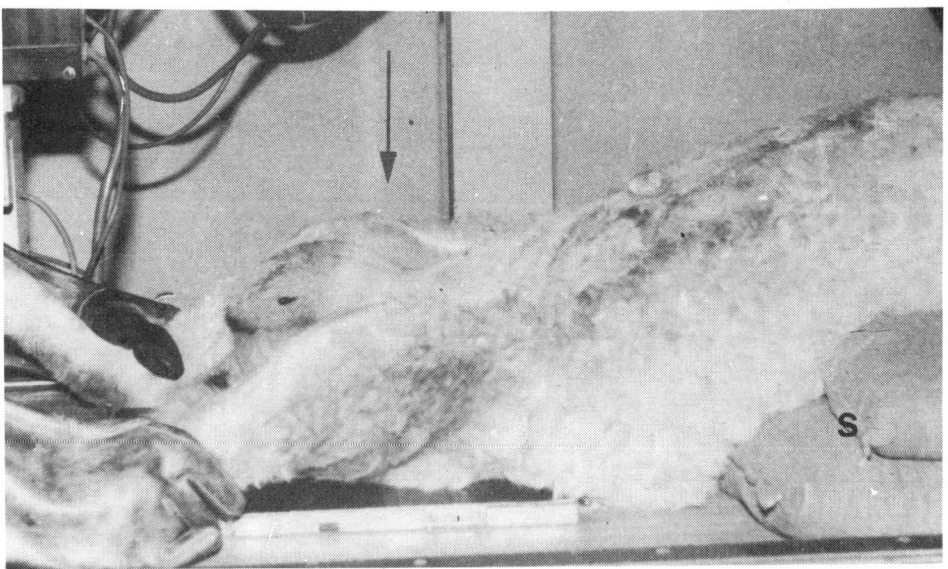

Fig. 13-79 : Sheep. Position for craniocaudal view of femur. S = Sand bags.

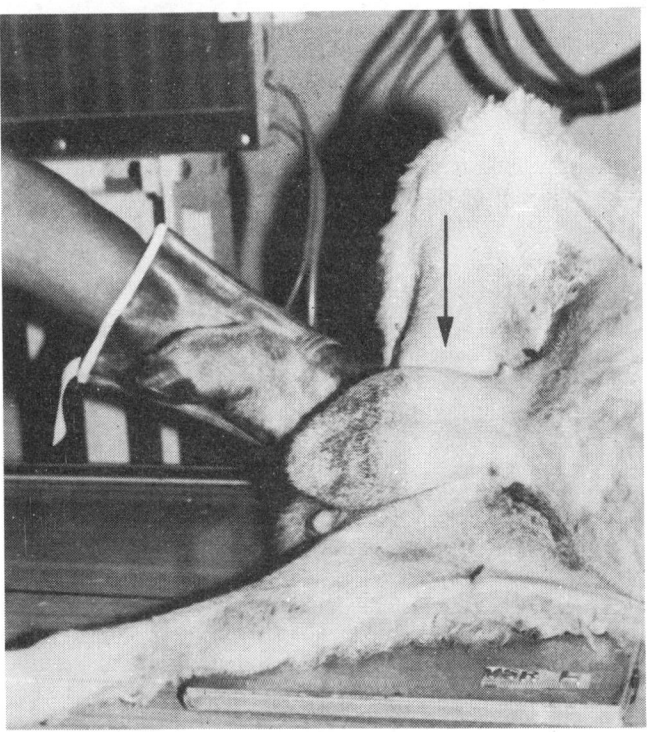

Fig. 13-80 : Sheep. Position for mediolateral view of femur.

THE STIFLE JOINT

Caudocranial View (CdCr)

Position: The sheep is controlled in lateral recumbency with the pelvic limb to be radiographed uppermost. The limb is now extended as much as possible and held in a position almost parallel to the table top. The cassette is held against the cranial aspect of the stifle joint (Fig. 13-81).

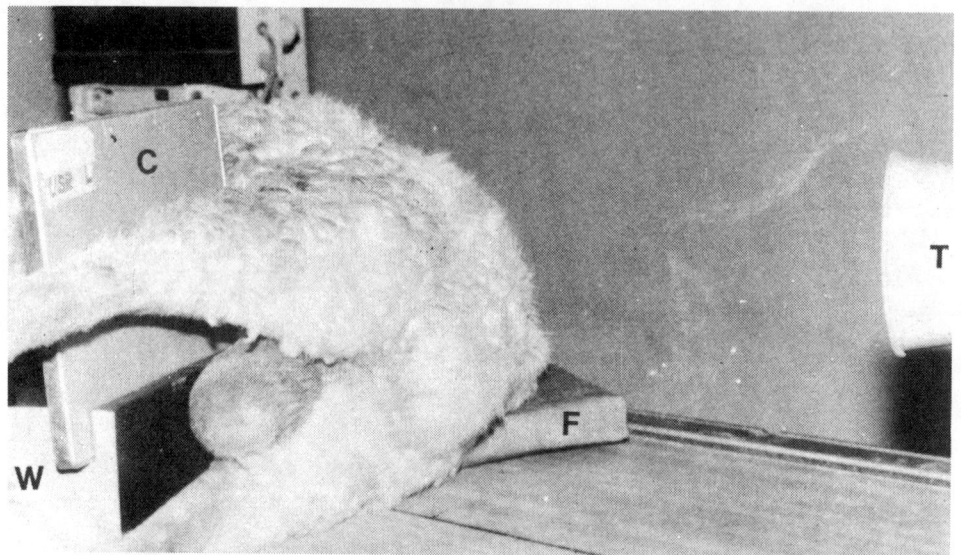

Fig. 13-81 : Sheep. Position for caudocranial view of stifle joint. W = Wooden block with a slot for cassette (C), F = Foam block, T = X-ray tube.

Beam centre: The X-ray tube is positioned on the caudal aspect of the stifle joint. The beam is directed parallel to the table top at right angle to the cassette and centered at the caudal aspect of the stifle joint.

Mediolateral View (ML)

Position: The animal is placed in lateral recumbency and positioned in the same way as described for the mediolateral radiograph of femur. However, in this case, the stifle joint is centered at the centre of the cassette.

Beam centre: The X-ray beam is directed vertically at right angle to the cassette and centered on the medial aspect of the stifle joint.

THE TIBIA

Caudocranial View (CdCr)

Position: The sheep is secured in lateral recumbency with the pelvic limb to be radiographed uppermost. The upper limb is extended and held in a position almost parallel to the table top to eliminate rotation. The cassette is placed against the cranial aspect of the tibia (Fig. 13-82A).

Beam centre: The X-ray tube is positioned caudal to the tibia. The beam is directed horizontally parallel to the table top and centered at the caudal aspect of the mid tibial region. The beam should be collimated to include both the stifle and tarsal joints.

Mediolateral View (ML)

Position: The sheep is placed in lateral recumbency with the pelvic limb to be radiographed lowermost. The limb is now extended and placed over the cassette. The upper limb is held away from the line of primary beam (Fig. 13-82B).

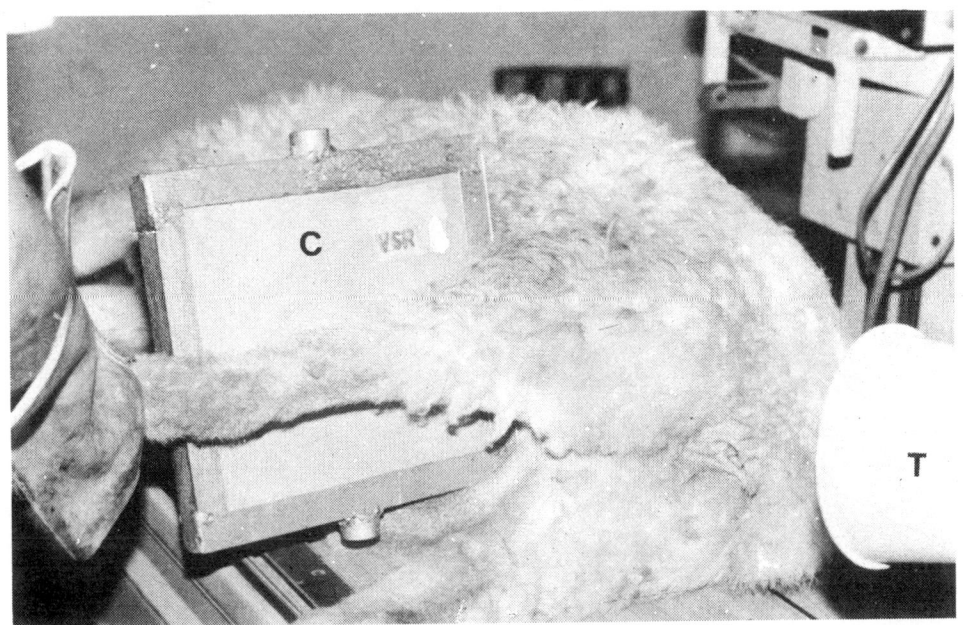

Fig. 13-82A : Sheep. Position for caudocranial view of tibia. C = Cassette in a holder, T = X-ray tube.

Beam centre: The beam is directed vertically at right angle to the cassette and is centered at the medial aspect of the mid tibial region. The beam should be collimated to include both the stifle and tarsal joints.

THE TARSUS

Plantodorsal View (PlD)

Position: The sheep is placed in lateral recumbency with the limb to be examined uppermost. The limb is extended and held parallel to table top. The cassette with a slot in a holder is placed on the dorsal aspect of the tarsus (Fig. 13-83).

Beam centre: The X-ray tube is positioned on the plantar aspect of the limb. The beam is directed horizontally parallel to the table top and centered on the plantar aspect of the tarsus.

Mediolateral View (ML)

Position: The sheep is placed in lateral recumbency with the limb to be radiographed lowermost. The tarsal joint of the limb is placed over the cassette. The upper limb is flexed and held away from the field of X-ray beam.

Beam centre: The beam is directed vertically at right angle to the cassette and centered at the medial aspect of the tarsal joint.

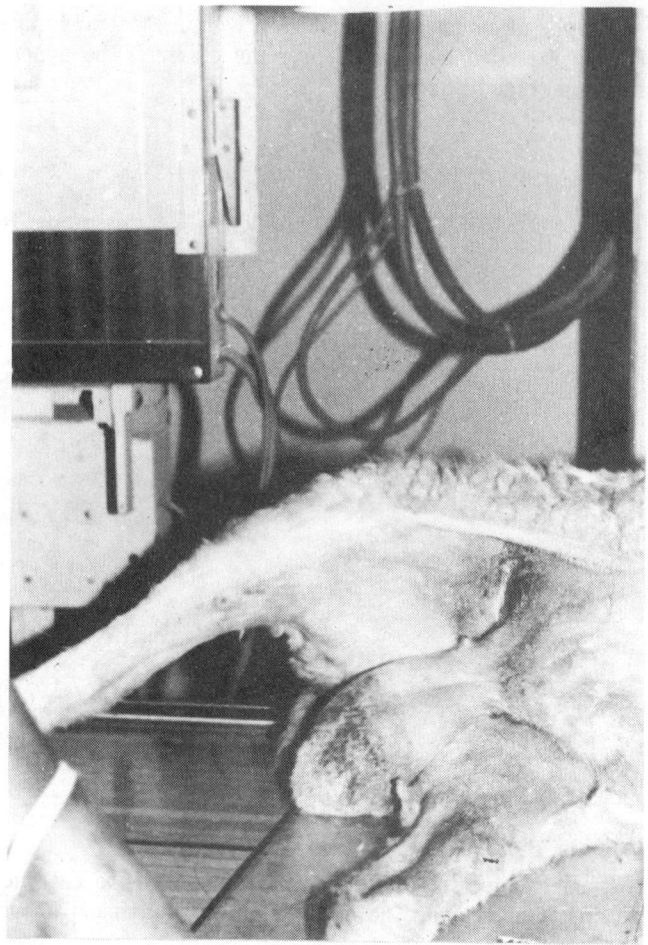

Fig. 13-82B : Sheep. Position for mediolateral view of tibia. T = X-ray tube.

Flexed mediolateral View (ML flexed)

The positioning of the animal and centering of the X-ray beam are same as described for ML view. The tarsal joint to be radiographed is, however, flexed and then placed over the cassette for the exposure (Fig.13-84).

THE METATARSUS

The procedures are same as described for the metacarpus.

THE VERTEBRAL COLUMN

A. THE CERVICAL VERTEBRAE

Ventrodorsal View (VD)

Position: The sheep is placed in dorsal recumbency under deep sedation and maintained in this position by using sand bags on both sides (Fig. 13-85). The thoracic limbs are drawn caudally and held lateral to the thoracic wall. The head and neck are extended. The cassette is placed beneath the cervical region to facilitate proper positioning and relaxation of the area. Because of the difference in the tissue thickness between the cranial and caudal parts of the cervical region, it is advisable to take two overlapping radiographs in well built sheep to eliminate over or under exposure of the vertebrae.

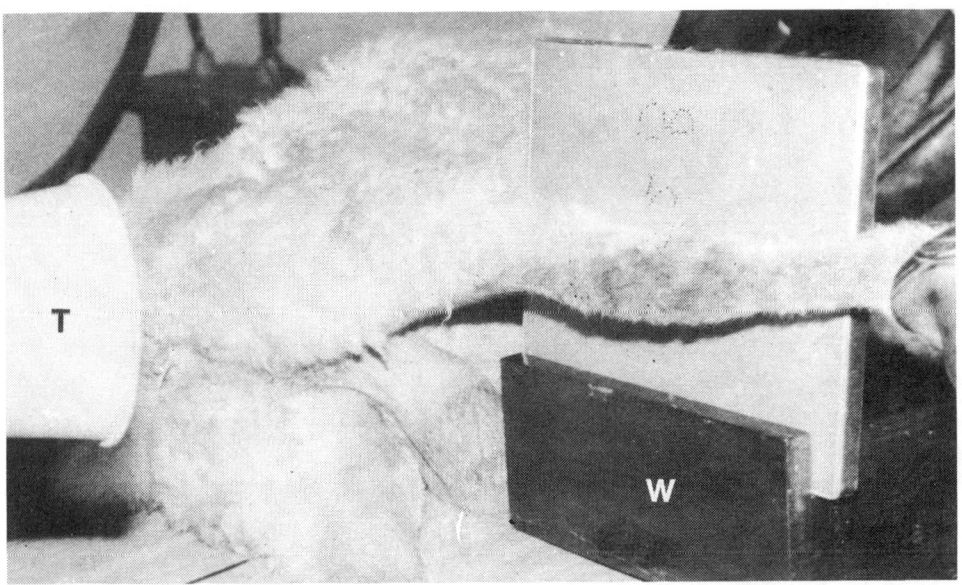

Fig. 13-83 : Sheep. Position for plantodorsal view of tarsus. W = Wooden block with slot for cassette T = X-ray tube.

Beam centre: The X-ray beam is directed vertically at right angle to the cassette and centered at the level of C_3-C_4 or at the area of specific interest. The X-ray beam should be collimated to exclude excessive soft tissues of the cervical region.

Lateral View

Position: The animal is controlled in lateral recumbency. The thoracic limbs are drawn caudally to eliminate superimposition of the shoulder joints and other structures. The head and neck should be extended. The cassette is placed under the cervical vertebral region.

Beam centre: The beam is directed vertically at right angle to the cassette to centre at the level of C_3-C_4 or over the area of specific interest.

B. THE THORACIC VERTEBRAE

Ventrodorsal View (VD)

Position: The sheep is controlled in dorsal recumbency. The thoracic and pelvic limbs are extended cranially and caudally, respectively and maintained in position with gloved hands or sand bags (Fig. 13-86). The thoracic vertebrae should be centered over a large cassette to include the entire thoracic vertebrae. A grid should be used.

Beam centre: The X-ray beam is directed vertically at right angle to the cassette and centered at the mid thoracic vertebrae or at the area of interest. The beam should be collimated to eliminate excessive soft tissues.

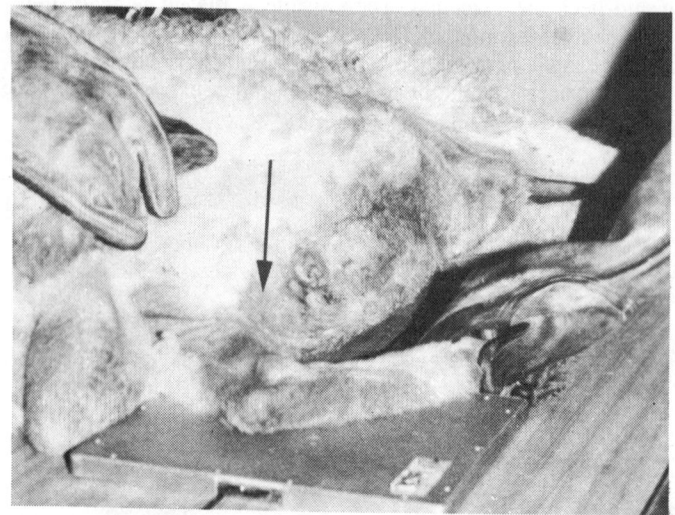

Fig. 13-84 : Sheep. Position for medio-lateral view of flexed tarsus.

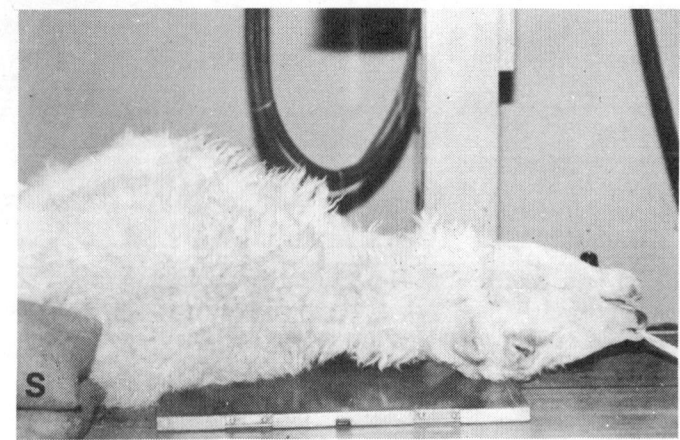

Fig. 13-85 : Sheep. Position for ventro-dorsal view of cervical region. Control the animal in lateral recumbency for a lateral view. S = Sand bags.

Fig. 13-86 : Sheep. Position for ventro dorsal view of thoracic vertebrae. S = Sand bags.

Lateral View

Position: After controlling the animal in lateral recumbency, the thoracic and pelvic limbs are extended moderately. A foam block may be placed beneath the sternum to raise its level parallel to thoracic vertebral bodies. The cassette is place under the thoracic vertebrae (Fig. 13-87). The cassette should be large enough to include a major part of the thoracic vertebral column. A grid should be used.

Beam centre: The X-ray beam is directed vertically at right anlge to the cassette and centered at the level of mid thoracic vertebral bodies or at the region of specific interest. The beam should be collimated to exclude excessive soft tissues.

C. THE LUMBAR VERTEBRAE

Ventrodorsal View (VD)

Position: The animal is suitably secured in true dorsal recumbency with the thoracic and pelvic limbs extended cranially and caudally, respectively. Positioning devices may be used on both sides of the lateral thoracic wall. The lumbar vertebrae are centered over a large sized cassette to include the entire lumbar vertebral region (Figs 13-88A). A grid should be used.

Beam centre: The X-ray beam is directed vertically at right angles to the cassette and centered at the level of L_3-L_4. The beam should be collimated to eliminate excessive soft tissues.

Lateral View

Position: The animal is placed in lateral recmbency with the thoracic and pelvic limbs extended moderately (Fig. 13-88B). This eliminates rotation of lumbar vertebrae. A foam block of appropriate thickness may be placed under the sternum to make the vertebral column in the same plane and parallel to the cassette. The lumbar vertebrae should be centered over the cassette. The cassette should be large enough to include the entire lumbar vertebral region. A grid should be used in well built animals.

Beam centre: The beam is directed at L_3-L_4 lumbar vertebrae or at the area of specific interest.

THE SKULL

Ventrodorsal View (VD)

Position: The sheep is secured in dorsal recumbency with the aid of suitable positioning devices preferably after deep sedation. The thoracic limbs are pulled caudally and maintained in position with gloved hands. A sand bag of appropriate thickness should be placed under the cervical region to maintain proper positioning of the skull. The head is held in extended position over the cassette with the help of a tape (Fig. 13-89).

Beam centre: The beam is directed vertically from above and centered at the level of the orbit.

Lateral View

Position: The sheep is placed in lateral recumbency on the X-ray table with the head resting on the cassette (Fig. 13-90). A foam block or cotton pad of appropriate thickness should be placed under the cranial cervical region to eliminate rotation and to permit a true lateral view of the head.

Beam centre: The beam is directed vertically at right angle to the cassette and centered at the level of the orbit.

Fig. 13-87 : Sheep. Position for lateral view of thoracic vertebrae.

Fig. 13-88A : Sheep. Position for ventro-dorsal view of lumbar vertebrae. S = Sand bags.

Fig. 13-88B : Sheep. Position for lateral view of lumbar vertebrae of sheep.

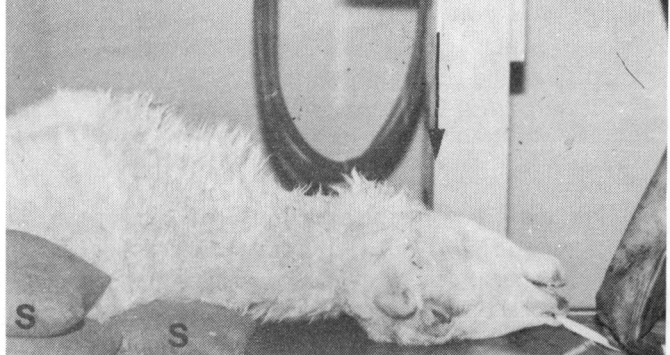

Fig. 13-89 : Sheep. Position for ventro-dorsal view of skull. S = Sand bags.

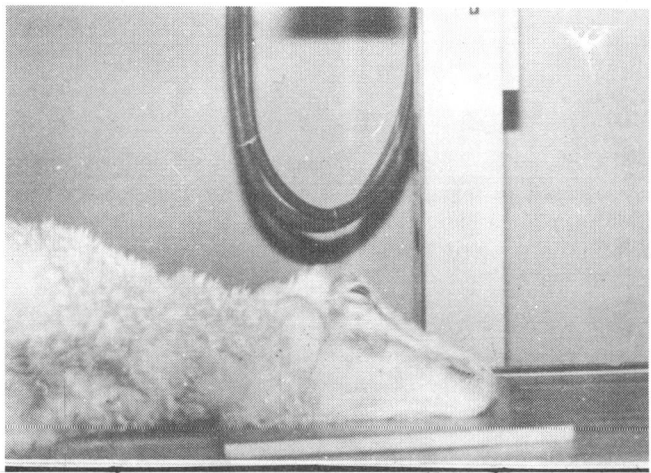

Fig. 13-90 : Sheep. Position for lateral view of skull.

FRONTAL SINUS REGION

Oblique View

Position: The animal is controlled in lateral recumbency with the thoracic limbs extended caudally. The cassette is placed beneath the head. A foam block of appropriate thickness is placed under the cassette to acehieve about 20⁰ elevation of the dorsal aspect of the skull.

Beam centre: The X-ray beam is directed vertically and centered at the site of interest.

THE MANDIBLE

Lateral View

Position: The sheep is secured in lateral recumbency with the head resting on the cassette. A foam block of appropriate thickness should be placed under the cranial cervical region to permit true lateral view of the mandible.

Beam centre: The X-ray beam is directed vertically at right angle to the cassette and centered at the mid horizontal ramus of the mandible.

TURBINATES

Lateral View

The protocol for radiography of the turbinates is same as described for the mandible except that the beam is centered at the nasal region.

LARYNGOPHARYNGEAL REGION

Lateral View

Position: The animal is secured in lateral recumbency with the thoracic limbs drawn caudally. The atlanto-occipital articulation is moderately flexed. The cassette is placed under the laryngopharyngeal region.

Beam centre: The X-ray beam is directed vertically at right angle to the cassette to centre at the laryngopharyngeal region.

THE THORAX

Ventrodorsal View (VD)

Position: The sheep is placed in dorsal recumbency with the thoracic limbs extended cranially. Positioning devices may be used to maintain a true ventrodorsal position. A large sized cassette is placed beneath the animal to include the entire thorax (Fig. 13-91A). A grid may be used.

Beam centre: The beam is directed vertically at right angle to the cassette and centered at the level of the fifth intercostal space.

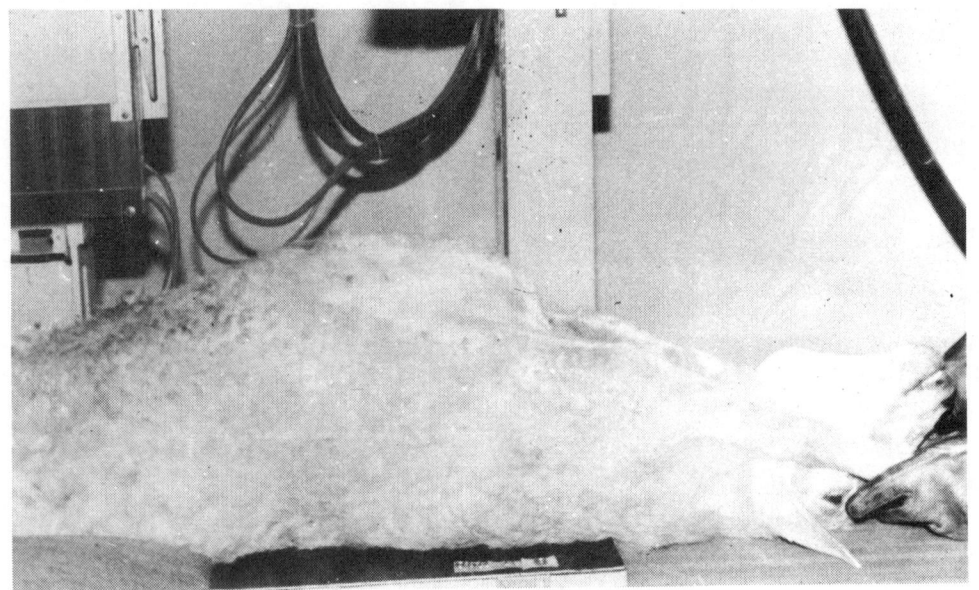

Fig. 13-91A : Sheep. Position for ventrodorsal view of thorax.

Lateral View

Position: After placing the animal in lateral recumbency, the thoracic limbs are extended cranially (Fig. 13-91B). The cassette is placed beneath the thoracic region. A large cassette should be used to include the entire thorax on a single film.

Beam centre: The X-ray beam is directed vertically at right angle to the cassette and is centered at the fifth intercostal space. A grid is not used.

ABDOMEN

Postion: The animal is placed in the left lateral recumbency with the thoracic and pelvic limbs extended cranially and caudally, respectively. A large sized cassette is positioned beneath the abdominal region (Fig. 13-92). Alternatively, the sheep may be positioned in dorsal recumbency with the thoracic and pelvic limbs extended cranially and caudally, respectively. The cassette is placed on the left side of the abdominal region (Fig. 13-93).

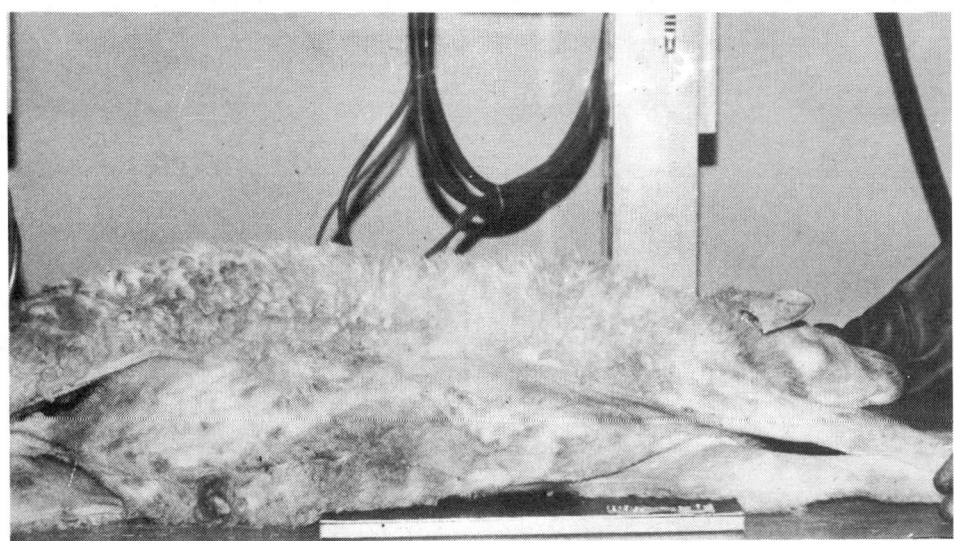

Fig. 13-91B : Sheep. Position for lateral view of thorax..

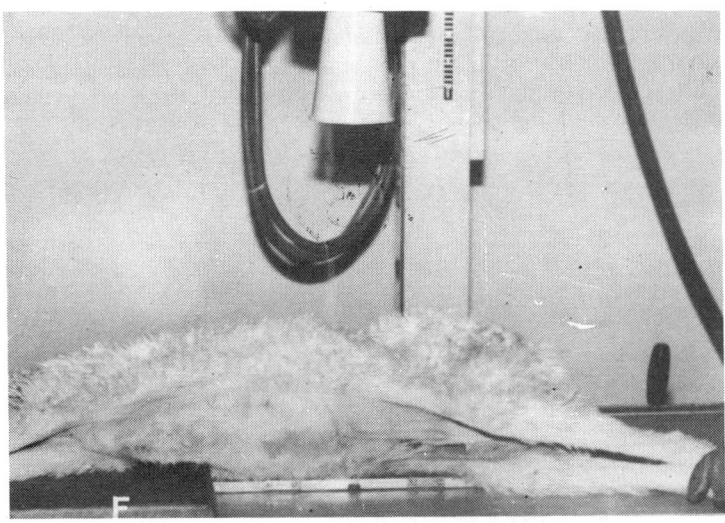

Fig. 13-92 : Sheep. Position for lateral view of abdomen. F = Foam block.

Beam centre: The X-ray beam is directed vertically at right angle to the face of the cassette, if the animal is placed in lateral recumbency. In case the animal is positioned in the dorsal recumbent position, the beam is directed horizontally at the cassette. The beam is centered at the middle of seventh rib. A grid should be used.

Fig. 13-93 : Sheep. Position for lateral view of abdomen with the animal in dosal recumbency. C = Cassette in a holder, X = Beam centring.

HORSE AND DOG

Radiographic positioning protocol for the horse is almost similar to that described for the bovine. For the dog, the protocol is almost similar to that of sheep. However, some variations exist. Some of the radiographic positioning techniques which are commonly required in day to day practice are illustrated for the horse (Figs 13-94 to 13-111) and dog (Fgs. 13-112 to 13-135).

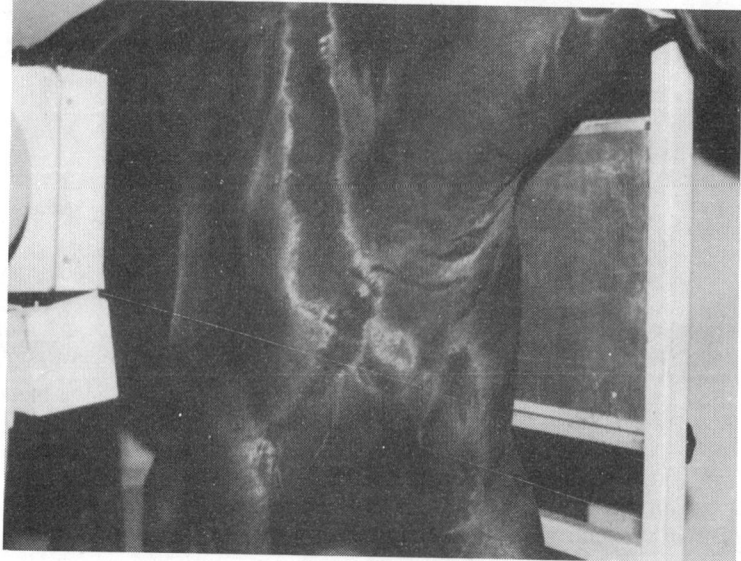

Fig. 13-94 : Horse. Position for mediolateral view of shoulder.

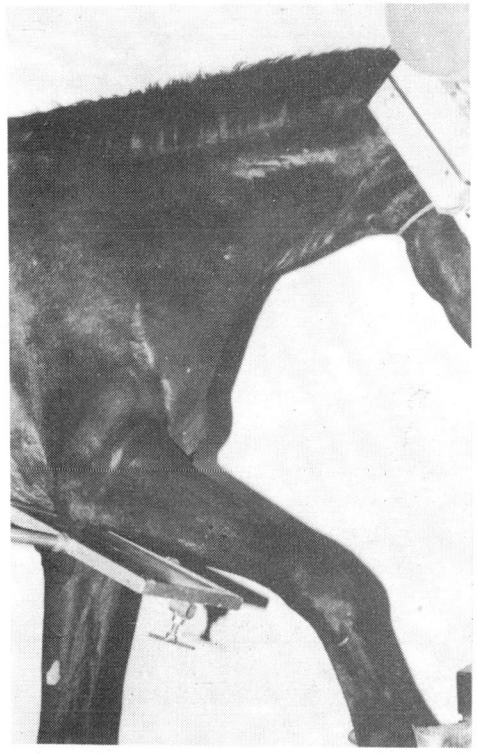

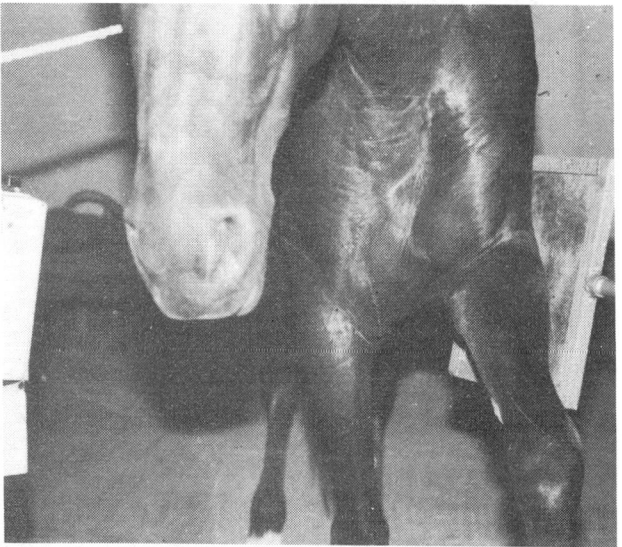

Fig. 13-96 : Horse. Position for mediolateral view of elbow joint.

Fig. 13-95 : Horse. Position for craniocaud view of elbow joint.

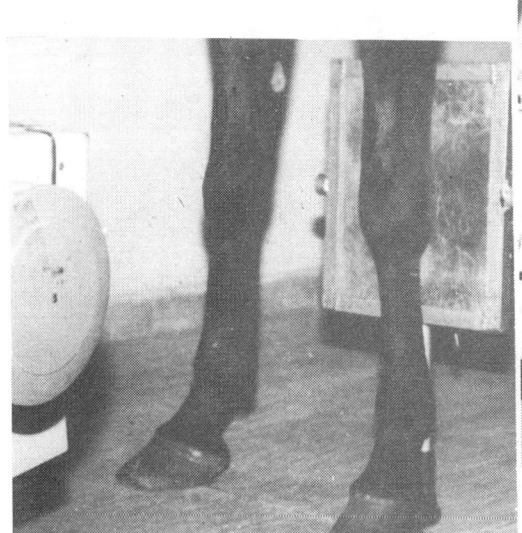

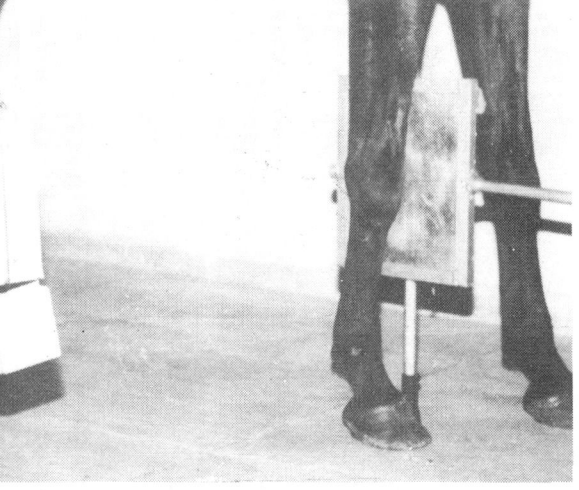

Fig. 13-98 : Horse. Position for latromedial view of carpus.

Fig. 13-97 : Horse. Position for dorsopalmar view of carpus.

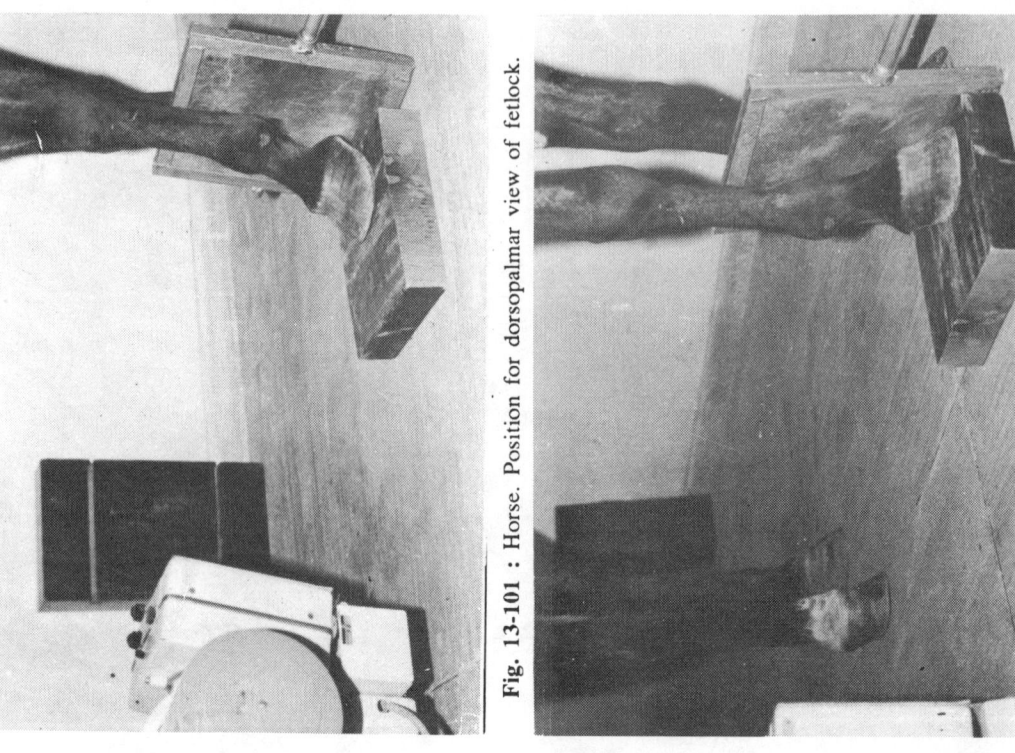

Fig. 13-101 : Horse. Position for dorsopalmar view of fetlock.

Fig. 13-102 : Horse. Position for lateromedial view of fetlock.

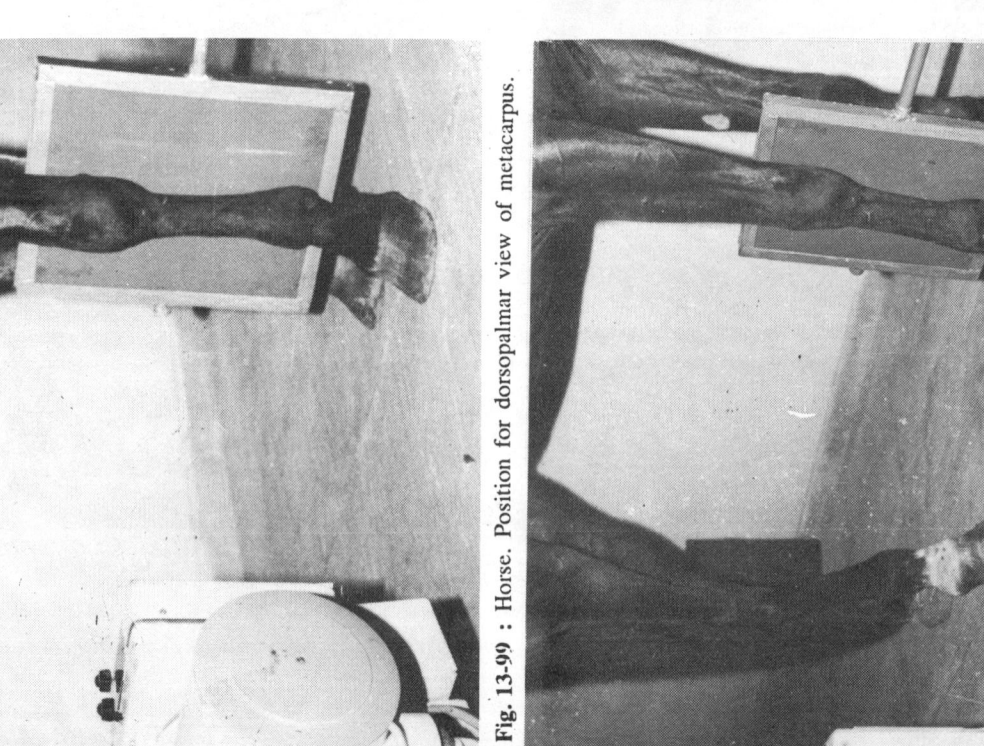

Fig. 13-99 : Horse. Position for dorsopalmar view of metacarpus.

Fig. 13-100 : Horse. Position for lateromedial view of metacarpus.

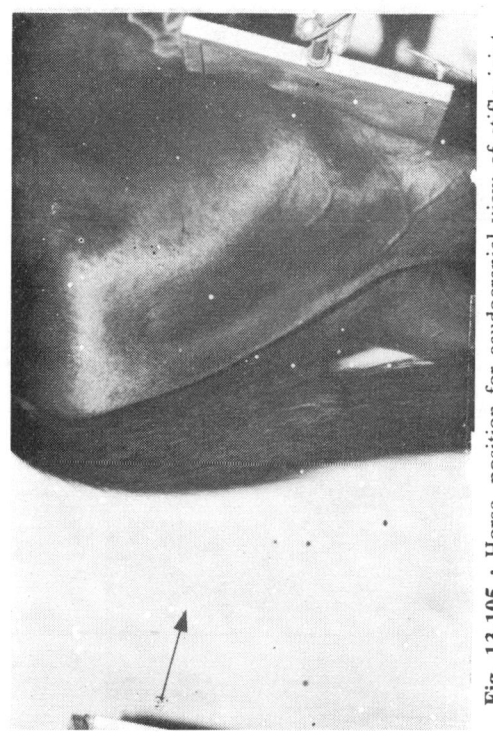

Fig. 13-105 : Horse. position for caudocranial view of stifle joint.

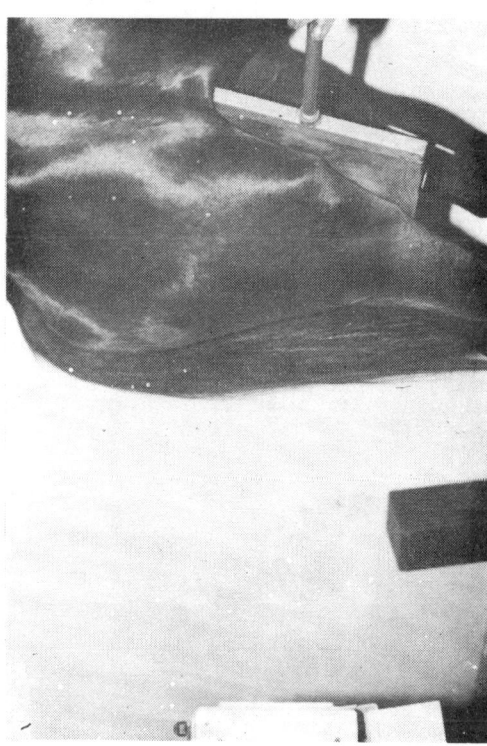

Fig. 13-106 : Horse. Position for lateromedial oblique view of stifle joint.

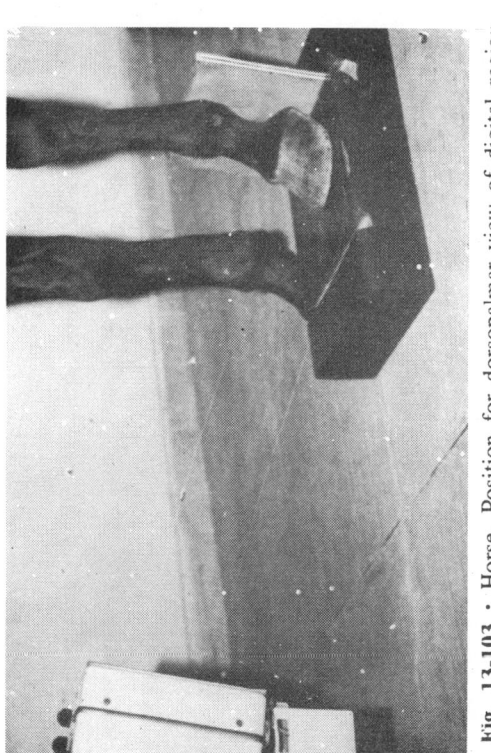

Fig. 13-103 : Horse. Position for dorsopalmar view of digital region using wooden block with slot for cassette.

Fig. 13-104 : Horse. Position for dorsopalmar view of distal phalanx and distal sesemoids using wooden block with slot for cassette.

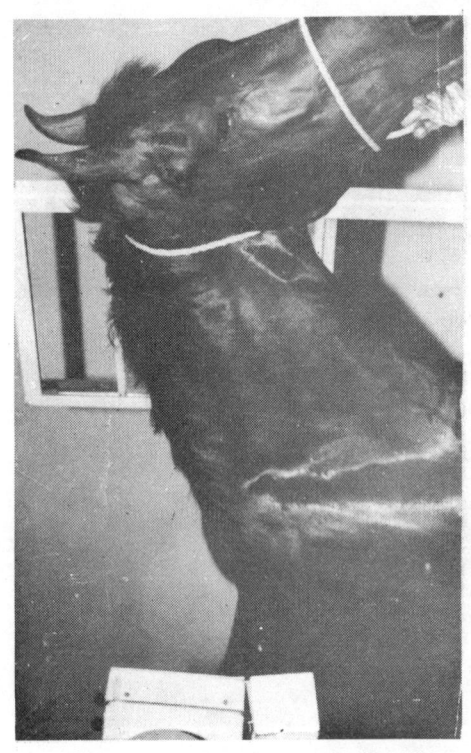

Fig. 13-109 : Horse. Position for lateral view of cervical vertebrae using a cassette stand.

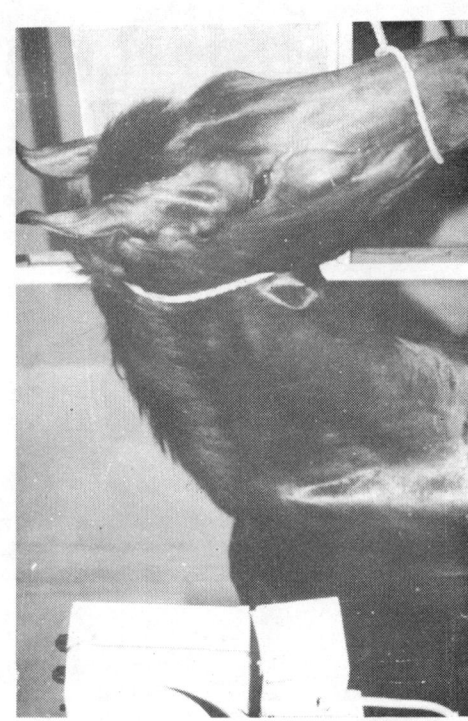

Fig. 13-110 : Horse. Position for lateral view of frontal region.

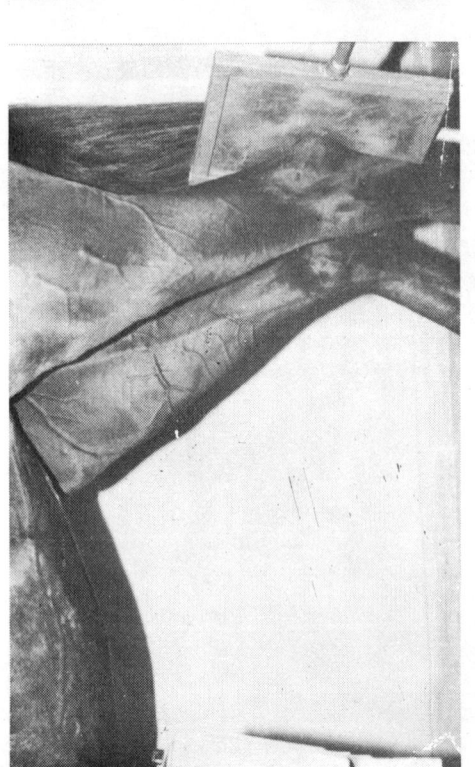

Fig. 13-107 : Horse. Position for dorsoplantar view of tarsus.

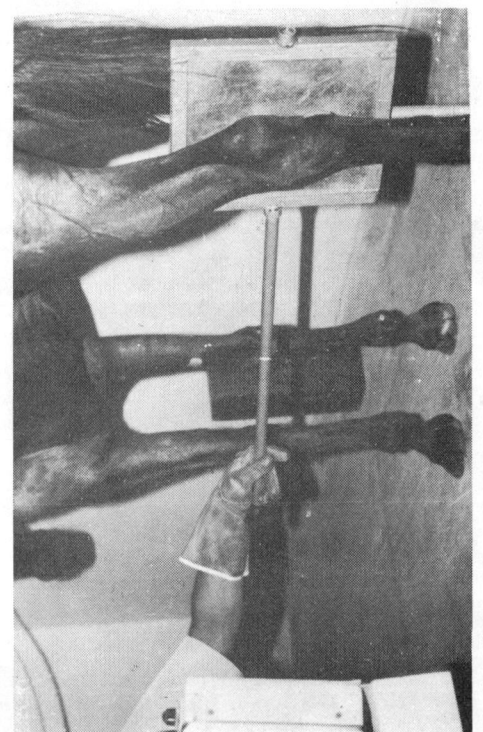

Fig. 13-108 : Horse. Position for latromedial view of tarsus.

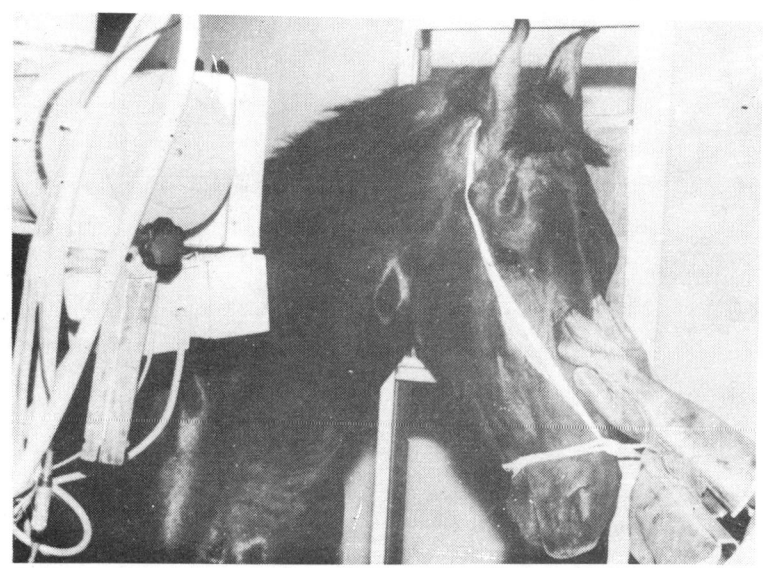

Fig. 13-111 : Horse. Position for lateral view of laryngopharyngeal region.

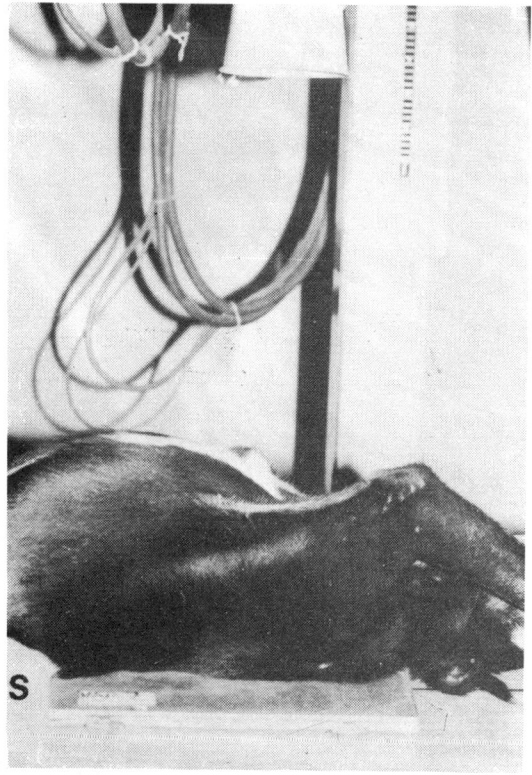

Fig. 13-112 : Dog. Position for caudocranial view of shoulder. S = Sand bags.

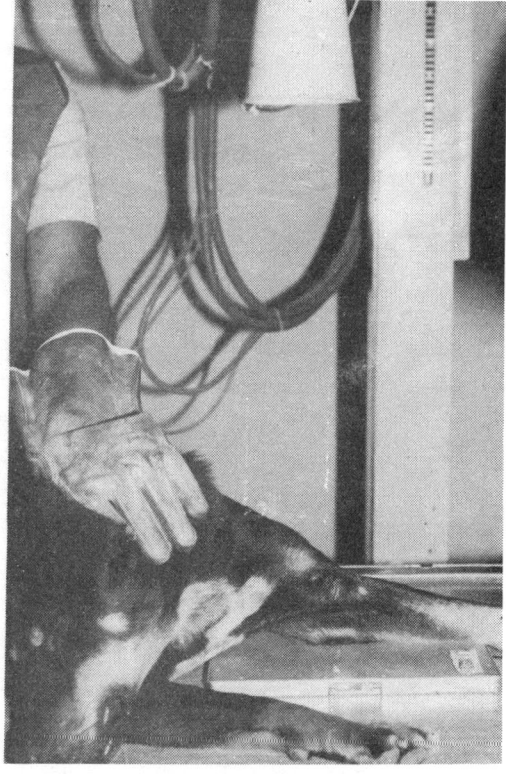

Fig. 13-113 : Dog. Position for craniocaudal view of elblow joint.

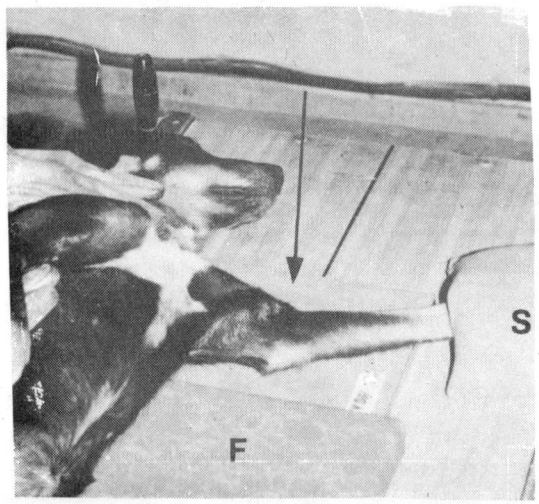

Fig. 13-114 : Dog. Position for mediolateral view of elbow joint. Arrow indicates X-ray beam direction. F = Foam block, S = Sand bag.

Fig. 13-116 : Dog. Position for mediolateral view of radius and ulna. Arrow indicates X-ray beam direction. S = Sand bag.

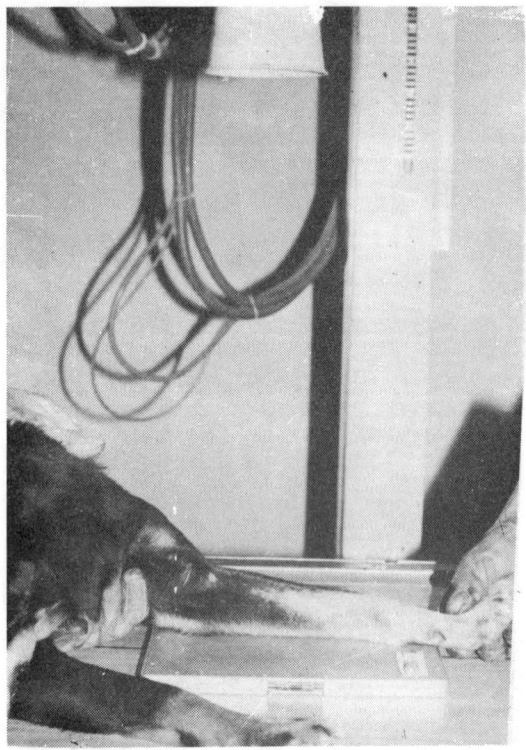

Fig. 13-115 : Dog. Position for craniocaudal view of radius and ulna.

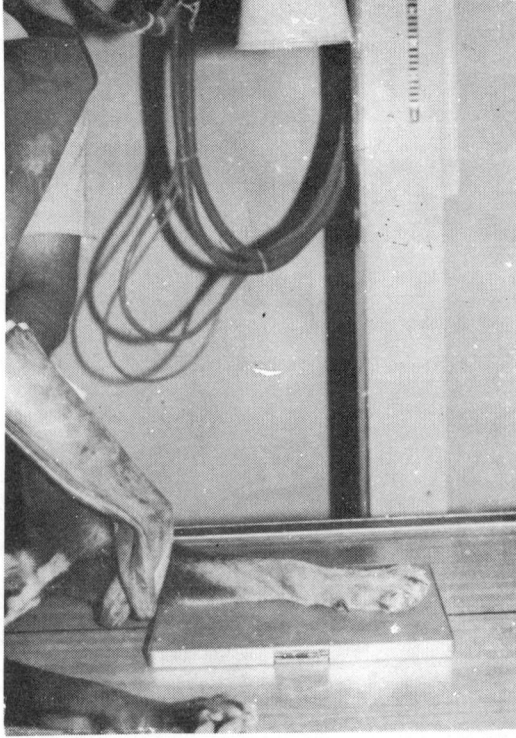

Fig. 13-117 : Dog. Position for dorsopalmar view of carpus and metacarpus.

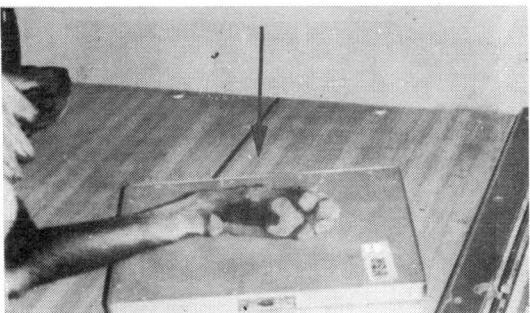

Fig. 13-118 : Dog. Position for mediolateral view of carpus and metacarpus. Arrow indicates direction of X-ray beam.

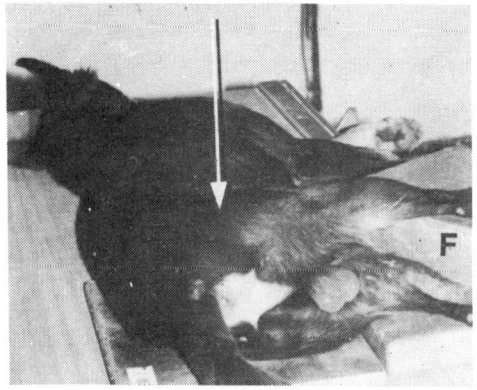

Fig. 13-120A : Dog. Position for lateral view of pelvis. Arrow indicates X-ray beam direction. F = Foam block placed between hind limbs.

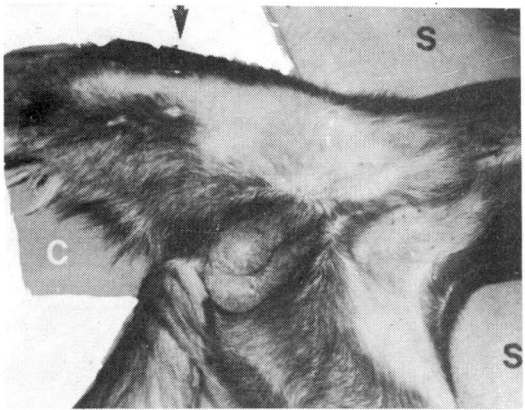

Fig. 13-120B : Dog. Position for craniocaudal view of femur. C = Cassette. S = Sand bags. For a lateral projection Position is same as shown in fig. 13-122 except that cassette is pushed more proximally to cover femoral area.

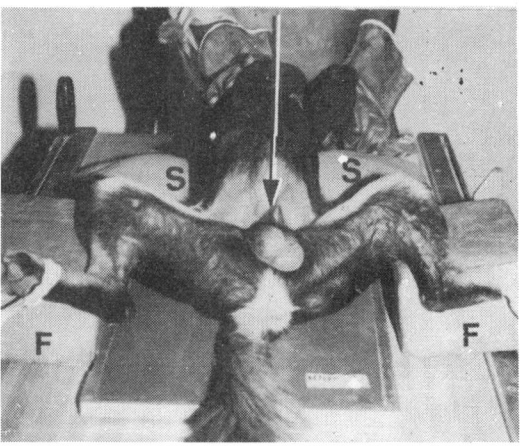

Fig. 13-119 : Dog. Position for ventrodorsal view of pelvis. F = Foam blocks. S = Sand bags. Arrow indicates X-ray beam direction.

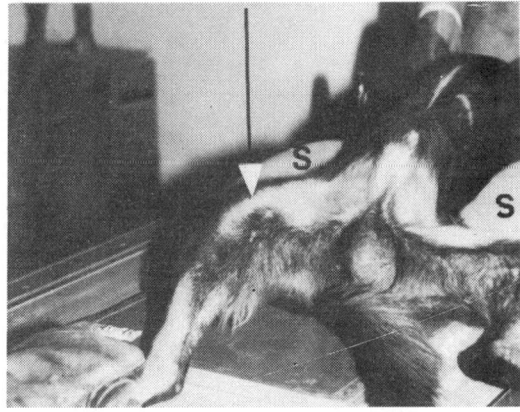

Fig. 13-121 : Dog. Position for craniocaudal view of stifle joint. Arrow indicates X-ray beam direction. S = Sand bags.

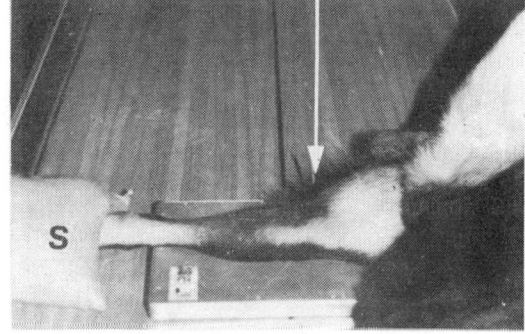

Fig. 13-122 : Dog. Position for mediolateral view of stifle joint. Arrow indicates X-ray beam direction. S= Sand bag.

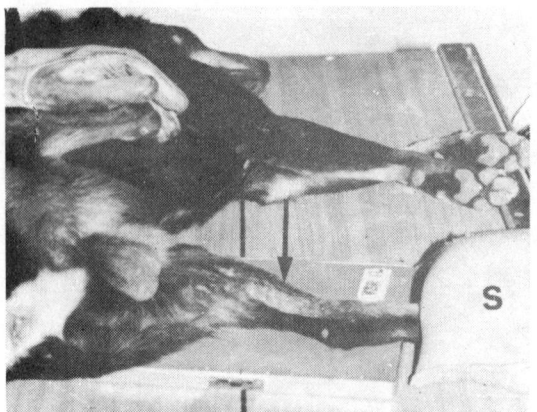

Fig. 13-123 : Dog. Position for mediolateral view of tibia. Arrow indicates X-ray beam direction. S = Sand bag.

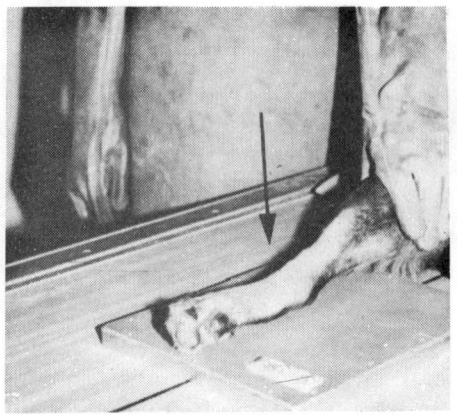

Fig. 13-124 : Dog. Position for dorsoplantar view of tarsus and metatarsus. Arrow indicates X-ray beam direction.

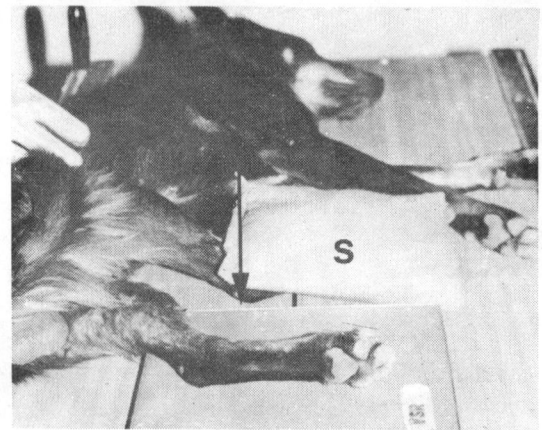

Fig. 13-125 : Dog. Position for lateromedial view of tarsus and metatarsus. Arrow indicates X-ray beam direction. S = Sand bag.

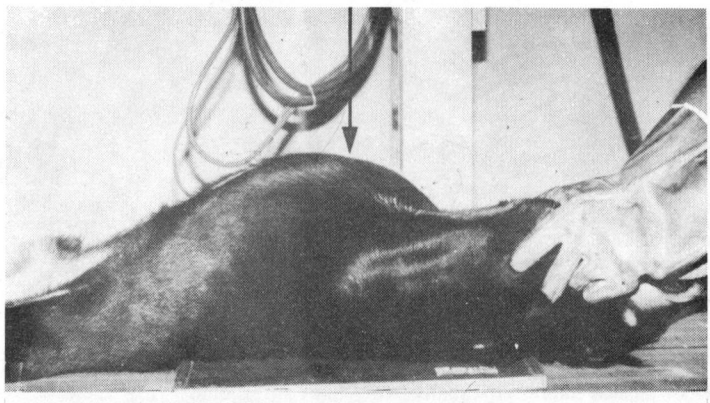

Fig. 13-126 : Dog. Position for ventrodorsal view of thorax.

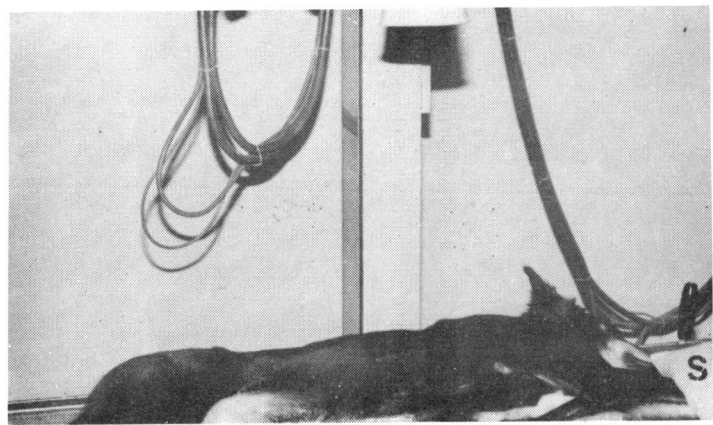

Fig. 13-127 : Dog. Position for lateral view of thorax. S = Sand bag.
F = Foam block

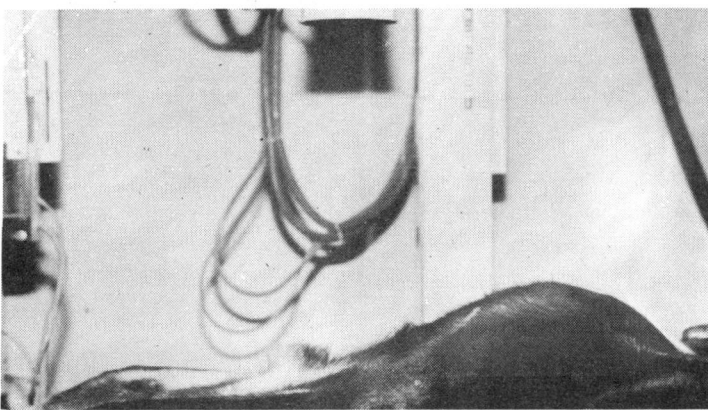

Fig. 13-128 : Dog. Position for ventrodorsal view of abdomen.
S = Sand bags.

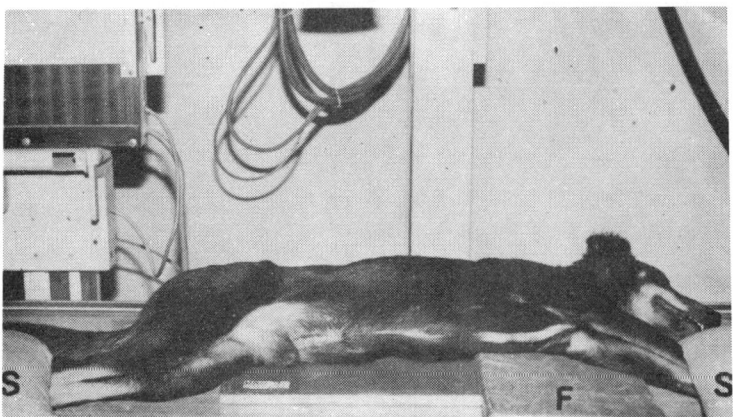

Fig. 13-129 : Dog. Position for lateral view of abdomen. S = Sand bags,
F = Foam block.

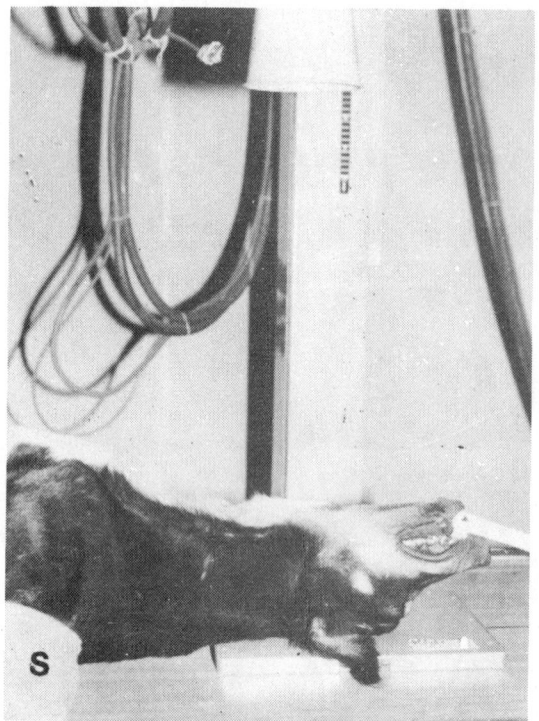

Fig. 13-130 : Dog. Position for ventrodorsal view of head. S = Sand bags.

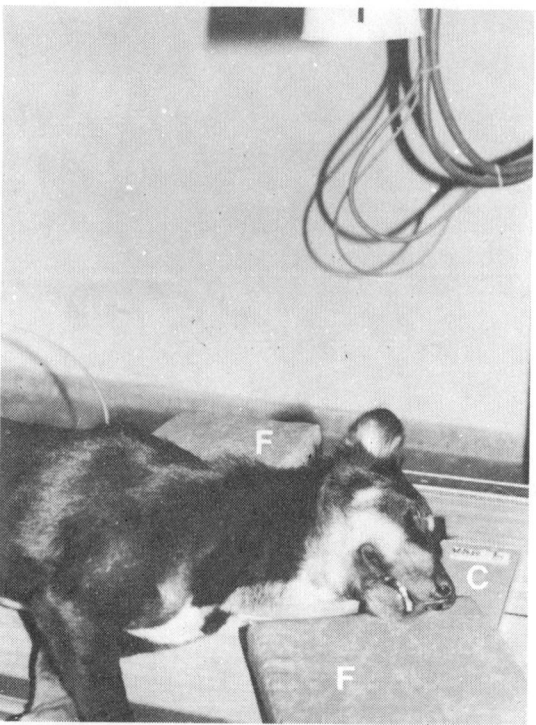

Fig. 13-131 : Dog. Position for lateral view of laryngopharyngeal region. F = Foam blocks, C = Cassette, T = X-ray tube.

Fig. 13-132 : Dog. Position for dorsoventral view of maxilla with intraoral placement of cassette. Arrow indicates X-ray beam direction.

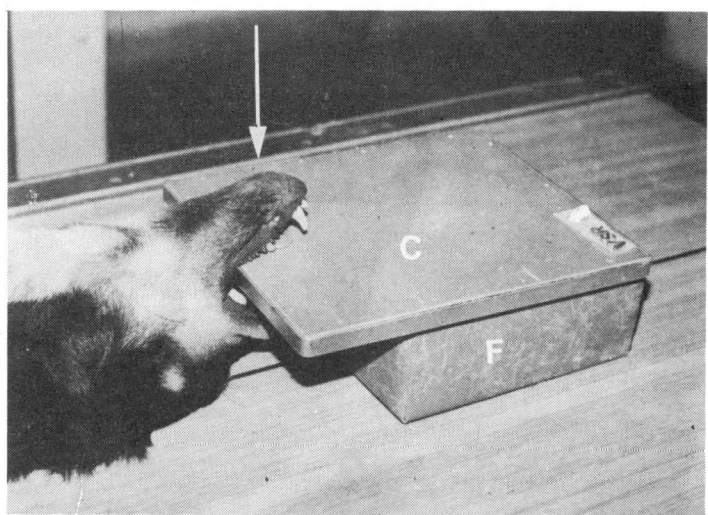

Fig. 13-133 : Dog. Position for ventrodorsal view of rostral part of mandible. Arrow indicates X-ray beam direction. C = Cassette, F = Foam block.

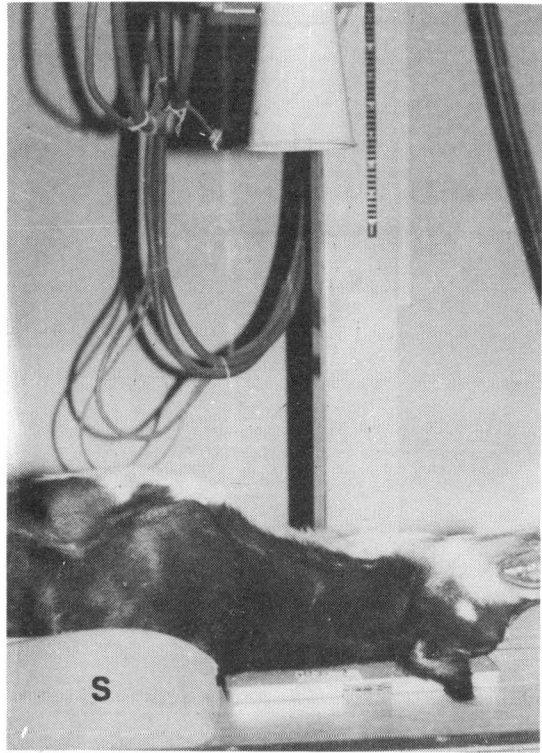

Fig. 13-134 : Dog. Position for ventrodorsal view of cervical region. S = Sand bags.

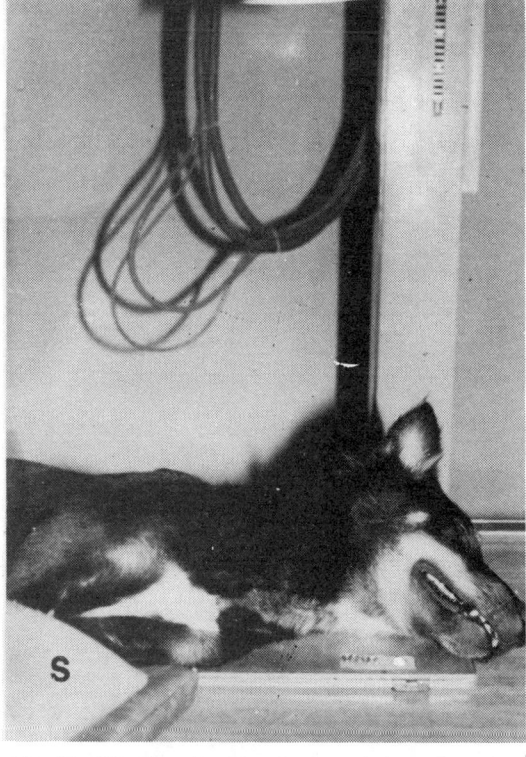

Fig. 13-135 : Dog. Position for lateral view of cervical region. S = Sand bags.

REFERENCES

Bargai, U., Pharr, J.W. and Morgan, J.P. 1989. Bovine Radiology. Ist edn., Iowa State University Press, Ames.

Douglas, S.W. and Williamson, H.D. 1980. Principles of Veterinary Radiography. 3rd edn., Williams and Wilkins Co., Baltimore.

Gillette, E.L., Thrall, D.E., and Lebel, J.L. 1977. Čarlson's Veterinary Radiology. 3rd edn., Lea and Febiger, Philadelhpia.

Hickman, J.A. 1954. A review of some technical aids in veterinary radiology. Vet. Rec. **66**, 805.

Morgan, J.P. and Silverman,S. 1990. Techniques of Vetrinary Radiography. 4th edn., Iowa State University Press, Ames.

Park, R.D. and Lebel, J.L. 1987. Equine radiology. In: Adam's Lamaness in Horses. Ed. Stashak, T.S., 4th edn, Lea and Febiger, Philadelphia.

Ryan, G.D. 1981. Radiographic Positioning of Small Animals. Lea and Febiger, Philadelphia.

Ticer, J.W. 1984. Radiographic Technique in Veterinary Practice. 2nd edn. W.B. Saunders Co., Phildelphia.

Williams, F. 1957. Cassette holders for large animal radiography. J. Am Vet. Med. Assoc. **130**, 28.

```
- RADIOGRAPHIC INTERPRETATION -
CONFUSION AND PRECONCEIVED IDEAS
          NOT ALLOWED
```

1 SPECIMEN EXPOSURE CHART FOR CATTLE AND BUFFALOES

X-ray Unit Siemens 'Heliophos'® 125 kV/500 mA

Calcium tungstate 'Hi-speed' intensifying screen. 'Indu'® medium speed film. Focal film distance = 90 cm.

Cd - Caudal, Cr - Cranial, D - Dorsal, L - Lateral, ML - Mediolateral, Pa - palmar, Pl -Plantar, V - Ventral

Part	View	Young animal		Medium sized animal		Heavy animal	
		mAs	kV	mAs	kV	mAs	kV
1	2	3	4	5	6	7	8
Fore limb							
Scapula or shoulder joint	ML	16	60	25	73	120	90*
Humerus	ML	12	60	20	73	80	85*
Elbow joint	ML	10	60	20	70	85	81*
	CdCr	12	63	20	77	-	-
Radius and ulna	CrCd	10	60	16	70	20	73
	L	10	60	12	66	20	73
Carpus	DPa	10	60	12	70	16	70
	L	10	60	12	70	16	70
Metacarpus	DPa	8	57	10	66	12	70
	L	10	57	10	66	12	70
Fetlock and digit	DPa	8	57	10	66	12	70
	L	10	57	10	66	12	70
Hind limb							
Hip joint	VD	20	70	-	-	-	-
	L	30	77	30	81	-	-
Femur	CrCd	12	70	-	-	-	-
	ML	12	70	25	81	40	90
Stifle joint	CrCd	12	66	20	77	-	-
	L	10	63	16	73	20	77
Tibia	CrCd	10	63	10	70	16	73
	L	10	60	10	70	12	73
Tarsus	DPl	10	63	10	70	16	73
	L	10	60	10	66	12	73

contd.

Appendix 1 contd.

1	2	3	4	5	6	7	8
Metatarsus	DPl	8	60	10	66	12	70
	L	8	60	10	63	10	70
Fetlock and digit	DPl	8	60	10	66	12	70
	L	8	60	10	63	10	70

Head

Skull	DV	20	70	40	81	80	81*
	L	12	66	20	73	80	90*
Nasal cavities	L	10	55	10	63	16	70
Lower jaw	L	10	60	12	66	12	70
Laryngeal region	L	10	63	12	70	20	73

Spine

Cervical spine	DV	16	70	25	77	-	-
	L	19	66	16	70	20	73
Thoracic spine	L	12	70	20	73	30	81
Lumbar spine	L	30	73	60	81*	-	-

Thorax

Lungs	L	16	70	25	82	40	90
Heart area (*recumbent position*)	L	20	70	60	85*	90	93*

Abdomen

Cranial abdomen	L	30	73	80	90*	120	102*

* With stationary grid 8:1

2 SPECIMEN EXPOSURE CHART FOR CAMELS

X-ray Unit Siemens 'Heliophos'® 125 kV/500 mA

Calcium tungstate 'Hi-speed' intensifying screen. focal film distance = 90 cm. 'Indu'® medium speed film.

Part	View	Adult animal	
		mAs	*kV*
1	2	3	4
Fore limb			
Elbow joint	CrCd	24	73
	ML	16	73
Radius and ulna	CrCd	12	73
	L	12	73
Carpus	DPa	16	73
	L	12	73
	ML Flexed	12	73
Metacarpus	DPa	12	70
	L	12	70
Fetlock and digit	DPa	10	70
	L	12	70
Hind limb			
Distal femur	L	30	85
Stifle joint	CrCd	30	81
	L	20	77
Tibia	CrCd	16	73
	L	12	73
Tarsus	DPl	20	73
	L	12	70
Metatarsus	DPl	12	70
	L	12	70
Fetlock and digit	DPl	12	66
	L	12	66

contd.

Appendix 2 contd.

1	2	3	4
Head			
Skull	DV	30	85
	L	30	81
Nasal cavity	L	16	70
Cranial region	VD	120	90*
	L	80	90*
Lower jaw	L	12	70
mandibular angle and molar teeth	L	12	70
Laryngeal region	L	16	70
Spine			
Cervical spine	DV	60	81*
	L	22	77*
Thorax			
Lungs	L	40	93

* With stationary grid 8 : 1

APPENDIX

3 SPECIMEN EXPOSURE CHART FOR SHEEP AND GOATS

X-ray Unit Siemens 'Heliophos'® 125 kV/500 mA
Calcium tungstate 'Hi-speed' intensifying screen. Focal film distance = 90 cm. 'Indu'® medium speed film.

Part	*View*	*Adult animal*	
		mAs	*kV*
1	2	3	4
Fore limb			
Shoulder joint and humerus	ML	12	70
Elbow joint	CrCd	10	60
	ML	8	60
Radius and ulna	CrCd	10	60
	L	8	60
Carpus, metacarpus, and digit	DPa	8	55
	L	8	55
Hind limb			
Femur	CrCd	10	70
	L	10	66
Stifle joint	CrCd	10	66
	L	8	63
Tibia	CrCd	8	60
	L	8	57
Tarsus	DPl	10	60
	L	8	55
Metatarsus and digit	DPl	8	52
	L	8	52
	Oblique	8	52
Head			
Skull	VD	12	63
	L	10	60
Nasal cavity	L	8	55

contd.

Appendix 3 contd.

1	2	3	4
Spine			
Cervical spine	VD	12	63
	L	10	60
Dorsal spine	L	12	70
Lumbar spine	VD	30	77
	L	16	73
Pelvis	VD	16	70
	L	24	73
Thorax	L	10	70
Abdomen	VD	24	73
	L	12	70

4 EQUINE

X-ray Unit Siemens 'Heliophos' 125/500 mA. Calcium tungstate 'Hi-speed' intensifying screen, medium speed film. Focal film distance = 90 cm.

Part	View	Young animal		Medium sized animal		Heavy animal	
		mAs	*kV*	*mAs*	*kV*	*mAs*	*kV*
1	2	3	4	5	6	7	8
Fore limb							
Shoulder joint & humerus	ML	40	70*	60	81*	120	90*
Elbow joint	ML	12	73	20	70*	30	77*
	CrCd	16	85	24	77*	40	77*
Radius and ulna	CrCd	8	63	10	66	12	70
	LM	8	63	10	66	12	70
Carpus	DPa	8	57	8	63	12	66
	LM	8	57	8	63	12	66
Metacarpus	DPa	8	57	8	63	10	66
	LM	8	55	8	60	10	63
Fetlock and digit	DPa	8	55	8	60	10	63
	LM	8	55	8	60	10	63
Hind limb	VD	24	70*	-	-	-	-
Hip joint	Lat	30	73*	60	81	-	-
Femur	CrCd	16	70	-	-	-	-
	ML	12	70	30	81*	40	90*
Stifle	CdCr	20	77	80	81*	80	90*
	LM	20	73	40	73*	60	77*
Tarsus	DPl	10	63	10	70	16	73
	LM	8	60	10	70	12	73
Metatarsus	DPl	8	57	8	63	10	66
	LM	8	55	8	63	10	63
Head							
Skull	DV	20	73	60	81*	80	81*
	Lat	16	66	40	73*	60	73*

contd.

Appendix 4 contd.

1	2	3	4	5	6	7	8
Nasal cavities	Lat	8	63	10	63	16	70
Lower jaw	Lat	10	63	12	66	12	70
Laryngeal region	Lat	10	63	12	70	16	73
Mandible	Lat	24	70*	40	81*	60	81*

Spine

Cervical	Lat	20	66	30	70*	40	81*
Thoracic	Lat	24	70	40	81*	60	81*
Lumbar	Lat	30	73*	40	81*	60	81*

* With stationary Grid (8 : 1).

5 SPECIMEN EXPOSURE CHART FOR DOGS

X-ray units Siemens Ergosphos® 125/200 mA
Calcium tungstate 'Hi-speed' intensifying screen. Indu® medium speed film. Focal film distance = 90 cm.

Part	View	Dog 10 kg		Dog 20 kg		Dog 30 kg	
		mAs	kV	mAs	kV	mAs	kV
1	2	3	4	5	6	7	8
Fore limb							
Shoulder joint	CrCd	8	50	8	55	10	65
	ML	8	45	8	50	10	55
Humerus	CrCd	6	50	8	50	8	60
	ML	6	50	8	50	10	55
Elbow joint	CrCd	8	50	8	60	10	65
	ML	8	50	8	55	8	60
Radius & ulna	CrCd	6	45	6	50	8	55
	ML	6	45	6	50	8	50
Carpus	DPa	6	45	6	50	6	55
	ML	6	45	6	50	6	55
Metacarpus & phalanges	DPa	6	45	6	50	6	55
	ML	6	50	6	50	6	55
Hind limb							
Hip joint	VD	10	60	12	65	16	70
	L	12	60	24	70*	30	70*
Femur	CrCd	10	55	10	60	12	65
	ML	8	50	8	60	10	65
Stifle joint	CdCr	8	55	8	60	10	65
	ML	8	50	8	55	10	55
Tibia	CrCd	6	50	6	55	8	55
	ML	6	45	6	50	8	50
Tarsus	DPl	6	45	6	50	6	55
	ML	6	45	6	50	6	55
Metatarsus and phalanges	DPl	6	45	6	50	6	55
	ML	6	45	6	50	6	55

contd.

Appendix 5 contd.

1	2	3	4	5	6	7	8
Spine							
Cervical	VD	8	55	10	60	12	65
	L	8	50	8	60	10	60
Thoracic	VD	12	70	16	70	30	75*
	L	10	55	12	60	12	70
Lumbar	VD	10	60	12	70	24	70*
	L	10	60	10	65	12	75
Sacrum	VD	12	65	16	70	20	75
	L	10	70	12	70	16	75
Thorax	VD	8	55	10	65	12	65
	L	8	50	8	60	10	65
Abdomen	VD	8	65	10	70	12	75
	L	8	65	8	70	10	75
Skull (General)	VD	8	60	8	65	10	70
	L	8	60	8	65	10	70
Turbinate	VD	6	50	8	50	8	55
	L	6	50	8	50	8	55
Teeth-upper Jaw (Intraoral)	DV	4	50	6	55	6	60
Teeth-lower Jaw (Intraoral)	VD	4	50	6	55	6	60
Laryngeal region	L	6	45	6	50	6	55

* With stationary grid (8 : 1)

6 SUPPLIERS OF X-RAY EQUIPMENT AND RELATED ITEMS IN INDIA

This list is not exhaustive. Mostly central office addresses are mentioned. Branch offices are only listed. For details, central offices can be contacted.

A. Equipment

(i) Siemens India Pvt. Ltd.,
Medical Engineering Division
130, Pandurang Budhakar Marg, Worli,
Bombay 400 018
New Delhi-4-A Ring Road, I.P. Estate, New Delhi - 110 002
(Regional Offices also at Madras and Calcutta, branch offices at Ahmedabad, Banglore and Hyderabad).

(ii) Toshniwal Bros. (Bombay) Pvt Ltd.,
198, Jamshedji Tata Road,
Bombay 400 020
(Delhi, Calcutta, Madras, Bangalore, Hyderabad; Indore, Ajmer).

(iii) International General Electric Co. (India) Ltd.
Nirmal, Nariman Point, Bombay 400 021
(Ahmedabad, Banglore, Bhopal, Calcutta, Chandigarh, Chinchwad, Delhi, Ernakulam, Hubli, Jaipur, Lucknow, Madras, Patna, Pune, Rajkot, Secunderabad, Trivandrum).

(iv) Philips Products
Peico Electronics and Electricals Ltd.
Medical Systems Division
Shivsagar Estate, Block A
Dr. Annie Besant Road, Worli, Bombay 400 018

(v) General Electric Co. India Ltd.
Magnet House, N Morarji Road, Bombay - 400 038

(vi) AGFA-GEVAERT India Ltd.,
Merchant Chamber, 41 New Marine Lines, Bombay 400 020
(New Delhi, Madras, Calcutta).

(vii) The Scientific Instruments Co. Ltd.
410/411, Pragati Towers, Rajender Place, New Delhi - 110 008
(Ahmedabad, Allahabad, Bombay, Calcutta, Madras, banglore, Bhopal, Guwahati, Hyderabad, Kanpur, Lucknow).

(viii) Electro-Medical and Allied Industries Ltd.,
4/2, B.T. Road, Calcutta 700 056

(ix) Medical Coordinators Pvt. Ltd.
Khaleel Shirazi Estate, 4th Floor,
Pantheon Road, Madras 600 008

B. Related Items (Films, Cassettes, darkroom equipment, safety equipment etc.)

- (i) Anita Enterprises, Belfer
  147, Water Field Road (Ext.)
  Bandra, Bombay 400 050

- (ii) CBO Engineering Co. Pvt. Ltd.,
  Halcyon Homes TPS III
  Jn. of 24 and 29th Roads,
  Bandra, Bombay 400 050

- (iii) Meditronics Corporation of India,
  6/9 Umarkhadi, Ist Road,
  Bombay 400 002

- (iv) Precision Electronic Instruments and .Components
  17, Ranvir building, 66/70 Princess Street,
  Bombay 400 002

- (v) Hindustan Photo Films Mfg. Co. Ltd
  (A Govt. of India Enterprise)
  Indunagar, Ootacomund 643 005
  (Sales Offices throughout India)

- (vi) X-ray Allied Film Co.
  XXXV/2453-A Karimpatta Road
  Pallimukku, Cochin 682016

- (vii) Wipro Ge Medical Systems Ltd
  Surya Towers, 6th floor
  Sardar Patel Road,
  Secunderabad 500 003

- (viii) Rege Cine Films
  263, Dr. Annie Besant Road,
  Worli, Bombay 400 018

- (ix) Kiran X-ray Screens Ltd.
  509, Dalamia Chambers,
  Marine Lines, Bombay 400 020

- (x) B.K. X-ray equipment Co.
  110, R.B. Marg, Ghodapdev,
  Bombay 400 033

C. Chemicals and Contrast Agents

- (i) Schering Division,
  German Remedies Ltd.
  P.O. Box 6570, Bombay 400 018

- (ii) Fairdeal Corporation Pvt. Ltd.
  66, Lakshmi Building,
  Sir P.M. Road, Bombay 400 001

- (iii) May and Baker India Ltd.
  Shastri Marg, Bombay 400 078
  (Bangalore, Calcutta, Guwahati, Hyderabad, Indore, Jaipur, Lucknow, Madras, New Delhi, Patna)

- (iv) AGFA-GEVAERT India Ltd.
  Address Listed earlier

- (v) Hindustan Photo Films
  Address Listed earlier

(vi)   M/S Anita Enterprises
        Address listed earlier
(vii)   International General Electric Co. (IGE)
        Address listed earlier.
(viii) Samrat Chemical Industries Pvt. Ltd.
        Halcyon Homes, TPS III,
        Jn. of 24th and 29th Roads,
        Bandra, Bombay 400 050
(ix)   Pharmed Ltd.
        25-31, Rope Walk Lane, Bombay 400 023
(x)   Nycomed Imaging
        P.O. Box 4283, Bangalore 560 042

INDEX

A

Abbreviations
    radiographic views, 6
Abdomen, radiographic view
    bovine, 240
    dog, 285
    sheep, 274
Absorber
    atomic number effect, 16
    density, 17
    thickness effect, 16
Absorption unsharpness
    image, 53,54
Acoustic enhancement, 146
Acoustic shadowing, 146
Air gap technique, 81,82
Amplitude, sine-wave, 9
Angiography
    cavernous sinus, 194
    general, 202
    orbital, 194
    renal, 197
    vertebral, 198
    see also osteomedullography
Anode
    angle, 23,25
    cooling curve, 46
    damage, 26,27
    failure, 26
    functions, 21
    heat cooling curve, 45
    prewarming, 27
    rotating, 23
    rotation, 24
    rotation, speed, 23
    stationary, 22
    thermal capacity, 45

Aperture diaphragm, 72
Arteriography, 197, 202
Artifacts
    radiographic, 96-104
    ultrasound, 144,147
Arthrography, 204
Atom, basic structure, 11
Atomic mass number, 172
Atomic number
    effective, 17
    elements, 16, 17, 172
    effect on contrast, 70
Attenuation, X-rays
    by anode, 25
    exponential, 14
    factors affecting, 16
    measurement, 14

Automatic processor, 96
Autotransformer, 36,37

B

Barium enema, 190
Barium series
    gastrointestinal tract, 190
Barium swallow
    see oesophagraphy
Basic density, of film, 56
Beam restricting devices, 71-74
Brachytherapy, 176
Bremstrahlung radiation, 30-31
Bronchography, 185
Bucky grid, 79

C

Caliper, 121,122